PROGRESS IN
PARKINSON RESEARCH

PROGRESS IN PARKINSON RESEARCH

Edited by
Franz Hefti and William J. Weiner

National Parkinson Foundation
Miami, Florida

PLENUM PRESS • NEW YORK AND LONDON

Library of Congress Cataloging in Publication Data

National Parkinson Foundation Symposium on Parkinson Research (1st: 1988: Miami
Beach, Fla.)
  Progress in Parkinson research / edited by Franz Hefti and William J. Weiner.
    p.   cm.
  Proceedings of the First National Parkinson Foundation Symposium on Parkinson
Research, held January 25–26, 1988 in Miami, Florida.
  Bibliography: p.
  Includes index.

  ISBN-13: 978-1-4612-8068-2   e-ISBN-13: 978-1-4613-0759-4
  DOI: 10.1007/978-1-4613-0759-4

  1. Parkinsonism — Pathophysiology — Congresses. 2. Methylphenyltetrahydropyridine
— Physiological effect — Congress. I. Hefti, Franz. II. Weiner, William J. III. Title.
RC382.N37 1988   88-30731
616.8′33 — dc19   CIP

Proceedings of the First National Parkinson Foundation Symposium
on Parkinson Research, held January 25–26, 1988,
in Miami, Florida

© 1988 Plenum Press, New York

Softcover reprint of the hardcover 1st edition 1988

A Division of Plenum Publishing Corporation
233 Spring Street, New York, N.Y. 10013

PREFACE

In the past five years significant progress has been made in our basic and clinical understanding of Parkinson's disease. The discovery that MPTP, a relatively simple molecule, is able to induce parkinsonism in otherwise healthy adult humans, and the recent interest in the possibility of "transplantation" procedures as a therapeutic modality in the treatment of Parkinson's disease have generated enormous interest in research related to Parkinson's disease. In this setting, the National Parkinson Foundation decided to organize a research meeting to bring together scientists actively engaged in research relevant to the study of Parkinson's disease, to accelerate its progress and to promote an exchange of ideas. This meeting took place in January 1988 at Key Biscayne, Florida. It was decided to publish the proceedings of this meeting to allow rapid documentation of the participants current findings and views regarding this rapidly evolving field. The structure of this volume follows the organization of the meeting and begins with a clinical and neuropathological review of current knowledge regarding Parkinson's disease. Since dopaminergic neurons play a major role in the pathophysiology of the disease, many of the contributions relate to some aspects of dopaminergic function including localization, regulation, and pharmacology of dopamine receptors. A special effort has been made to provide a summary of the present knowledge of the cellular biology of the dopaminergic neurons. There is extensive discussion of MPTP with particular emphasis on the role of potential other neurotoxins which might relate to Parkinson's disease. The topic of "transplantation" is addressed in several contributions. We thank the National Parkinson Foundation for making the meeting possible, its staff for excellent organization, and Merck, Sharp and Dohme for their generous grant which supported the meeting.

Franz Hefti and William J. Weiner

Miami, July 1988

CONTENTS

1

THE CURRENT CLINICAL PICTURE OF PARKINSON'S DISEASE

Stewart A. Factor * and William J. Weiner +

* Albany Medical College, Dept of Neurology
  Albany, New York 12208
+University of Miami, Dept of Neurology
  Miami, Florida 33101

The following chapters in this book are very specifically directed at individual research topics which relate to Parkinson's disease (PD). Often in our quest for answers to specific questions we lose sight of the patient and their disease. It is our intent to provide a clinical description of PD and to discuss the current symptomatic treatment. We will also review new evidence related to the etiology of PD and briefly discuss new directions in attempting to treat this debilitating disease.

CLINICAL SYNDROME OF PARKINSON'S DISEASE

Parkinson's disease is a progressive neurodegenerative disorder with well recognized clinical and pathological features. The clinical syndrome was first described in detail by James Parkinson (1817) and was referred to as "the shaking palsy". It was Charcot in the late 1800's who first referred to this disorder as "Parkinson's disease".

PD is a slowly progressive disease with an insidious onset. It most commonly occurs between the ages of 58 and 61 but may present anywhere from 30 to 80 years of age. The clinical syndrome is comprised of four cardinal signs and symptoms; tremor, rigidity, bradykinesia, gait and postural abnormalities. The tremor is classically referred to as a "resting" tremor because it occurs while the patient is at rest and disappears for the most part with goal directed movement. Sixty to 70% of patients present with tremor as the initial symptom and approximately 90% experience this symptom at some time during the illness. The tremor most frequently occurs in the hands, however, the legs, jaw, and tongue are often involved. It is commonly unilateral when it begins but it almost invariably becomes bilateral with time. The hand typically assumes a posture of flexion of the metacarpo-phalangeal joint with extension of the more distal joints. The presence of tremor in the fingers make it appear as if the patient were rolling a pill between the thumb and index finger thus resulting in the term "pill-rolling tremor". The tremor is frequently more embarrassing to the patient than disabling and it generally worsens with stress or heightened emotion. A minority of the patients also have a postural and kinetic hand tremor. Rigidity can manifest in the patient as muscle aches and stiffness. This frequently occurs in the neck and back and may be accompanied by occipital headache. With passive movement of the involved extremity the examiner will find increased muscle tone which has a ratchetty quality to it and thus the term "cogwheel rigidity". Bradykinesia is defined as slowness or poverty of movement with loss of automatic stereotyped movements. The slowness is perceived by the patient as a slowing down of the their ability to perform the usual activities of daily living such as bathing, dressing and turning over in bed. There is an associated loss of dexterity making buttoning, shaving and other activities quite difficult. Hesitation on initiation of movement and early fatigue are also features of bradykinesia. The loss of automatic movement includes loss of arm swing when walking, decrease in the frequency of swallowing resulting in an accumulation of saliva in the mouth and drooling, a decrease in blinking frequency and loss of facial expression resulting in the so called masked face, and loss of expressive gestures of the hands. The posture of the parkinsonian patient involves flexion of the head, trunk and extremities. Thus the patients appear stooped with the head bent forward, legs bent at the knees, and arms bent at the elbows. The trunkal flexion is often severe and the patient will usually lean to one side even when sitting. Loss of postural reflexes often results in poor balance. If the patient encounters even minor postural perturbation (slight push) they will fall. The normal protective mechanisms associated with falling are also lost so that it is unlikely that the person will do anything to break their fall. It is not uncommon to see patients with multiple contusions, some quite severe, and even fractures. The gait is also abnormal. In addition to the postural abnormalities and loss of arm swing the patients generally take small shuffling steps. The forward leaning of the trunk moves the center of gravity forward causing the patient to hasten his pace in order to catch up with it (propulsion). Patients are often unable to lift their feet off the floor as if they are stuck to it exhibiting another aspect of the gait abnormality, start hesitation or freezing.

Other clinical features which occur with variable frequency in PD patients include micrographia (small handwriting), hypophonia and dysarthria. The voice volume diminishes in some cases to a whisper and may be associated with a dysarthria. Constipation, pedal edema, seborrhea of the scalp and forehead are also frequently observed in PD. Less commonly, autonomic dysfunction with orthostatic hypotension, urinary incontinence, and impotence is seen (Goetz et al, 1986). Sensory symptoms such as numbness, coldness, burning, akathisia, and pain, which may be quite severe have also been reported to be more prevalent than previously realized (Lang and Johnson, 1987; Quinn et al, 1986; Goetz et al, 1986a; Koller, 1984). Olfactory impairment is also a feature of PD (Ward et al, 1983a). Many forms of dystonia have been reported (Kidron and Melamed, 1987; Weiner and Nausieda, 1982; Nausieda et al, 1980). Although PD is considered to be primarily a disorder of movement, psychiatric abnormalities have also been observed. Dementia and depression are frequently observed but the exact frequency with which they both occur is unknown (Brown and Marsden, 1984; Santamaria et al, 1986). Whether dementia and depression are independent disorders occurring in a patient with PD or part of the parkinsonian syndrome itself remains to be elucidated.

Although PD is a progressive degenerative disorder the exact prognosis for an individual patient is difficult to precisely predict. This wide variability in prognosis was noted by Schwab (1960). He presented evidence of patients whose symptomatology became worse only very slowly and in whom disease progression was almost imperceptible from year to year. In some of his most dramatic examples patients were still independent and working after 25 years of PD. On the other hand Mjones (1949) found that 50% of patients were completely disabled or forced to retire after only 4 years of PD. In 1967, Hoehn and Yahr (1967) reviewed evidence from 866 patients regarding age of onset, progression of disease, and mortality and this report published just prior to the introduction of levodopa in the treatment of PD stands as the most authoritative statement regarding disease progression in the prelevodopa era. This study reported that 28% of all parkinsonian patients were disabled or dead within 5 years of disease onset and that by 10 years of disease this figure had risen to 61%. Life span was shortened and the mortality rate was 3 times that of the general population.

The introduction of levodopa therapy has altered these grim statistics and several 5 year followup reports of levodopa in these patients revealed that disability onset was delayed and mortality was decreased (Sweet and McDowell, 1975; Yahr, 1975; Markham et al, 1974). There was such initial enthusiasm that it was hoped that levodopa might actually halt disease progression. This unfortunately was not the case. In fact, recent long term studies of patients taking levodopa have suggested that the drug improves life expectancy for the first 6 years of treatment but that after 12 years of continuous therapy the protective effect may be lost (Stern, 1987). It is the continued inexorable progression of PD which leads to severe disability and it is hoped that either a specific etiology will be discovered which can be treated or successful approaches will be found to alter the rate of progression.

SYMPTOMATIC THERAPY

The degeneration of the zona compacta and loss of dopaminergic input to the striatum results in the motor symptomatology of PD. It is the loss of this dopaminergic influence which forms the basis of symptomatic therapy in PD. The introduction of levodopa in clinically useful doses in the late 1960's represents a remarkable therapeutic achievement (Cotzias et al, 1969) in the treatment of this disorder. The use of a peripheral dopa decarboxylase inhibitor (PDI) in association with levodopa prevented the occurrence of frequent gastro-intestinal side effects and this combination is now standard therapy for symptomatic treatment of disabling PD. Although reasonably effective in relieving many of the major motor symptoms of PD the chronic use of these agents is associated with well recognized adverse effects. These adverse effects include loss of efficacy, dyskinesia (peak dose, diphasic, and early morning dystonia) psychiatric side effects (hallucinations and confusion) and motor fluctuations. The most commonly recognized motor fluctuations include end of dose failure, sudden on-off phenomenon, and loss of efficacy of individual doses. The mechanisms behind these complications have not been entirely elucidated, however, pharmacokinetic (reflecting alterations of bioavailability of levodopa in the striatum), pharmacodynamics (alterations in postsynaptic dopamine receptors), and the natural progression of PD appear to be involved (Fahn, 1982).

These complications have led to the formulation of new strategies in the treatment of PD. One such strategy involves the utilization of dopamine agonists which act directly on dopamine receptors without requiring metabolism by the presynaptic nigrostriatal neurons. There are two types of dopamine receptors; D-1 and D-2. These will be discussed in detail in later chapters. The dopamine agonists have variable activity at these receptor sites but it appears that D-2 activity is primarily related to clinical response in PD. Bromocriptine is an ergot derivative which appears to have its major activity at the D-2 receptor and it has been widely and effectively used to treat PD. It has also been used alone in denovo patients with limited efficacy and, more commonly, as an adjunct to levodopa/PDI therapy (Leiberman, 1985). When used as a late addition to levodopa/PDI it improved parkinsonian symptoms and, in addition, improves motor fluctuations. It has been suggested that early combination therapy of levodopa/PDI and bromocriptine may actually prevent the onset of fluctuations while having the same early efficacy as levodopa/PDI therapy alone (Rinne, 1985). Double blind studies of this strategy are currently in progress (Weiner et al, 1988). A number of other ergot derivative dopamine agonists with varying properties (e.g. lergotrile, pergolide, mesulergine, lisuride) have been studied and found to be effective in PD. Pergolide is primarily a D-2 dopamine receptor agonist with D-1 agonist properties at low doses and antagonist activity at high doses. It too has been found to have anti-parkinsonian activity and it improves motor fluctuations. Adverse effects are similar to bromocriptine and include hallucinations, confusion, dyskinesia, orthostatic faintness, and nausea/vomiting (Factor et al, 1988). Lisuride is another dopamine agonist which can be given intravenously (Obeso et al, 1986). The usefulness of these drugs is limited both by toxic side effects and eventual loss of efficacy with chronic use. New novel compounds with dopamine agonist properties which are non-ergot derivatives are being studied. Ciladopa demonstrated efficacy in patients but was withdrawn from clinical trials because of tumorogenesis in rodents (Weiner et al, 1987). Another of these compounds is 4-propyl-9-hydroxynaphthoxazine (PHNO). This is a powerful D-2 receptor agonist with anti-parkinsonian properties which has been efficacious in the treatment of PD significantly improving tremor, rigidity and bradykinesia at low doses (Weiner et al, 1988a; Stoessl et al, 1985). The potency and solubility characteristics of PHNO may allow development of parenteral delivery systems. Each dopamine agonist will probably be efficacious in select subgroups of PD patients. It appears impossible at this time to predict a specific therapeutic response to a dopamine agonist in patients already on levodopa/PDI. Furthermore, the response or lack of response to a particular dopamine agonist does not preclude response or failure of response to another agonist (Factor et al, 1988). The more agonists that become available for trial in patients, the better the chance for continuing success in treating the parkinsonian patient.

Another therapeutic strategy has been to alter the delivery of levodopa in an attempt to maintain constant levodopa blood levels, improve its bioavailability, and alter certain levodopa induced side effects. The administration of frequent, small doses of levodopa/PDI can be useful in this regard. Patients, however, find this plan inconvenient and often of limited usefulness. In addition, this strategy may play a role in the formation of more unpredictable responses which occur as the result of a longer duration of therapy (Nutt, 1987). Alterations in dietary protein levels (Pincus and Barry, 1987), direct duodenal infusion of levodopa (Kurlan et al, 1986), and intravenous forms of levodopa (Nutt et al, 1984) have been utilized with some success. We and others have studied a controlled release form of Sinemet (Sinemet CR4). In our trials CR4 was found to increase "on" time in a subgroup of patients with an older age of onset of PD, a shorter duration of disease and motor fluctuations than nonresponders (Factor et al, 1988a).

ETIOLOGIC CLUES IN PARKINSON'S DISEASE

The etiology of PD remains unknown despite intensive recent interest in the subject. Obviously, if a specific etiology is discovered PD might become preventable. However, even if PD were not preventable, if the cause or mechanism of dopamine cell death in PD were known, therapeutic measures might be developed which could interfere with the progressive cell loss and therefore the progression of PD might be altered and slowed. Many attempts have been made to associate PD with various factors in patients personal and/or social histories but these have not yielded any clear picture of a patient who is more "at risk" for the development of PD. There is no association between the development of PD and HLA type, blood group, serological markers, cancer, coronary heart disease, cerebro-atherosclerosis, common viral infections, diet peculiarities, alcohol consumption, vaccinations, exposure to drugs, fumes, gases, radioactive materials, or animal contacts (Bharucha et al, 1986; Martilla and Rinne, 1986).

The role of heredity in the pathogenesis of PD has been, and continues to be, somewhat controversial. Twin studies (Ward et al, 1983), for the most part have proven that heredity plays virtually no role in the development of this disorder. In the study by Ward et al (1983), only 1 of 43 monozygotic pairs were concordant. A review of this and two other studies combined showed that the percentage of concordant monozygotic pairs was less than concordant dizygotic pairs (Duvoisen, 1986). Despite the more recent reports of two other concordant pairs of monozygotic twins (Koller et al, 1986; Jankovic et al, 1986) it remains unlikely that heredity plays a major role. Some have suggested the presence of familial subsets of PD thus making heredity important in a subgroup of patients (Barbeau and Roy, 1984; Zetusky et al, 1985; Geraghty et al, 1985). This too remains controversial (Duvoisin, 1986).

It has been proposed that PD is etiologically related to an accelerated aging process (Barbeau, 1973). This will be further discussed by Finch in a later chapter. This concept implies that dopaminergic neurons undergo an age related deterioration which depends primarily on predetermined intrinsic properties of the cell. There is evidence that cell bodies in the substantia nigra decrease in number with age, and that striatal dopamine, dopamine receptors, and dopamine synthesizing enzymes decrease with age. However these changes are not profound and come nowhere near the proposed 80% decrease in dopaminergic systems which is required for a patient to manifest parkinsonian symptoms. The age related changes also do not take into account compensating changes in the dopaminergic system even in the aging brain. If PD were simply age related then all individuals who reach a certain age should develop PD and this is not the case. In addition, evidence has been presented that the age of onset peak is in the early 60's and that after 75 years of age the incidence of new cases falls off dramatically (Koller et al, 1986). Since there is no linear increase in incidence with age (for example as seen with cerebral vascular disease) a strong argument against aging as an etiology can be made. Calne and Langston (1983) have also argued that in twin pairs aging of the central nervous system should occur at a similar rate and since there was no concordance in the previously cited twin study this is also evidence against aging as an etiologic factor. However, aging is often still considered as an additional risk factor in the development of PD since normal aging changes in the substantia nigra and striatum are added to a yet unknown factor which also produces either a static injury or a progressive injury to nigral and striatal neurons.

The final area of current research related to the etiology of PD relates to environmental causes and toxins. This has been a very exciting area of research in which several new facts have had startling implications for the etiology of PD. The most important recent finding of an environmental toxin which produces clinical symptoms and perhaps pathologic features very similar to PD was the discovery of MPTP induced parkinsonism (Langston et al, 1983, Davis et al, 1979). Dr. Langston will further discuss his most recent evidence linking MPTP and PD in a later chapter. MPTP was accidently synthesized in an attempt to produce a cogener of demoral, 1-methyl-4-phenyl-4 proprionoxy-pyridine (MPPP), for recreational use and profit. The eventual result was a series of drug abusers with a pure, permanent parkinsonian syndrome (Langston et al, 1983; Davis et al, 1979). In addition, since this discovery a chemist who worked with MPTP developed parkinsonism and it has been suggest that inhalation or transcutaneous intoxication may occur (Langston et al, 1983a). Clinically, the syndrome resulting from MPTP toxicity mimics PD more closely than any previous toxin in both man and nonhuman primates (Ballard et al, 1985; Burns et al, 1983). The aspects of the clinical syndrome which differ from PD in patients include a young age of onset, a rapid onset with rapid progress to severe disability and the onset of clinical fluctuations in response to levodopa after a relatively short exposure to the drug. Response to levodopa/carbidopa is quite dramatic as in PD. In addition, there is some question as to whether or not MPTP induced parkinsonism progresses. MPTP induced parkinsonism results in significant reduction of the metabolites of dopamine in CSF just as PD however, the metabolites of norepinephrine, which are also depleted in PD, are normal in the MPTP patients (Burns et al, 1985). Pathologically, in the one patient examined, MPTP causes a significant isolated substantia nigra lesion with neuronal loss and gliosis. In addition, a single inclusion body similar to a Lewy body was found (Davis et al, 1979). More recently, in older monkeys, more extensive pathologic lesions, similar to that seen in PD, including involvement of the locus ceruleus and ventral tegmental tract and with the presence of eosinophilic inclusion bodies have been reported (Forno et al, 1986; Mitchell et al, 1985). Dr. Forno will discuss her latest findings regarding Lewy bodies in MPTP parkinsonism and PD in the next chapter.

The mechanism of action by which MPTP induces selective dopaminergic cell death in the substantia nigra is not entirely known. It is proposed that MPTP enters the brain and interacts with mao B to form its toxic by-product MPP+ (Glover et al, 1986: Reznikoff et al, 1985; Uh1 et al, 1985). This step occurs in glial cells, not neurons, and is blocked by mao B inhibitor deprenyl but not mao A inhibitor clorgiline. MPP+ enters the neurons via a specific dopamine uptake system and this can be blocked by dopamine uptake inhibitors such as mazindol and benztropine (Sanchez-Ramos et al, 1988: Sanchez-Ramos et al, 1986; Bradbury et al, 1986, Snyder and D"Amato, 1986). The toxic levels of MPP+ in nigral neurons are believed to affect the mitochondria, interfere with cellular respiration, and result in cell death (Sanchez-Ramos et al, 1988a: Vyas et al, 1986).

Interest in MPTP-induced dopamine neuronal toxicity has led to renewed interest in the concept that free radical-induced dopamine cell damage plays an etiologic role in PD. Oxidative deamination of dopamine can result in the formation of hydrogen peroxide, hydroxy radicals, and superoxide . These agents interact with polyunsaturated lipids and initiate further reactions which then lead to membrane lipid peroxidation and cellular damage (Halliwell and Gutteridge, 1985; Cohen, 1983; Freeman and Crapo, 1982). Superoxide dismutase, glutathione peroxidase, catalase, and other soluble reducing substances are the neurons self defense which promote low free radical levels. When free radical formation is either accelerated or scavenging enzymes are reduced, as has been reported in PD, then cell damage occurs (Perry and Yong, 1986; Kish et al, 1985; Perry et al, 1982). Kopin (1986) has proposed that accumulated MPP+ in the neuron generates free radicals by its interaction (redox cycling) with neuromelanin, dopamine, and other reducing agents. This excessive production of free radicals exceeds the ability of the neuron to contain them and cell death ensues. This proposal is very interesting because it not only accounts for the susceptibility of dopamine neurons to MPTP but it also accounts for the selectivity of cell damage to the substantia nigra (neuromelanin).

Additional possible environmental clues have come from the work of Rajput et al (1986, 1984) who have observed that patients with younger onset of PD are more likely to have been exposed to a rural environment. In addition, in these patients childhood drinking water supply was almost exclusively from well water. The investigators postulated that drinking water may be the source of a common toxic etiologic agent which could have access to all nationalities and geographic areas. This work remains to be confirmed.

Another environmental toxin is the cycad seed from the false sago palm cycas circinalis L. In the 1950's the Chamorro population of the western Pacific islands Guam and Roto suffered from a high incidence of a disorder which presented as ALS, parkinsonism, dementia or a combination of all three. There were no demonstrable hereditary factors or viruses found to cause this syndrome. Attention was placed on the cycad seed because of its frequent use in foods and medicines during and after WWII and its declining use in association with a decline in incidence of the disorder after 1955 and the Americanization of the people. Spencer et al (1987) demonstrated in monkeys that the inciting toxin in the cycad seed was B-N-methyl-amino-alanine (BMAA). BMAA is an excitotoxin, a compound with excitatory properties at the synapse and neurotoxic properties at higher than physiologic levels (for review of excitotoxins refer to Schwarcz and Meldrum, 1985). Although Spencer et al (1987) found that the toxin effected motor neurons most severely the fact that Parkinsonism was also noted in the chamorro population and in the laboratory animals, is significant and suggests that multiple types of toxins may be implicated in the pathogenesis of PD. In addition, it was observed that monkey's with parkinsonian features responded to some extent to levodopa/PDI suggesting that a dopaminergic lesion is present. Although this discovery will not provide the etiologic answer to PD it does demonstrate the potential implications of environmental toxins producing neurologic disease.

An etiologic hypothesis which combines genetics and environmental toxins has been put forward by Barbeau et al (1986, 1985). This theory states that PD is multifactorial in origin and that the disease occurs as a result of the interaction between potential environmental toxins and genetically suseptible individuals. They present evidence from Quebec for a geographic distribution to the number of PD cases; a higher prevalence of PD corresponding with regions in which the highest level of various pollutants including pesticides and herbicides are used. Hepatic cytochrome P450 monooxygenases are a broad class of enzymes which serve as "detoxifying" enzymes. Using the metabolism of the compound debrisoquin PD patients were demonstrated to have deficiencies in P450 activity. This combination of high incidence of PD in rural Quebec in those regions with high pesticide use and a deficiency in the P450 detoxifying system in PD patients suggests that individuals in these areas are more suseptible to environmental toxins. There is no confirmed relationship at this time between pesticide use and PD.

NEW DIRECTIONS IN THERAPY

New symptomatic drugs for the treatment of PD continue to be developed. These agents both ergot derivatives and non-ergot derivatives are for the most part D-2 receptor agonists. These drugs will play an important role in the symptomatic treatment of PD because if future D-2 agonists resemble past D-2 agonists there will be a subgroup of PD patients who will respond positively to each new drug. Since agonist induced response in individual patients cannot be predicted at this time the availability of a wide range of D-2 agonists will provide the greatest chance for continued successful symptomatic treatment in these patients. In addition to new agonists, there will also shortly be available novel delivery systems so that dopaminergic agents can provide steady plasma levels of drug and perhaps avoid some of the drug induced complications.

In the most exciting departure from symptomatic therapy of PD the Parkinson Study Group has undertaken a very large double blind study to evaluate the usefulness of deprenyl and tocopherol in altering the progression of PD. The selection of these two drugs for trial in this study is based largely on the suspected role of environmental neurotoxins and the potential role of oxidative mechanisms within dopaminergic cell that may induce cell death. This clinical trial will test the hypothesis that interference with oxidative mechanisms will slow or halt the progression of PD. The results of this study may alter future treatment of PD.

Another interesting departure from symptomatic pharmacologic treatment is the use of tissue transplants (adrenal medulla or fetal cell) in the treatment of PD. This is a very controversial topic at this time since it is unclear if the original success reported by the Madrazo group (1987) can be replicated. In addition, there remain many significant questions regarding this procedure. For example, is the adrenal medulla the optimum tissue to transplant or should fetal substantia nigra cell be used? If fetal cells are to be used what ethical considerations are to be considered? If fetal cells are utilized will patients have to be immunosuppressed? What evidence exists to demonstrate that the adrenal transplants survive and function? If tissue transplantation is shown to be successful in lessening the symptoms of PD, should patients who have early disease or late disease be subjected to this procedure? Would it be more advantageous for patients to undergo surgery first and then be treated with drugs or would the reverse order be most effective? What effects will surgery have on pre-existing complications of dopaminergic drugs such as dyskinesia and hallucinations? If tissue transplantation can alleviate parkinsonian symptoms, one of the most essential questions remains whether or not surgery is any better than currently available medical treatment. After all, if tissue transplantation is successful simply because dopamine is produced which then acts on dopamine receptors will there be any advantage over the oral administration of levodopa. On the other hand, if there are other benefits from tissue transplantation (e.g. the release of specific trophic factors) there might be additional reasons to believe that transplantation might have advantages in the long term care of PD patients. Obviously these questions require a major research effort to establish the efficacy and possible long term benefit of this procedure (Joynt and Gash, 1987).

REFERENCES

Ballard PA, Tetrud JW, Langston JW., 1985, Permanent human parkinsonism due to l-methyl-4-phenyl-1,2,3,6-tetrahydropyridine (MPTP): seven cases. Neurology 35:949-956.

Barbeau A., 1973, Aging and the extrapyramidal system. J AM Geriatrics Soc 21:145-149.

Barbeau A, Roy M, Cloutier T, Plasse L, Parls S., !986, Environmental and genetic factors in the etiology of Parkinson's disease. Adv Neurol 45:299-306.

Barbeau A, Roy M, Paris S, Cloutier T, Plasse L, Poirier J., 1985, Ecogenetics of Parkinson's disease: 4 hydroxylation of debrisoquin. Lancet 2:1213-1216.

Barbeau A and Roy M., 1984, Familial subsets in idiopathic Parkinson's disease. Can J Neurol Sci 11:144-150.

Bharucha NE, Stokes L, Schoenberg BS, Ward C, Ince S, Nutt JG, Eldridge R, Calne DB, Mantel N, Duvoisin RC., 1986, A case control study of twin pairs discordant for Parkinson's disease: a search for environmental risk factors. Neurology 36:284-288.

Bradbury AJ, Kelly ME, Costall B, Naylor RJ, Jenner P, Marsden CD., 1985, Benztropine inhibits toxicity of MPTP in mice. Lancet 1:1444-1445.

Brown RG and Marsden CD., 1984, How common is dementia in Parkinson's disease? Lancet 2:1262-1265.

Burns RS, Lewitt PA, Ebert MH, Pakkenberg H, Kopin IJ., 1985, The clinical syndrome of striatal dopamine deficiency: Parkinsonism induced by 1-methyl-4-phenyl-1,2,3,6-tetrahydropyridine (MPTP). N Engl J Med 312:1418-1421.

Burns RS, Chiueh CC, Markey SP, Ebert MH, Jacobowitz DM, Kopin JJ., 1983, A primate model of parkinsonism: selective destruction of dopaminergic neurons in the pars compacta of the substantia nigra by N-methyl-4-phenyl-1,2,3,6-tetrahydropyridine. Proc Natl Acad Sci USA 80:4546-4550.

Calne DB and Langston JW., 1983, Aetiology of Parkinson's disease. Lancet 2:1457-1459.

Cohen G., 1983, The pathobiology of Parkinson's disease: biochemical aspects of dopamine neuron senescence. J Neural Trans 19(suppl):89-103.

Cotzias GC, Papavasiliou PS, Gellene R., 1969, Modification of parkinsonism: chronic treatment with l-dopa. N Engl J Med 280:337-345.

Davis GC, Williams AC, Markey SP, Ebert MH, Caine ED, Reichert CM, Kopin IJ., 1979, Chronic parkinsonism secondary to intravenous injection of meperidine analogs. Psychiatry Res 1:249-254.

Duvoisin RC., 1986, On heredity, twins and Parkinson's disease. Ann Neurol 19:409-411.

Factor SA, Sanchez-Ramos JR, Weiner WJ., 1988, Parkinson's disease: an open label trial of pergolide in patients failing bromocriptine therapy. J Neurol Neurosurg Psychiatry 51:529-33.

Factor SA, Sanchez-Ramos JR, Ingenito AM, Weiner WJ., 1988a, Efficacy of Sinemet CR4 in subgroups of patients with Parkinson's disease. J. Neurol Neurosurg Psychiarty submitted.

Fahn S., 1982, Fluctuations of disability in Parkinson's disease: pathophysiologic aspects (Chap 8). In: Marsden CD and Fahn S, eds. Movement disorders. London: Butterworth Scientific pp123-145.

Forno LS, Langston JW, DeLanney LE, Irwin I, Ricaurte GA., 1986, Locus ceruleus lesions and eosinophilic inclusions in MPTP-treated monkeys. Ann Neurol 20:449-455.

Freeman BA and Crapo JD., 1982, Biology of disease: free radicals and tissue injury. Lab invest 47:412.

Geraghty JJ, Jankovic J, Zetusky WJ., 1985, Association between essential tremor and Parkinson's disease. Ann Neurol 17:329-333.

Glover V, Gibb C, Sandler M., 1986, Monoamine oxidase B (MAO-B) is the major catalylst for 1-methyl-4-phenyl-1,2,3,6-tetrahydropyridine (MPTP) oxidation in human brain and other tissues. Neurosci let 64:216-220.

Goetz CG, Lutge W, Tanner CM., 1986, Autonomic dysfunction in Parkinson's disease. Neurology 36:73-75.

Goetz CG, Tanner CM, Levy M, Wilson RS, Garron DC., 1986a, Pain in Parkinson's disease. Movement Disorders 1:45-49.

Halliwell B and Gutteridge JMC., 1985, Oxygen radicals and the nervous system. Trends In Neuroscience 8:22-26.

Hoehn MD and Yahr MM., 1967, Parkinsonism: onset, progression and mortality. Neurology 17:427-442.

Jankovic J and Reches A., 1986, Parkinson's disease in monozygotic twins. Ann Neurol 19:405-408.

Joynt RJ and Gash DM., 1987, Neural tranplants: are we ready ? Ann Neurol 22:455-456.

Kidron D and Melamed E., 1987, Forms of dystonia in patients with Parkinson's disease. Neurology 37:1009-1011.

Kish SJ, Morito C, Hornykiewicz 0,, 1985, Glutathione peroxidase activity in Parkinson's disease brain. Neuroscience letters 58:343-346.

Koller W, O'Hara R, Nutt J, Young J, Rubino F., 1986, Monozygotic twins with Parkinson's disease. Ann Neurol 19:402-405.

Koller W, O'Hara R, Weiner W, Lang A, Nutt J, Agid Y, Bunnet AM, Jankovic J., 1986, Relationship of aging to Parkinson's disease. Adv Neurol 45:317-321.

Koller WC., 1984, Sensory symptoms in Parkinson's disease. Neurology 34:957-959.

Kopin IJ., 1986, Toxins and Parkinson's disease: MPTP parkinsonism in humans and animals. Adv Neurol 45:137-144.

Kurlan R, Rubin AJ, Miller C, Rivera-Calimlim L, Clarke A, Shoulson I., 1986, Duodenal delivery of levodopa for on-off fluctuations in parkinsonism: preliminary observations. Ann Neurol 20:262-265.

Lang AE and Johnson K., 1987, Akathesia in idiopathic Parkinson's disease. Neurology 37:477-481.

Langston JW, Ballard P, Tetrud JW, Irwin I., 1983, Chronic parkinsonism in humans due to a product of meperidine synthesis, Science 219:979-980.

Langston JW and Ballard PA., 1983a, Parkinson's disease in a chemist working with 1-methyl-4-phenyl-1,2,3,6-tetrahydropyridine. N Engl J Med 309:310.

Leiberman AN and Goldstein M., 1985, Bromocriptine in Parkinson's disease. Pharmacologic Rev 37:217-227.

Madrazo I, Drucker-Colin R, Diaz V, Martinez-Mata J, Torres C, Becerril JJ., 1987, Open microsurgical autograft of adrenal medulla to right caudate nucleas in two patients with intactable Parkinson's disease. N Engl J Med 316:831-834.

Markham CH, Treciokas LJ, Diamond SG., 1974, Parkinson's disease and levodopa - a five year followup. West J Med 121:188-206.

Martilla RJ and Rinne UK., 1986, Clues from epidemiology of Parkinson's disease. Adv Neurol 45:285-288.

Mitchell IJ, Cross AJ, Sambrook MA, Crossman AR., 1985, Sites of neurotoxic action of 1-methyl-4-phenyl-1,2,3,6-tetrahydropyridine in the macque monkey include the ventral tegmental area and the locus coeruleus. Neuroscience letters 61:195-200.

Mjones H., 1949, Paralysis agitans: a clinical and genetical study. Acta Psychiatrica Scand Supplementum 54:1.

Nausieda PA, Weiner WJ, Klawans HL., 1980, Dystonic foot response of parkinsonism. Arch Neurol 37:132-136.

Nutt JG, Woodward WR, Hammerstad JP, Carter JT, Anderson JL., 1984, The "on-off" phenomenon in Parkinson's disease: relation to levodopa absorption and transport. N Engl J Med 310:483-488.

Nutt JG., 1987, On-off phenomenon: relationship to levodopa pharmacokinetics and pharmacodynamics. Ann Neurol 22:535-540.

Obeso JA, Luguin MR, Martinez Lage JM., 1986, Intravenous lisuride corrects oscillations of motor performance. Ann Neurol 19:31-35.

Parkinson J., 1817, Essay on the shaking palsy. London: Whittingham and Rowland, for Sherwood, Neely and Jones.

Perry TL and Yong VW., 1986, Idiopathic Parkinson's disease, progressive supranuclear palsy and glutathione metabolism in substantia nigra of patients. Neuroscience letters 67:269-274.

Perry TL, Godin DV, Hansen S., 1982, Parkinson's disease: a disorder due to nigral glutathione deficiency ? Neuroscience letters 33:305-310.

Pincus JH and Barry KB., 1987, Influence of dietary protein on motor fluctuations in Parkinson's disease. Arch Neurol 44:270-272.

Quinn NP, Lang AE, Koller WC, Marsden CD., 1986, Painful Parkinson's disease. Lancet 1:1366-1369.

Rajput AH., 1984, Etiology of Parkinson's disease: environmental factor(s). Neurology 34(suppl 1):207.

Rajput AH, Uitti RJ, Stern W, Laverty W., 1986, Early onset Parkinson's disease and childhood environment. Adv Neurol 45:295297.

Reznikoff G, Manaker S, Parsons B, Rhodes CH, Rainbow TC., 1985, Similar distribution of monoamine oxidase (MAO) and parkinsonian toxin (MPTP) binding sites in human brain. Neurology 35:1415-1419.

Rinne UK., 1985, Combined bromocriptine - levodopa therapy early in Parkinson's disease. Neurology 35:1196-1198.

Sanchez-Ramos JR, Michel P, Weiner WJ, Hefti F., 1988, Selective destruction of cultured dopaminergic neurons from embryonic rat mesencephalon: cytochemical and morphological evidence. J Neurochem in press.

Sanchez-Ramos JR, Hollinden CE, Sick TJ, Rosenthal M., 1988a, l-methyl-4-phenylpyridinium (MPP+) increases oxidation of cytochrome b in rat striatal slices. Brain Res 443:183-189.

Sanchez-Ramos J, Barret JN, Goldstein M, Weiner WJ, Hefti F., 1986, l-methyl-4-phenylpyridinium (MPP+) but not l-methyl-4-phenyl-1,2,3,6-tetrahydropyridine (MPTP) selectively destroys dopaminergic neurons in cultures of dissociated rat mesencephalic neurons. Neuroscience letters 72:215-220.

Santamaria J, Tolosa E, Valles A., 1986, Parkinson's disease with depression: a possible subgroup of idiopathic parkinsonism. Neurology 36:1130-1133.

Schwab RS., 1960, Progression and prognosis of Parkinson's disease. J nervous & Mental Disease 130:556-566.

Schwarcz R and Meldrum B,, 1985, Excitatory aminoacid antagonists provide a therapeutic approach to neurological disorders. Lancet 2:140-143.

Snyder SH and D'Amato RJ., 1986, MPTP: a neurotoxin relevant to the pathophysiology of Parkinsin's disease. Neurology 36:250-258.

Spencer PS, Nunn PB, Hugon J, Ludolph AC, Ross SM, Roy DN, Robertson RC., 1987, Guam amyotrophic lateral sclerosis-parkinsonism-dementia linked to plant excitant neurotoxin. Science 237:517-522.

Stern G., 1987, Prognosis in Parkinson's disease (Chap 7). In: Marsden CD and Fahn S eds. Movement disorders 2. Butterworth. London: pp 91-98.

Stoessl AJ, Mak E, Calne BD., 1985, (+)-4-propyl-9-hydroxynaphthoxazine (PHNO). a new dopaminemimetic, in treatment of parkinsonism. Lancet 2:1330-1331.

Sweet RD and McDowell FH., 1975, Five years treatment of Parkinson's disease with levodopa. Ann Int Med 83:456-463.

Uhl GR, Javitch JA, Snyder SH., 1985, Normal MPTP binding in parkinsonian substantia nigra: evidence for extraneuronal toxin conversion in human brain. Lancet 1:956-957.

Vyas I, Heikkila RE, Nicklas WJ., 1986, Studies on the neurotoxicity of l-methyl-4-phenyl-1,2,3,6-tetrahydropyridine: inhibition of NAD-linked substrate oxidation by its metabolite, l-methyl-4-phenylpyridinium. J Neurochem 46:1501-1507.

Ward CD, Duvoisin RC, Ince SE, Nutt JD, Eldridge R, Calne DB., 1983, Parkinson's disease in 65 pairs of twins and in a set of quadruplets. Neurology 33:815-824.

Ward CD, Hess WA, Calne DB., 1983a, Olfactory impairment in Parkinson's disease. Neurology 33:943-946.

Weiner WJ, Factor SA, Sanchez-Ramos JR., 1988, A comparison of bromocriptine and levodopa/carbidopa both alone and in combination in de novo patients with Parkinson's disease. Arch Neurol 45:207.

Weiner WJ, Factor SA, Sanchez-Ramos JR., 1987, The efficacy of (+)-4-propyl-9-hydroxynaphthoxazine (PHNO) as adjunctive therapy in Parkinson's disease. Neurology 37 (Suppl 1):276,1987.

Weiner WJ, Factor SA, Sanchez-Ramos JR, Berger J., 1987, A double blind evaluation of Ciladopa in Parkinson's disease. Movement Disorders 2:211-217.

Weiner WJ and Nausieda PA., 1982, Meiges syndrome during long term dopaminergic therapy in Parkinson's disease. Arch Neurol 39:451-452.

Yahr MD., 1975, Levodopa. Ann Int Med 83:677-682.

Zetusky WJ, Jankovic J, Pirozzolo FJ., 1985, The heterogeneity of Parkinson's disease: clinical and prognostic implications. Neurology 35:522-526.

2

THE NEUROPATHOLOGY OF PARKINSON'S DISEASE
(The Lewy body as a clue to the nerve cell degeneration)

Lysia S. Forno

From the Departments of Pathology, Veterans Administration
Medical Center, Palo Alto, CA 94304 and Stanford University
School of Medicine, Stanford, CA 94305

INTRODUCTION

The etiology and pathogenesis of Parkinson's disease have remained a mystery since the first description of the disease, but the neuropathology of the disorder, as we now know it, is fairly simple and straightforward. The most important lesion is in the substantia nigra pars compacta, where nerve cells degenerate together with their nigro-striatal fiber tract, causing a severe depletion of dopamine in the basal ganglia.

A very similar nerve cell degeneration takes place in the noradrenergic locus ceruleus, located in the dorso-lateral rostral tegmental pons. Nerve cell degeneration in both locations are characteristically accompanied by the formation of eosinophilic intraneuronal inclusions, called Lewy bodies. It is generally accepted that neuropathology is present also outside the substantia nigra and locus ceruleus, in highly selective areas and manifested by the presence of Lewy body formation, with or without conspicuous nerve cell loss. The sites of predilection are the dorsal motor nucleus of the vagus, the innominate substance with the nucleus basalis of Meynert and certain portions of the hypothalamus, all areas where Lewy (1912, 1923) first described them. Other common sites include the serotonergic raphe nuclei, especially the dorsal raphe nucleus and the superior central nucleus, the sympathetic ganglia, and the amygdala and cerebral cortex, with limbic structures more susceptible than neocortical regions. The spinal cord also may display Lewy body inclusions, not only in autonomic nuclei, but also in the anterior and posterior horns. Other locations such as neostriatum and globus pallidus, dentate fascia and inferior olives are much less common, and some locations appear totally spared, for example the lateral geniculate body, Betz cells, and perhaps Purkinje cells.

From this it will already be obvious that, if Lewy bodies are an indication, Parkinson's disease is not a disease of one or a few neurotransmitter systems. What then is it? This is a question that can not be answered at the present time, but in the following some of the available information will be reviewed and assessed.

Because the most important question to be asked concerns the nature of the nerve cell degeneration, and the only available clue to this is the presence of the Lewy body inclusions, these peculiar structures will be dealt with in some detail. First, however, a couple of other concerns regarding the neuropathology will be considered.

PARKINSON'S DISEASE: ONE DISEASE OR SEVERAL?

Until proven otherwise, a clinical syndrome with a consistent pathological substrate from one case to another, can justifiably be classified as one disease. A good example of this is the use of the designation "Alzheimer's disease" for both presenile and senile variants of this disease. In the same way the neuropathologist may elect to reserve the term Parkinson's disease or idiopathic parkinsonism for cases with clinical parkinsonism during life and the neuropathological picture just outlined, with the substantia nigra degeneration and the Lewy bodies as the cardinal features. For the clinician, who deals with the living patient, diagnosis may be far more difficult, and the possibility that the patient may be suffering from another form of parkinsonism, can not be dismissed.

In the definition of Parkinson's disease used here, the presence of Lewy bodies has been included as one important criterion, raising the question: Does Parkinson's disease without Lewy bodies exist? This question can not be answered, if the presence of the Lewy body phenomenon is included among the criteria for diagnosis. There is of course no way of knowing, whether or not the patients that Parkinson described, fulfilled these criteria. Perhaps one might look for cases that in all other respects fulfill clinical and neuropathological criteria for the disease. To some degree this is being done, when for example Bernheimer et al. (1973) found Lewy bodies in 36 out of 39 patients with idiopathic parkinsonism.

The question is not a trivial one, since the Lewy body at present is our only structural clue to the nerve cell degeneration in Parkinson's disease. If we choose to ignore this inclusion and to regard the many other conditions associated with parkinsonism as different parts of a spectrum, we loose the only guiding light available to us. Therefore, while it is possible that Parkinson's disease without Lewy bodies exists, such cases will not help us in understanding Parkinson,s disease, so for practical reasons they will be excluded from further discussion in this review. One final comment about Lewy bodies: It is very likely that they only represent the morphological expression of some special derangement in the biochemistry or the molecular events in nerve cells, that can occur under several different circumstances, but since this particular form of nerve cell degeneration is highly characteristic for Parkinson's disease, this does not detract from their usefulness as a target for investigation.

Viewing Parkinson's disease as one clinico-pathological entity doe not rule out that several variants or a spectrum of the disease process may exist. In most Veterans Administration Medical Centers, for example in the Palo Alto VA Medical Center, the most commonly encountered parkinsonian patient is elderly and often demented, and the neuropathological case material reflects this. Approximately 3 out of 4 brains from patients that fulfill the neuropathological criteria for Parkinson's disease are from such elderly demented patients. Many of them also fulfill the criteria for the diagnosis of Alzheimer's disease. Others have a variable number of Alzheimer type senile changes, and some can be classified as having Lewy Body Dementia (Kosaka et al., 1984). In these elderly patients, with late onset of parkinsonism, it is difficult to establish the main cause of the dementia. Nerve cell loss in the nucleus basalis of Meynert is often severe, and cortical Lewy bodies, especially in limbic structures, are common. It is certainly possible that the Lewy body-related cortical nerve cell degeneration may play a part in the mental deterioration. The parkinsonism in these patients is often mild, and also of brief duration, because of the old age and the concurrent existence of other somatic disease processes.

The other variant of Parkinson's disease, seen in approximately one fourth of our Parkinson material, is actually the classical Parkinson's disease, with onset, most often in the 50'es, long duration and typical symptoms, that often are severe and have required treatment with levodopa. Before levodopa, these patients may have had stereotaxic surgery. Mentally, the majority are well preserved. The neuropathology differs only slightly between the two groups. If the substantia nigra degeneration is severe, Lewy bodies may be relatively infrequent here, but can sometimes be found in much greater numbers in extranigral sites. Cortical Lewy bodies, especially Lewy bodies in the amygdala-parahippocampal region, are perhaps less common. Other variants of Parkinson's disease such as young onset parkinsonism (Quinn et al., 1987) and juvenile Parkinson's disease (Yokochi et al., 1984) have not been represented in our case material.

The impression gained is that Parkinson's disease and Alzheimer's disease exist as two quite separate diseases when the onset is before age 65. After that age the two diseases appear to gradually blend, so that many patients with Alzheimer's disease also have evidence neuropathologically and often also clinically of Parkinson's disease, and patients with Parkinson's disease have variable degrees of Alzheimer changes. This overlap between the two diseases could be due to increased susceptibility to either one with age, but it is also possible that each disease process might in some way predispose towards the other. The phenomenon must be age-related, since the same overlap rarely is seen in the classical presenile variants of the two diseases.

NEUROPATHOLOGY. THE MAIN FEATURES

The paucity of abnormalities present on gross examination of the brain is striking. Unless there is coexistent Alzheimer's disease, the weight of the brain is not decreased (Jellinger and Riederer, 1984). In demented patients with Lewy Body Disease, a condition that probably represents one part of the spectrum of Parkinson's disease variants, one may be tipped off to the possibility of this disorder by the normal weight and normal general external appearance of the brain. A varying degree of pallor of the substantia nigra and often also of the locus ceruleus is usually the only abnormality to be found before cell groups, with ventral cell groups more severely involved than dorsal ones. The best documentation of this distribution to date remains the studies by Rolf Hassler (1938). He demonstrated the marked difference in distribution of lesions between Parkinson's disease and postencephalitic parkinsonism with the more severe lesions in all cell groups in the latter condition. Likewise, other conditions with parkinsonism, such as Striatonigral Degeneration and Progressive Supranuclear Palsy, often have much more marked involvement of the substantia nigra reticulata and the ventral tegmental area. The latter area is not completely spared in Parkinson's disease, and Lewy bodies can be found here. It has recently been suggested (Sima et al., 1886) that the ventral tegmental area is markedly involved in Lewy Body Dementia, so the distribution of the lesions can not be used as an absolute criterion for diagnosis of Parkinson's disease. Also, when substantia nigra degeneration is very severe, the nerve cell loss can be expected to be more diffuse. The focal nerve cell loss is often accompanied by distinct, although rather delicate glial scars (Hallervorden, 1957), as if the insult to the cell group in question had been especially severe, or as if damage to more than just the nerve cells in the area, had taken place. Since the disease is progressive, there is usually evidence of active nerve cell degeneration, and because the neuromelanin in the nerve cells is not easily digested, the nerve cell degeneration is highlighted by the presence of increased neuromelanin granules in the neuropil. Macrophages are quick to take up this pigment and may form clusters that mimic the shape of the nerve cell.

The Locus Ceruleus

Although it is not clear what role the disease process in the locus ceruleus plays in Parkinson's disease and it does not rival the substantia nigra in importance for Parkinson's disease, it is nevertheless affected in some way in nearly all typical cases. It has a high prevalence for Lewy bodies, which in preclinical cases may be seen here, when they are not present in the substantia nigra. On the other hand, Lewy bodies may be absent in the locus ceruleus in a few typical cases of Parkinson's disease. The changes in the locus ceruleus are essentially the same as in the substantia nigra, including the inclusion bodies and the increase of extraneuronal neuromelanin pigment. With marked nerve cell loss there may be some fibrillary gliosis in the nucleus, but it does not present as a focal scar, and all portions of the nucleus appear almost equally involved. If the caudal locus ceruleus appears to be more severely affected, this may be attributed to the fact that nerve cells are more closely packed and more numerous in the caudal half of the nucleus, making it an easier portion of the nucleus to evaluate. Because the locus ceruleus is much more prone to develop neurofibrillary tangles with age than the substantia nigra, it is not uncommon to find both Lewy bodies and neurofibrillary tangles in this nucleus, especially in elderly individuals. Nerve cell loss is usually more severe in the presenile variant of Alzheimer's disease than in Parkinson's disease. It is therefore not surprising to find that in cases with combination of Parkinson's disease or Lewy body dementia, the nerve cell loss may become extreme. Only a few nerve cells may remain, and on gross examination, the locus ceruleus is practically invisible. Reliable studies of actual nerve cell loss in the classical variant of Parkinson's disease would be of some interest, but are, as far as I am aware, sadly lacking.

Other extranigral sites

The main locations for the disease process in Parkinson's disease have already been listed, and for reasons of space, and because so little is known about the importance of these sites or why they should be preferentially affected, they will not be further discussed here, except to say that the involvement of the nucleus basalis of Meynert may well play a role in the dementia, and that the question of the cortical Lewy bodies, especially those in the amygdala-parahippocampal region also deserves further investigation (Forno and Langston, 1988).

THE LEWY BODIES

General remarks and staining reactions

Whenever characteristic alterations can be found in nerve cells, it is a reasonable assumption that such changes are the result of molecular and biochemical events that have taken place in the cell. If, as in Parkinson's disease, this abnormality is a well-defined structure, such as the Lewy body, it opens up possibilities for study of the nerve cell degeneration that are not available in diseases that undergo "simple atrophy". Therefore the emphasis on the Lewy bodies in this chapter.

The general appearance of the Lewy bodies as eosinophilic, sometimes concentrically laminated, often multiple, globular inclusions with a central dense core and a paler periphery in the cytoplasm of nerve cells, is well known and have been described repeatedly by many authors (see Greenfield and Bosanquet, 1953; Bethlem and den Hartog Jager, 1960; Forno, 1987) (Figure 1 A). Perhaps because of the pattern of immunoreaction with antibodies to neurofilaments (see later) three different zones have been described: a dense inner zone, a less dense adjacent band, and a peripheral halo, but many Lewy bodies consist only of one eosinophilic body, with variable intensity of staining. Frequently, pale, fairly homogenous areas are seen in nerve cells adjacent to Lewy bodies (Figure 1 B). These are probably early stages in Lewy body formation ("pre-Lewy bodies"), but should not be diagnosed as Lewy bodies, since they occasionally can be found in cases without fully developed Lewy bodies. Of the staining reactions commonly used in neuropathology the routine hematoxylin-eosin stain is probably the most satisfactory for demonstration of Lewy bodies, but any acidophilic stain will demonstrate the inclusion. The periodic acid Schiff reaction (PAS) is negative, since the Lewy bodies do not contain carbohydrates. Lewy bodies also have much less affinity for silver stains than the markedly argyrophilic Pick bodies, but the staining can vary from unstained over brown to black (Tiller-Borcich and Forno, 1988. The Lewy bodies are composed mainly of protein (Bethlem and den Hartog Jager, 1960), and from ultrastructural studies, first performed by Duffy and Tennyson over 20 years ago (1965), as well as from later immunocytochemical studies, to be discussed shortly, it has been clear that abnormalities in cytoskeletal elements are largely responsible for the appearance of the Lewy bodies. If it can be established what cytoskeletal ele-

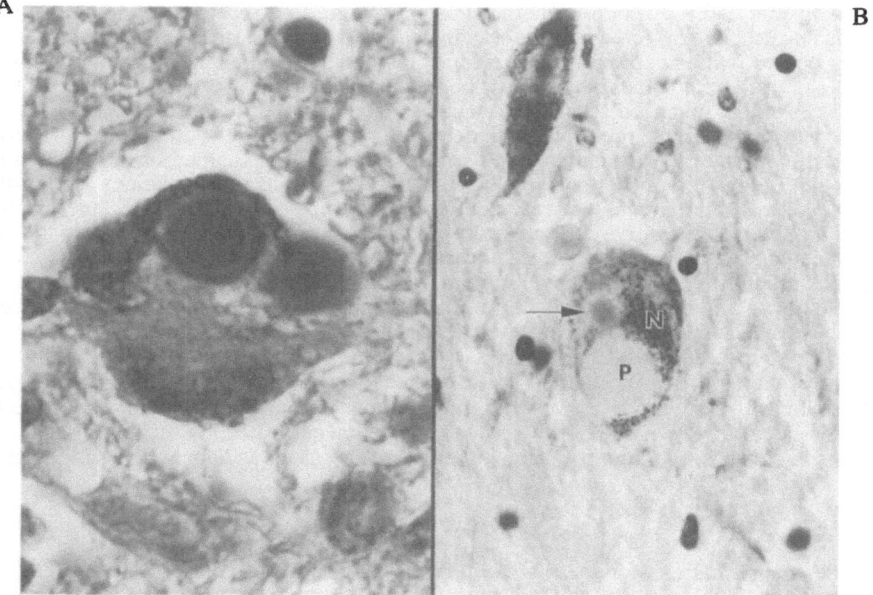

Figure 1 A. Lewy body in pigmented nerve cell in the locus ceruleus. Note the central core and the concentric lamination. Hematoxylin-eosin. X 1480.

Figure 1 B. Pigmented nerve cell in the substantia nigra with a small Lewy body (arrow) and a pale, depigmented area, the pre-Lewy body (P). N = neuromelanin. Hematoxylin-eosin. X 550

ments (normal or abnormal) go into the formation of the inclusions, we might be on our way to a better understanding of the nature of the nerve cell degeneration. Two methods, electron microscopy and immunocytochemistry, with a combination of the two in the form of immuno-electron microscopy, have been employed in trying to get to the root of this problem. As will be seen, results from different laboratories so far have not been in complete agreement. It has also been pointed out (Forno, 1987) that because of the variable composition and appearance of Lewy body inclusions in different locations in the nervous system, the task of unravelling the pathogenesis of these peculiar structures is particularly difficult. A structure with a much simpler and more consistent appearance, the paired helical filament (PHF) of the neurofibrillary tangle, has not as yet yielded conclusive results in spite of the many investigators who have been working on it. Here the ultrastructure of the classical Lewy bodies in the substantia nigra and locus ceruleus will first be considered and then the immunocytochemical studies will be reviewed.

First, a few remarks will be addressed to the question of the less typical Lewy bodies (see den Hartog Jager and Bethlem, 1960, and Forno, 1982 and 1987). It must be remembered that the inclusions that Lewy described (1912, 1923) were not what we now regard as classical Lewy bodies, nor were they found in the substantia nigra and locus ceruleus. In these extranigral sites, Lewy bodies often have bizarre shapes and are situated in nerve cell processes. All forms of Lewy bodies do, however, appear to be closely related, and usually both intraneuritic and perikaryal Lewy bodies tend to be present In the same anatomical location. As a general rule perikaryal Lewy bodies are the more common form in the substantia nigra, locus ceruleus, raphe nuclei, amygdala and cerebral cortex. Bizarre forms, often in nerve cell processes, sometimes brilliantly red and with concentric laminations, are particularly common in the nucleus basalis of Meynert, the hypothalamus (especially the tuberomammillary nucleus), the dorsal motor nucleus of the vagus (the three locations where Lewy described them), and the sympathetic ganglia.

The perikaryal Lewy bodies in the amygdala and cerebral cortex have other differences from the classical Lewy bodies, in that they more often have a less pronounced central core. Often they are homogenous pink swellings that may fill the cytoplasm and displace the nucleus.

The Lewy bodies we will consider here are the classical Lewy bodies in the substantia nigra and locus ceruleus. Because of their characteristic and consistent appearance, they are more likely to yield valuable information about their cytoskeletal composition and the stages in their formation.

Ultrastructure of classical Lewy bodies

The examination of the Lewy bodies by the aid of the electron microscopy might seem to be the obvious and clearest way to find out what elements these inclusions consist of, but as for the neurofibrillary tangles results have not been unequivocal. Duffy and Tennyson (1965) described the typical dense core and radiating filaments, surrounded by neuromelanin granules, as well as the lack of a confining membrane. In the center of the inclusion linear structures or ringformed elements of identical width (7-8 nm) could be found in some of the Lewy bodies. In others the core was extremely electron dense and only a slight granularity could be discerned. The radiating filaments were more widely spaced and less regular, with diameters varying in width from 7.5 to 20 nm. This description is being quoted in some detail, because it still is the most accurate description published of the Lewy bodies. Studies from our own laboratory are in complete agreement with those by Duffy and Tennyson (Figure 2).

These authors considered the possibility that neurofilaments might contribute to the formation of Lewy bodies, but noted that the arrangement and perhaps the nature of the filaments were different from that seen in the neurofibrillary tangle, which when that was written was thought to be made up of neurofilaments. Duffy and Tennyson also mentioned the presence of circular profiles and their fusion into a granular mass as distinctive features, representing an alteration of neurofilaments.

Because of the morphological similarity to neurofilaments, and the immunocytochemical reaction with antibodies to neurofilaments (see later), a close relationship to neurofilaments has been suspected, but so far definite proof is lacking. Neurofilamentous hyperplasia is a fairly common occurrence and can be seen in experimentally induced neurofibrillary tangles, for example after exposure to aluminum salts, as well as in axonal spheroids in Amyotrophic Lateral Sclerosis (see Hirano, 1985). These neurofilamentous accumulations do not assume the characteristic sunflower formation seen in the Lewy body. Also, since the diameter of the Lewy body filaments shows more variation than normal neurofilaments, and normal neurofila-

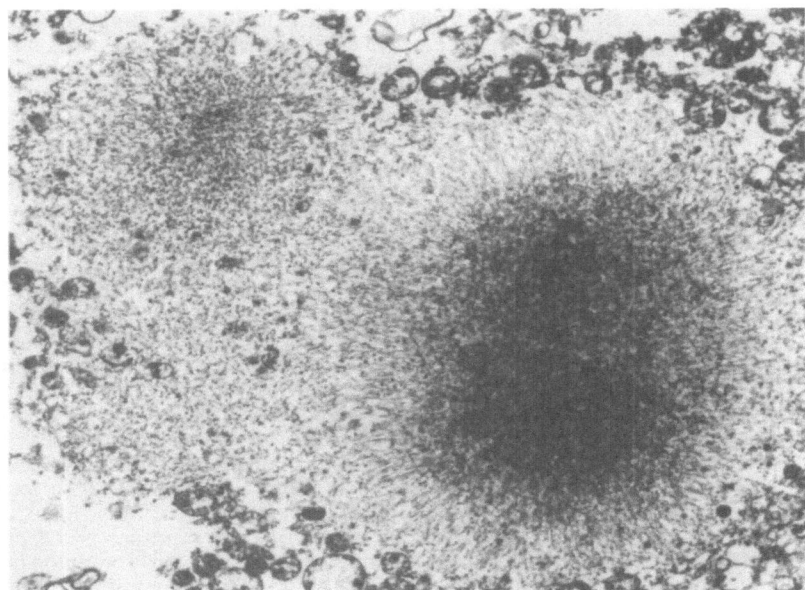

Figure 2. Electron micrograph of a Lewy body in the substantia nigra. Note the dense core and the radiating fila-
    ments. A smaller Lewy body, perhaps sectioned through its peripheral portion,can be seen in the upper left
    corner. Uranyl acetate and lead citrate. X 7,000.

ments occasionally can be seen to be displaced to the periphery in Lewy bodies in nerve cell
processes (Forno, 1982), one must assume that if neurofilaments are involved in formation of
Lewy bodies, it is in the form of altered neurofilaments or filaments formed from protofila-
ments. For the PHF of the neurofibrillary tangles, it has been suggested (Wischik and
Crowther, 1986) that they are formed by de novo assembly. This could also be the case for the filaments in the Lewy body.

    In the neuropathology laboratory in the Palo Alto VA Medical Center we have recently
observed that the filaments of Lewy bodies tend to have fuzzy deposits along their course. In
most cases these deposits look more granular than the sidearms seen on neurofilaments (Figure 3).
Our attention was drawn to this characteristic by Munoz-Garcia and Ludwin (1984), when they
described similar deposits on filaments in inclusion bodies in the generalized variant of Pick's

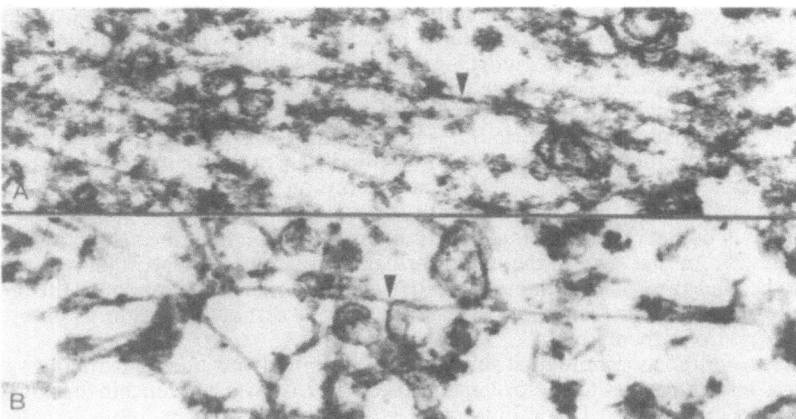

Figure 3 A. Periphery of a Lewy body in the substantia nigra to show the individual filaments (arrowhead) with
    fuzzy deposits along their course. X 87,000. Compare with Figure 3 B.

Figure 3 B. Filament (arrowhead) from a Pick body in the temporal cortex in Pick's disease. The contour of the
    filament is fairly smooth, without sidearms or fuzzy deposits. X 87,000.

disease. In contrast the filaments of classical Pick's disease had a smooth contour (Figure 3 B). The authors mentioned that the inclusions with the fuzzy filaments resembled those seen in filaments in cortical Lewy-like bodies. We have recently confirmed this in a case where some difficulty existed in separating Pick bodies from Lewy bodies in the temporal cortex (Tiller-Borcich and Forno, in press). Review of Lewy bodies in the substantia nigra and locus ceruleus (unpublished observations) have revealed that similar granular deposits on filaments in classical Lewy bodies can be observed (Figure 3 A), the main difference being that the deposits tend to be coarser, and the diameter of the filaments more difficult to measure, in the cortical Lewy bodies than in those of the brainstem.

To sum up the ultrastructural characteristics of classical Lewy bodies, they consist of: a core of variable density, sometimes with circular elements; filaments of 8 to 20 nm diameter radiating out from the core; fuzzy deposits of electron dense material on the filaments; lack of a limiting membrane; and neuromelanin granules in the periphery. The location is in the perikaryon, and mitochondria and other organelles, but not Nissl substance, can be found within the Lewy body. Pale areas devoid of Nissl substance and usually also of neuromelanin can be found in the cytoplasm, probably representing the pre-Lewy body seen by light microscopy.

Immunocytochemical studies

Goldman and colleagues (1983) were the first to observe that antibodies to neurofilaments had antigenic sites in common with Lewy bodies. They used four different polyclonal antibodies, which did not react with neurofibrillary tangles. Several later studies have confirmed that Lewy bodies do indeed in many instances share epitopes with Lewy bodies (Forno et al., 1986; Sima et al., 1986; Kahn et al., 1985, 1988; Galloway et al., 1987). In most of these studies, monoclonal antibodies were used, and it was found that antibodies against phosphorylated neurofilaments gave the strongest and most consistent reaction (Figure 4) (Forno et al., 1986). Since phosphorylated neurofilaments are mainly present in axons and not in perikarya or dendrites, this led to the speculation that abnormal posttranslational phosphorylation of proteins in the Lewy bodies had occurred. a situation already described for neurofibrillary tangles (Sternberger et al., 1985; Cork et al., 1986). The reaction of the Lewy bodies with antibodies to neurofilaments has mainly been directed against the 200 kDa portion of the neurofilament triplet. Some of the antibodies that reacted with Lewy bodies also immunostained neurofibrillary tangles, whereas others did not, indicating that not all antibodies were directed against the same antigenic sites.

A puzzling feature of the antibody reaction in the Lewy bodies was the presence of immunostaining mainly in a peripheral band that left the most peripheral halo unstained. Although a lack of staining of the core might be explained either on the basis of such severe alterations that the antigen was lost or because of difficult access to this portion of the Lewy body, this explanation falls short for the peripheral halo, and this reaction pattern is still a mystery. Pappola (1986) demonstrated a weak immunoreaction in both core and periphery of the Lewy body by immuno-electron microscopy, using neurofilament antibody. Interestingly, cortical

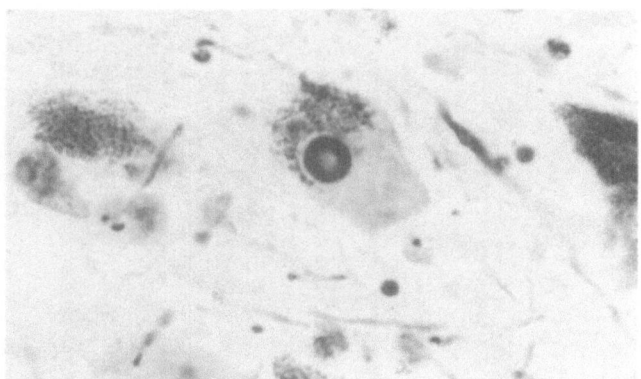

Figure 4. Lewy body in the substantia nigra, reacted with Sternberger 07-5 monoclonal antibody to phosphorylated neurofilaments. Only the most central portion of the Lewy body and a narrow peripheral rim remain unstained. Positive immunostaining of axons can be seen in the neuropil. Hematoxylin counter stain. X 550.

Lewy bodies often show a more intense staining of the center of the Lewy body than of the periphery.

Immunocytochemical studies using antibodies to other neuroskeletal elements have now been performed. Galloway and associates (1987) performed a battery of immuno-reactions and found some positive reaction with antibodies against neurofilaments, against tubulin or micro-tubules, and with antibodies against neurofibrillary tangles, but not against the microtubule-associated protein tau. Rasool and Selkoe (1985) found that their antibodies against neurofibrillary tangles reacted with neurofibrillary tangles and Pick bodies, but not with Lewy bodies. In our own laboratory, antibody against alpha tubulin (Miles Laboratory) does not react with Lewy bodies, but gives an intense staining reaction with neurofibrillary tangles and especially with Pick bodies (unpublished observations).

It seems as if Lewy bodies can be made to react with most of the antibodies to cytoskeletal proteins tried except to actin (Goldman et al., 1983; Kahn et al., 1987) and to tau. Since tau antibody decorates neurofibrillary tangles heavily, this is one of the clear differences in antigenicity between the neurofibrillary tangle and the Lewy body. Although it is possible that by using different modifications of fixation and preparation of the tissue, a positive reaction of Lewy bodies with the antibody against tau might be obtained, a comparison of reactions when tissues are treated in the same way, would seem permissible and valid.

Most interesting is the recent finding (Kuzuhara et al., 1988), that Lewy bodies also react with ubiquitin. Ubiquitin is a 76 amino acid protein, highly conserved during evolution, that is involved in ATP-dependent proteolysis (see Rechsteiner, 1987 for review). It has been found to be an important component of PHF (Mori et al., 1987), and it is possible that the positive reaction of Lewy bodies with antibodies to neurofibrillary tangles in some laboratories may be due to the fact that these antibodies recognize ubiquitin. In cytoplasmic proteolysis, ubiquitin appears to be conjugated with short-lived proteins that are earmarked for destruction.

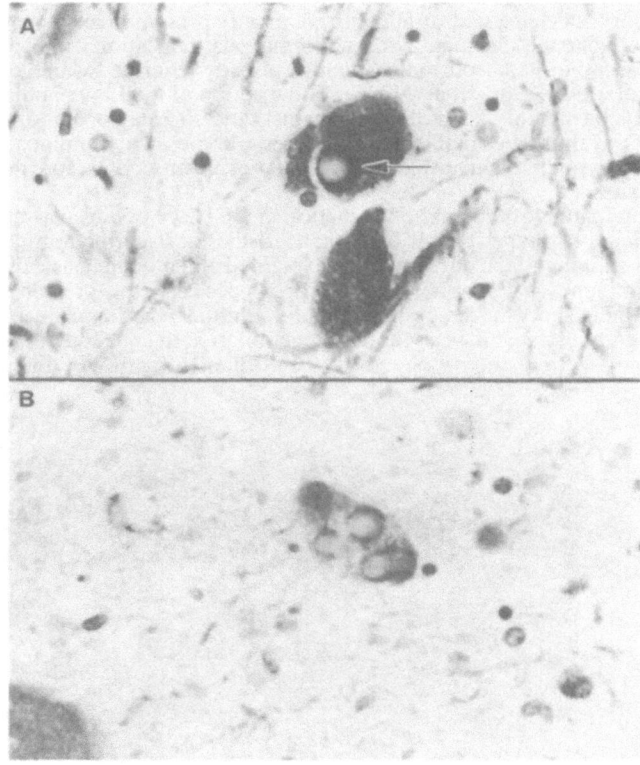

Fig. 5 A and B. Examples of staining of Lewy bodies in the substantia nigra (A) and locus ceruleus (B) with antibody to tyrosine hydroxylase (Eugene Tech). The reaction is present in the periphery of a single Lewy body (arrow) in A, and surrounds 3 Lewy bodies in B. Note the vigorous reaction in the cytoplasm of the inclusion-bearing Lewy body in A. Hematoxylin counterstain. X 550.

Obviously, neither PHF or Lewy body filaments are broken down as a result of their ubiquitination, and the positive reaction with ubiquitin for the Lewy bodies might be related to other functions of ubiquitin. At present it is not clear, if ubiquitin becomes attached to Lewy bodies because they contain abnormal or altered proteins. An intriguing aspect of the reaction of Lewy bodies with ubiquitin, is that the staining is located as a peripheral ring, reminiscent of the reaction of Lewy bodies with antibodies to neurofilaments.

This peripheral reaction is found again, when Lewy bodies are reacted with an antibody to tyrosine hydroxylase (Figure 5). Tyrosine hydroxylase antibody staining of Lewy bodies was first demonstrated by Nakashima and Ikuta (1984) in the locus ceruleus and substantia nigra, but not in Lewy bodies in non-catecholaminergic neurons. The inclusion-bearing nerve cells in the substantia nigra and locus ceruleus show a normal reaction for tyrosine hydroxylase. Since tyrosine hydroxylase protein easily aggregates and becomes insoluble in vitro, according to Nagatsu (cited by Nakashima and Ikuta, 1984), this does not necessarily mean that these nerve cells function normally.

The only attempt to produce antibodies raised against Lewy bodies has come from Dr. Agid's laboratory (Hirsch et al. (1985). They used the hybridoma technique to obtain monoclonal antibodies against the inclusions. Their two antibodies, which have no relationship to neurofilament antibodies, react with Lewy bodies, but also against antibodies normally present in the substantia nigra. This suggests that the Lewy bodies may contain normal constituents of the cell, but does not as yet clarify the nature of the Lewy bodies or their constituent of filaments. Because Lewy bodies are relatively scarce even in the substantia nigra, compared with the large number of neurofibrillary tangles than can be found in the hippocampal-parahippocampal region, the production of antibodies to Lewy bodies is not an easy task, but further developments from Dr. Agid's laboratory can probably be expected.

CONCLUDING REMARKS

As we have seen a considerable amount of information about the structure and composition of Lewy bodies have begun to be accumulated, but a clear picture has not as yet emerged. When it does, it is the hope that it will also illuminate some of the mysteries surrounding the nerve cell degeneration in Parkinson's disease.

ACKNOWLEDGMENTS

This review is based on studies supported by the Veterans Administration Medical Research Program. I thank Lawrence F. Eng, Ph.D. for help and advice with the immunocytochemical studies, and Ruth Grajcer, Roxana Norville and Chi Yin Choy for technical assistance.

REFERENCES

Bernheimer H, Birkmayer W, Hornykiewicz O, Jellinger K, Seitelberger F. Brain dopamine and the syndromes of Parkinson and Huntington. Clinical, morphological and neurochemical correlations. J Neurol Sci 1973; 20:415-455

Bethlem J, Jager WAdH. The incidence and characteristics of Lewy bodies in idiopathic paralysis agitans (Parkinson's disease). J Neurol Neurosurg Psychiatry 1960; 23:74-80

Cork LC, Sternberger NH, Sternberger LA, Casanova MF, Struble RG, Price DL. Phosphorylated neurofilament antigens in neurofibrillary tangles in Alzheimer's disease. J Neuropathol Exp Neurol 1986; 45:56-64

Duffy PE, Tennyson VM. Phase and electron microscopic observations of Lewy bodies and melanin granules in the substantia nigra and locus caeruleus in Parkinson's disease. J Neuropathol Exp Neurol 1965; 24:398-414

Forno LS. Pathology of Parkinson's disease. In: Marsden CD, Fahn S, eds. Movement Disorders. London, Butterworth Scientific. 1982:25-40

Forno LS. The Lewy body In Parkinson's disease. In: Yahr MD, Bergmann KJ, eds. Advances in Neurology, New York, Raven Press 1987; 45:35-43

Forno LS, Langston JW. The amygdala-parahippocampal region. A predilection site for Lewy bodies. J Neuropathol Exp Neurol 1988 (abstract; in press)

Forno LS, Sternberger LA, Sternberger NH, Strefling AM, Swanson K, Eng LF. Reaction of Lewy bodies with antibodies to phosphorylated non-phosphorylated neurofilaments. Neurosci Lett 1986; 64:253-258

Galloway P, Grundke-Iqbal I, Autilio-Gambetti L, Gambetti P, Perry G. Neuronal cytoskeletal involvement in Lewy body formation. J Neuropathol Exp Neurol 1987; 46:374

Goldman JE, Yen S-H, Chiu FC, Peress NS. Lewy bodies of Parkinson's disease contain neurofilament antigen. Science 1983; 221:1082-1084

Greenfield JG, Bosanquet FD. The brain-stem lesions in parkinsonism. J Neurol Neurosurg Psychiatry 1953; 16:213-226

Hallervorden J. Paralysis agitans (Anhang: Essentieller Tremor). In: Lubarsch O, Henke F, Rossle, eds. Handbuch der speziellen pathologischen Anatomie und Histologie. Berlin, Springer 1957; 13 (1A):900-924

Hassler R. Zur Pathologie der Paralysis agitans und des postenzephalitischen Parkinsonismus. J Psychol Neurol 1938; 48:387-476

Hirano A. Neurons, astrocytes and ependyma. In Davis RL, Robertson, DM, eds. Textbook of Neuropathology. Baltimore, Williams & Wilkins 1985: 1-91

Hirsch E, Ruberg M, Dardenne M, Portier M-M, Javoy-Agid F, Bach J-F, Agid Y. Monoclonal antibodies raised against Lewy bodies in brains from subjects with Parkinson's disease. Brain Research, 1985; 345:374-378

Jager WAdH, Bethlem J. The distribution of Lewy bodies in the central and autonomic nervous systems in idiopathic paralysis agitans. J Neurol Neurosurg Psychiatry 1960; 283-290

Jellinger K, Riederer P. Dementia in Parkinson's disease and (pre)senile dementia of Alzheimer type: morphological aspects and changes in the intracerebral MAO activity. In: Hassler RG, Christ JF, eds. Advances in Neurology. New York, Raven Press. 1984; 40:199-210

Kahn J, Anderton BH. Gibb WRG, Lees AJ, Wells FR, Marsden CD. Neuronal filaments in Alzheimer's, Pick's, and Parkinson's diseases. N Engl J Med 1985; 313:520-521

Kahn J, Anderton BH, Brion, J-P, Cowell I, Dale G, Kilford L. Marsden CD, Parke J, Robinson P. Immunocytochemical staining of Lewy bodies with antibodies to cytoskeletal proteins. In: Marsden, CD, Goldstein M, Calne DB, eds. Recent Developments in Parkinson's disease. Florham, New Jersey, Macmillan Healthcare Information 1987; volume 2:15-23

Kosaka K, Yoshimura M, Ikeda K, Budka H. Diffuse type of Lewy body disease: progressive dementia with abundant cortical Lewy bodies and senile changes of varying degree - a new disease? Clin Neuropathol 1984; 3:185-192

Kuzuhara S., Mori H, Izumiyama N, Yoshimura M, Ihara Y. Lewy bodies are ubiquitinated. A light and electron-microscopic immunocytochemical study. Acta Neuropathol 1988 (in press)

Lewy FH. Paralysis agitans. I. Pathologische Anatomie. In: Lewandowsky M, ed. Handbuch der Neurologie. Berlin, Springer; 1912:920-933

Lewy FH. Die Lehre vom Tonus und der Bewegung. Zugleich systematische Untersuchungen zur Klinik, Physiologie, Pathologie und Pathogenese der Paralysis Agitans. Berlin, Springer 1923

Mori H. Kondo J, Ihara Y. Ubiquitin is a component of paired helical filaments in Alzheimer's disease. Science 1987;235:1641-1444

Munoz-Garcia D, Ludwin SK. Classic and generalized variants of Pick's disease. A clinicopathological, ultrastructural and immunocytochemical comparative study. Ann Neurol 1984; 16:467-480

Nakashima S, Ikuta F. Tyrosine hydroxylase protein in Lewy bodies of parkinsonian und senile brains. J Neurol Sci 1984; 66:91-96

Pappola MA. Lewy bodies of Parkinson's disease. Immune electron microscopic demonstration of neurofilament antigens in constituent filaments. Arch Pathol Lab Med 1986; 110:1160-1163

Quinn N, Critchley P, Marsden CD. Young onset Parkinson's disease. Movement Disorders 1987; 2:73-91

Rasool CG. Selkoe DJ. Sharing of specific antigens by degenerating neurons in Pick's disease and Alzheimer's disease. N Engl J Med 1985; 312:700-705

Rechsteiner M. Ubiquitin-mediated pathways for intracellular proteolysis. Ann Rev Cell Biol 19S7; 3:1-30

Sima AAF, Clark AW, Sternberger NH, Sternberger LA, Lewy body dementia without Alzheimer changes. Can J Neurol Sci 1986; 13:490-497

Sternberger NH, Sternberger LA, Ulrich J. Aberrant neurofilament phosphorylation in Alzheimer's disease. Proc Natl Acad Sci USA 1985; 82:4274-4276

Tiller-Borcich JK, Forno LS. Parkinson's disease and dementia with neuronal inclusions in the cerebral cortex: Lewy bodies or Pick bodies? J Neuropathol Exp Neurol 1988 (in press)

Wischik CM, Crowther RA. Subunit structure of the Alzheimer tangle. British Medical Bulletin 1986; 42:51-56

Yokochi M, Narabayashi H, Iizuka R, Nagatsu T. Juvenile parkinsonism - some clinical, pharmacological and neuropathological aspects. In Hassler RG, Christ JF, eds. Advances in Neurology, New York, Raven Press 1984; 40:407-413

3

ANATOMICAL AND PHARMACOLOGICAL COMPARISONS BETWEEN DOPAMINE D-1 AND D-2 RECEPTORS IN MAMMALIAN CNS

Eric K. Richfield[1], Anne B. Young[2], and John B. Penney[2]

[1]Unit of Functional Neuroanatomy
National Institute of Mental Health
Bethesda, Maryland 20892   U.S.A.
[2]Department of Neurology
University of Michigan
Ann Arbor, Michigan 40814

INTRODUCTION

The dopamine neurotransmitter system has been of interest to neuroscientists since its discovery in the central nervous system (CNS) in the 1960s using catecholamine histofluroescence techniques (Dahlström and Fuxe, 1964, 1965). A variety of subsequent research techniques, including receptor autoradiography, have been developed and their application to the study of the dopamine (DA) system has provided us with significant information about how DA functions in normal and pathological conditions.

Dopamine functions as a neurotransmitter in both the autonomic nervous system (Greengard et al., 1972) and CNS (Moore and Bloom, 1978). In the CNS, several nuclei and their projection pathways have been characterized (Ungerstedt, 1971; Thierry et al., 1973; Lindvall and Björklund, 1974; Berger et al., 1976). The cells of origin for the major telencephalic projections are located in three midbrain nuclei designated A8 (retrorubral nucleus), A9 (substantia nigra pars compacta, SNC), and A10 (ventral tegmental area of Tsai, VTA) by Dahlström and Fuxe (1964). Together, they provide innervation to telencephalic structures via the mesocortical, mesolimbic and nigrostriatal pathways. Additional localized DA pathways exist in the hypothalamus and retina, as well as descending systems to the spinal cord (Björklund et al., 1975; Lindvall and Bjöjklund, 1974).

Classification of DA Receptor Subtypes

The initial link between DA and the enzyme responsible for its biochemical actions, adenylate cyclase (AC), was made by Greengard and coworkers in the autonomic nervous system, and subsequently extended to the CNS (Greengard et al., 1972; Kebabian and Greengard, 1972). The demonstration that antipsychotic drugs exerted their effect by interacting with DA receptors was supported by biochemical (Clement-Cormier et al., 1974; Inverson et al., 1976) and radioligand binding data (Creese et al., 1976). However, descrepancies existed between the in vivo and in vitro potency for some antipsychotics, notably the class known as butyrophenones, which were clinically quite potent, but in vitro were weak in antagonizing DA stimulated AC. This discrepancy eventually led to the classification of DA receptors into two types, designated D-1 and D-2 (Kebabian and Calne, 1979, Table 1).

Two important characteristics of the DA receptor subtypes serve to differentiate them. The D-1 receptor is stimulatory to AC (Stoof and Kebabian, 1984) and has high affinity for benzazepines (Kebabian et al., 1986), whereas the D-2 receptor is inhibitory to AC (Stoof and Kebabian, 1984) and has high affinity for butyrophenones and substituted benzamides (Creese et al., 1983). Currently, selective agonists and antagonists are available for both the D-1 and D-2 receptors (Table 1).

Table 1
Characteristics of DA D-1 and D-2 Receptors

	Receptor Subtypes	
	D-1	D-2

UNIQUE SUBTYPE CHARACTERISTICS

Biochemistry		
Effect on Adenylate Cyclase	stimulatory	inhibitory
G-Protein Linkage	G_S	G_I
Pharmacology		
Affinity for Butyrophenones	nM to μM	pM to nM
Affinity for Benzamides	nM to μM	pM to nM
Affinity for Benzazepines	pM	nM to μM
Selective Antagonists	SCH 23390	spiroperidol sulpiride
Selective Agonists RU 24213	SKF 38393	LY 17155
Anatomy		
Location	Parathyroid	Pituitary

SIMILAR SUBTYPE CHARACTERISTICS

Pharmacology		
Affinity for Phenothiazines	nM	nM
Affinity for Thioxanthenes	nM	nM
Affinity for Dopamine	nM	nM
Anatomy		
Location	CNS ANS Retina	CNS ANS Retina

Anatomical Distributions

The anatomical location of DA receptors was initially determined using homogenate preparations (for reviews, see Creese, 1982 and 1983; Seeman, 1980). Areas of the basal ganglia including the striatum, nucleus accumbens, and substantia nigra were found to contain high densities of DA receptors. Subsequently, autoradiographic techniques were applied to the DA system (Hollt and Schubert, 1978; Kuhar et al., 1978) and ultimately led to the use of quantitative autoradiographic techniques for both the D-1 and D-2 receptors (Altar et al., 1985; Dawson et al., 1985; Jastrow et al., 1984; Neve et al., 1984). Autoradiographic techniques allow for the determination and quantification of DA receptor subtypes in small anatomical regions throughout the CNS. They also allow the determination of different types of receptor heterogeneities such as receptor patches or gradients (Herkenham and Pert, 1981; Joyce et al., 1985; Nastuk and Graybiel, 1985).

Distributions of DA D-1 (Dawson et al., 1986; Dubois et al., 1986) and D-2 (Bouthenet el al., 1987) receptors have been determined in the rat. Reports of the widespread distribution of D-1 and D-2 receptors in most telencephalic regions were reported. Lack of correspondence to the known DA innervation of some of these regions has also been documented.

Species Comparisons

Because of their availability, ease of use, and cost, laboratory rats have been the best characterized species in terms of DA receptors and other measures of DA function. Despite species differences in many neurochemical markers (Graybiel, 1984; Graybiel and Ragsdale, 1978), few studies have been performed comparing the distribution and density of DA receptors in rats to those seen in other species.

One region where studying the distribution and density of DA receptors in other mammals (such as primates and humans) would be very helpful is the basal ganglia which differs from that of rats in several respects. Although rat striatum is heterogeneous for some markers, notably opiate receptors (Herkenham and Pert, 1981), is homogeneous for other markers including acetylcholinesterase staining (AChE), cholinergic receptors, and DA receptors. The striatum of other mammals (cats, monkeys and humans) is heterogeneous for many neurochemical markers and is considered to be divided into intermingled areas. Small areas staining weakly for AChE have been termed striosomes, while the area surrounding the striosomes stain more heavily for AChE and is termed the matrix (Graybiel and Ragsdale, 1987). During development, the characteristics of the striosome and matrix are inverted for AChE staining in that the striosomes stain darkly for AChE and the matrix stains weakly (Graybiel, 1984). The DA system has been reported to be heterogeneous in the developing striatum of rats and cats with denser innervation to the patch area, but homogeneous in adult rats and cats (Graybiel, 1984; Olson et al., 1972). The degree of matching between pre- and postsynaptic DA markers would be helpful in understanding the functional implications of dopaminergic activity in the basal ganglia during development and in mature animals.

Recent report have suggested that DA innervation of the cerebral cortex of monkeys may be more widespread than was previously suspected, and possibly unlike the innervation seen in rats (Berger et al., 1986; Lewis et al., 1987). Verification of DA innervation of the cerebral cortex using DA receptor autoradiography would be useful both to confirm and extend these findings. The role of DA in the cerebral cortex has not been studied to the degree that its role in the basal ganglia has been studied, apparently due to the low amount of innervation present in the cortex. The possible role that the DA system plays in the cerebral cortex, particularly as it relates to the symptoms and side effects in the treatment of Parkinson's disease, will be of interest.

Pathological implications of DA receptors

The pathological hallmark of Parkinson's disease is the loss of dopaminergic cell bodies in the substantia nigra pars compacta (Hornykiewicz, 1963) resulting in dopamine deficiency in projection areas including the basal ganglia and cerebral cortex (Scatton et al., 1982). Between 20 and 30% of patients with Parkinson's disease (PD) develop a dementia and a significant percentage also have cognitive side effects from the treatment of PD (Brown et al., 1984). The pathophysiology of these problems is not currently known. Changes in D-1 and D-2 receptors in PD have been contradictory, with different groups reporting increases, decreases or no change in receptor numbers (Lee et al., 1978; Pimoule et al., 1985). Data on the densities and distributions of D-1 and D-2 receptors in mammals may provide information on how to evaluate these issues.

QUANTITATIVE DA RECEPTOR AUTORADIOGRAPHY

Quantitative receptor autoradiography of DA D-1 and D-2 receptors was performed in a variety of mammals including shrews, opposums, rats, cats, monkeys, and humans (Richfield et al., 1986, 1987a,b,c). Frozen sections were cut on a cryostat, mounted onto gelatin-coated slides, dehydrated, and stored at $-20°C$ until used in assays. [^{3}H]-SCH 23390 and [^{3}H]-spiroperidol (in the presence of mianserin) were used to label the D-1 and D-2 receptors, respectively. Both the D-1 and D-2 receptor assays were performed in a 25 mM Tris-HCl buffer (pH 7.5) containing 100 mM NaCl, 1 mM $MgCl_2$, 1 μM pargyline, and 0.001% ascorbate. All distribution and competition experiments were performed at a [^{3}H]-ligand concentration equal to their respective K_Ds for the receptor.

The distributions of D-1 and D-2 receptors have been determined in the rat CNS by several investigators and appear to be similar to each other (Boyson et al., 1986; Dubois et al., 1986). A Pearson correlation between the densities of D-1 and D-2 receptors in 21 brain regions was 0.80, suggesting that there is not only a qualitative similarity between the distributions, but also a quantitative relationship. How D-1 and D-2 receptors might interact is not known. If both receptors were localized to the same neurons, an opportunity for their interaction would be possible. One way to ascertain this is to determine if one receptor subtype could alter the affinity state of the other subtype.

Dopamine receptors are analgous to β-adrenergic receptors in desplaying multiple affinity states when binding agonists (De Lean et al., 1982; Hancock et al., 1979; Kent et al., 1979; Sibley et al. 1982). The affinity states for the D-1 and D-2 receptors were found to be quite different from each other in terms of the percentage of high and low affinity state, and from the results obtained by others using homogenate preparations (Richfield et al., 1986 and 1987c). Both the D-1 and D-2 receptors were best modeled using a two state model. In the absence of exogenous guanine nucleotides, the D-1 receptor was primarily in a low affinity state ($R_H = 21 \pm 6\%$). In the presence of 10 μM guanyl-imidodiphosphate (GmP-PnP), the D-1 receptor was completely in a low affinity state ($R_L = 100\%$). In the absence of exogenous guanine nucleotides, the D-2 receptor was primarily in the high affinity state ($R_H = 77 \pm 3\%$). In the presence of 10 μM (GmP-PnP) the D-2 receptor was completely in a low affinity state ($R_L = 100\%$). The guanine nucleotide guanosine-5'-0-(2-thiophosphate (GDP-βS) had effects similar to GmP-PnP in shifting the affinity states for both the D-1 and D-2 receptors. Similar affinity states for the D-1 and D-2 receptors were found in the nucleus accumbens and olfactory tubercle.

Receptor occupancy of the D-2 receptor with either an agonist or antagonist did not alter the affinity states of the D-1 receptor, and vice versa, receptor occupancy of the D-1 receptor did not alter the affinity states of the D-2 receptor. This suggests that D-1 and D-2 receptors do not interact via changes in their affinity states.

SPECIES COMPARISONS

Basal ganglia

DA is a neurotransmitter in the basal ganglia of a diverse group of vertebrates, including reptiles, birds, and mammals (Parent, 1986). The presence of DA receptors would, therefore, not be unexpected in mammals. The percentage of D-1 and D-2 receptors was found to be quite similar among the mammals examined in areas of the basal ganglia including the striatum (caudate nucleus and putamen), pallidum (medial and lateral globus pallidus), and substantia nigra (pars compacta and reticulata). The number of D-1 receptors exceeded the number of D-2 receptors in the regions examined. However, there were some differences among the mammals. In the adult rat striatum, both the D-1 and D-2 receptors were "non-patchy." Kittens demonstrated patchy D-1 receptors that corresponded to AChE rich patches, whereas the D-2 receptor was homogeneous in its distribution. In adult cat, both D-1 and D-2 receptors were slightly heterogeneous in their distributions, but in patterns that did not match those seen with AChE staining. The degree of heterogeneity was slight, with increases of 10-20% above the average number of receptors. There were areas where the D-1 and D-2 receptor heterogeneity matched each other, as well as areas where they did not. The adult monkey and human striatum demonstrated less heterogeneous DA receptor binding than did the cat. Most regions of the basal ganglia demonstrated heterogeneous binding, including the presence of receptor gradients (Joyce et al., 1985).

Cerebral Cortex

Dopamine was identified as a neurotransmitter in the cerebral cortex of rats in the 1970's (Thierry et al., 1973; Berger et al., 1974; Fuxe et al., 1974). Dopaminergic innervation of the cortex was initially thought to be restricted to certain regions, unlike the pattern seen with other ascending catecholamine systems (Levitt and Moore, 1978). However, recent studies suggest that DA may have a much greater role in cerebral cortical function than was presumed (Lewis et al., 1987).

The D-1 receptor is present in widespread areas of the cerebral cortex of rats, cats, and monkeys (Boyson, et al., 1986; Dawson et al., 1986; Richfield et al., 1987a and 1987c). The number of D-1 receptors is greater than the number of D-2 receptors in most laminae and regions of cerebral cortex in all three species examined. The rat D-1 receptor displayed a relatively homogeneous laminar pattern in most regions except that the deeper laminae (V and VI) contain more receptors than the superficial layers. The cat and monkey, however, have distinct heterogeneous laminar paterns in all regions of cortex that varied from one region to another and were quite different from that seen in the rat. The cerebral cortical density of D-1 receptors was also much higher in the cats and monkeys than in the rats.

The D-2 receptor is also distributed in most regions of the cerebral cortex of the rats, cats, and monkeys (Boyson et al., 1986; Bouthenet et al., 1987; Richfield, et al., 1987a and 1987c). The D-2 receptor was more homogeneous in its laminar pattern and regional distribution than the D-1 receptor in all 3 species. The D-2 receptor is denser in the superficial layers (I and II) of cortex than in the deeper layers in the rat, but more homogeneous in the cat and monkey cerebral cortex.

These findings support and extend the notion that the dopamine system has a more widespread role in cerebral cortical function than was previously suspected. The high density, variability in laminar density and regional heterogeneity of the D-1 receptor in cat and monkey suggest that the dopamine system, especially the D-1 receptor, may be more involved in higher cortical processing in certain mammals than was presumed.

SUMMARY

Despite the many recent advances in our understanding of Parkinson's disease many questions regarding different aspects of the disorder remain. The widespread and heterogeneous distribution of dopamine D-1 and D-2 receptors in the CNS of mammals suggest that some of the symptoms and side effects of the treatment of Parkinson's disease may be due to factors outside the basal ganglia. Whether the dementia associated with PD is due to DA deficiency in areas of the cerebral cortex or other regions awaits further investigation. The long-term benefit of neural transplants in PD may depend on the ability of the transplants to improve the DA deficiency in multiple areas of the CNS.

REFERENCES

Altar, C.A., O'Neil, S., Walter, R.J., Marshall, J.F. (1985) Dopamine and serotonin receptor sites revealed by digital subtraction autoradiography. Science 228: 597-600.

Berger, B., Trottier, S., Gaspar, P., Verney, C., and Alvarez, C. (1986) Major dopamine innervation of the cortical motor areas in the Cynomolgus monkey. A radioautographic study with comparative assessment of serotonergic afferents. Neurosc. Lett. 72:121-127.

Berger, B., Thierry, A.M., Tassin, J.P. and Moyne, M.A. (1976) Dopaminergic innervation of the rat prefrontal cortex: A fluorescence histochemical study. Brain Res. 106:133-145.

Björklund, A., Lindvall, O., Nobin, A. (1975) Evidence of an incertohypothalamic dopamine neurone system in the rat brain. Brain Res. 89:29-42.

Björklund, A. and Lindvall, O. (1984) Dopamine-containing systems in CNS. In, Björklund, A. and Hökfelt, T. (eds). Handbook of Chemical Neuroanatomy, Vol. 2: Classical Transmitter in the CNS, Part I. New York: Elsevier Sciences Publishers.

Boyson, S.J., McGonigle, P., and Molinoff, P.B. (1986) Quantitative autoradiographic localization of the D₁ and D₂ subtypes of dopamine receptors in rat brain. J. Neurosci. 6:3177-3188.

Bouthenet, M.L., Matres, M.-P., Sales, N., Schwartz, J.-C. (1987) A detailed mapping of dopamine D-2 receptors in rat central nervous system by autoradiography with [^{125}I] iodosulpiride. Neurosciencce 20:115-155.

Brown, R.G. and Marsden, C.D. (1984) How common is dementia in Parkinson's disease? The Lancet i:1262-1265.

Clement-Cormier, Y.C., Kebabian, J.W., Petzold, G.L., and Greengard, P. (1974) Dopamine-sensitive adenylate cyclase in mammalian brain: A possible action of antipsychotic drugs. Proc. Natl. Acad. Sci. USA 29:1113-1117.

Creese, I., Burt, D.R., and Snyder, S.H. (1976) Dopamine receptor binding predicts clinical and pharmacological potencies of antischizophrenic drugs. Science 192:481-483.

Creese, I. (1982) Dopamine receptors explained. Trends in Neurosciences 3:40-43.

Creese, I., Sibley, D.R., Hamblin, M.W., and Leff, S.E. (1983) The classification of dopamine receptors: Relationship to radioligand binding. Ann. Rev. Neurosci. 6:43-71.

Dahlström, A. and Fuxe, K. (1964) Evidence for the existence of monoamine-containing neurons in the central nervous system. I. Demonstration of monoamines in the cell bodies of brainstem neurons. Acta. Physiol. Scand., (suppl.232), 62:1-55.

Dahlström, A. and Fuxe, K. (1965) Evidence for the existence of monoamine-containing neurons in the central nervous system. II. Experimentally induced changes in the interneuronal amine levels of the bulbospinal neuron system. Acta. Physiol. Scand., (suppl.247), 1-36.

Dawson, T.M., Gehlart, D.R., Yamamura, H.I., Barnett, A., and Wamsley, J.K. D-1 dopamine receptors in the rat brain: Autoradiographic localization. Eur. J. Pharm., 108 (1985) 323-325.

Dawson, T.M., Gehlert, D.R., McCabe, R.T, Barnett, A. and Wamsley, J.K. (1986) D-1 dopamine receptors in the rat brain: A quantitative autoradiographic analysis. J. Neurosci. 6:2352-2365.

DeLean, A., Kilpatrick, B.F., Caron, M.B. (1982) Dopamine receptors of the porcine anterior pituitary gland; Evidence for two affinity states discriminated by both agonists and antagonists. Molec. Pharm. 22:290-297.

Dubois, A., Savasta, M., Curet, A., and Scatton, B. (1986) Autoradiographic distribution of the D_1 agonist [^{3}H]SKF 38393, in the rat brain and spinal cord. Comparison with the distribution of D_2 dopamine receptors. Neuroscience 19:125-137.

Fuxe, K., Hökfelt, T., Johansson, D., Jonsson, G., Linbrink, P., Ljungdahl, A., (1974) The origin of the dopamine nerve terminals in limbic and frontal cortex. Evidence for mesocortical-dopamine neurons. Brain Res. 82:349-355.

Graybiel, A.M. and Ragsdale Jr., C.W. (1978) Histochemically distinct compartments in the striatum of human, monkey, and cat demonstrated by acetylthiocholinesterase staining. Proc. Natl. Acad. Sci. USA 75:5723-5726.

Graybiel, A.M. (1984) Correspondence between the dopamine islands and striosomes of the mammalian striatum. Neurosci. 13:1157-1187.

Greengard, P., McAfee, D.A., and Kebabian, J.W. (1972) On the mechanism of action of cyclic AMP and its role in synaptic transmission. In, Greengard, P. and Robinson,G.A. (ed) Advances in Cyclic Nucleotide Research, Vol. 1. New York: Raven Press

Hancock, A.A., DeLean, A.L., Lefkowitz, R.J. (1979) Quantitative resolution of beta-adrenergic receptor subtypes by selective legand binding: Application of a computerized model fitting technique. Mol. Pharm. 16:1-9.

Herkenham, M. and Pert, C.B. (1981) Mosaic distribution of opiate receptors, parafascicular projections and acetycholinesterase in rat striatum. Nature 291:415-418.

Hollt, V. and Schubert, P. (1978) Demonstration of neuroleptic receptor sites in mouse brain by autoradiography. Brain Res. 151:149-153.

Hornykiewicz, O. (1963) Die topische Lokalisation und das Verhalten von Noradrenalin and Dopamin (3-hydroxy-tyramin in der Substantia nigra des normalen und parkinsonkranken Menschen. Wiener Klinische Wochenschrift 75:309-312.

Iverson, L.L. Rogawski, M.A., Miller, R.J., (1976) Comparison of the effects of neuroleptic drugs on pre- and postsynaptic dopaminergic mechanisms in the rat striatum. Mol. Pharm. 12:251-262.

Jastrow, T.R., Richfield, E., Gnegy, M.E. (1984) Quantitative autoradiography of [³H]sulpiride binding sites in rat brain. Neurosci. Lett. 51:47-53.

Joyce, J.N. and Marshall, J.F. Striatal topography of D-2 receptors correlates with indexes of cholinergic neuron localization. Neurosc. Lett, 53 (1985) 127-131.

Kebabian, J.W. and Greengard, P. (1971) Dopamine-sensitive adenyl cyclases: possible role in synaptic transmission. Science 174:1346-1349.

Kebabian, J.W. and Calne, D.B. (1979) Multiple receptors for dopamine. Nature 277:93-96.

Kebabian, J.W., Agui, T., van Oene, J.C., Shigematsu, K., and Saavedra, J.M. (1986) The D₁ dopamine receptor: New perspectives. Trends in Pharm. Sci. 5:96-99.

Kent, R.S., DeLean, A.L., Lefkowitz, R.J. (1979) A quantitative analysis of beta adrenergic receptor interactions: Resolution of high and low affinity states of the receptor by computer modeling of ligand binding data. Mol. Pharm. 17:14-23.

Kuhar, M.J., Murrin, C., Malout, A.T., and Klemm, N. (1978) Dopamine receptor binding in vivo: The feasibility of autoradiographic studies. Life Sci. 22:203-210.

Lee, T., Seeman, P., Rajput, A., Farlye, I.K., and Hornykiewicz, O. (1978) Receptor basis for dopamnergic supersensitivity in Parkinson's disease. Nature 273:59-61.

Levitt, P. and Moore, R.Y. (1978) Noradrenaline neuron innervation of the neocortex in the rat. Brain Res. 139:219-231.

Lewis, D.A., Campbell, M.J., Foote, S., Goldstein, M., Morrison, J.H. (1987) The distribution of tyrosine hydroxylase-immunoreactive fibers in primate neocortex is widespread but regionally specific. J. Neurosci. 7:279-290.

Lindvall, O. and Björklund A. (1974) The organization of the ascending catecholamine neuron systems in the rat brain as revealed by the glyoxylic acid fluorescence method. Acta Physiol. Scand (Suppl.) 412:1-48.

Matres, M.-P., Bouthenet, M.-L., Sales, N., Sokoloff, P., and Schwartz, J.-C. (1985) Widespread distribution of brain dopamine receptors evidenced with [¹²⁵I]iodosulpiride, a highly selective ligand. Science 228:752-755.

Moore, R.Y. and Bloom, F.E. (1978) Central catecholamine neuron systems. Ann. Rev. Neurosci. 1:129-169.

Nastuk, M.A. and Graybiel, A.M. (1985) Patterns of muscarinic cholinergic binding in the striatum and their relation to dopamine islands and striosomes. J. Comp. Neurol. 237:176-194.

Neve, K.A., Altar, A., Wong, C.A. Marshall, J.F. (1984) Quantitative analysis of [³H]spiroperidol binding to rat forebrain sections. Plasticity of neostriatal dopamine receptors after nigrostriatal injury. Brian Res. 302:9-18.

Olson, L., Seiger, A. and Fuxe, K. (1972) Heterogeneity of striatal and limbic innervation: Highly fluorescent islands in developing and adult rats. Brain Res. 44:283-288.

Parent, A. (1986) Comparative Neurobiology of the Basal Ganglia. New York: Wiley and Sons.

Pimoule, C., Schoemaker, H., Reynolds, G.P., and Langer, S.Z. (1985) [³H]-SCH 23390 labeled D₁ dopamine receptors are unchanged in schizophrenia and Parkinson's disease. Eur. J. Pharm. 114:235-237.

Richfield, E.K., Young, A.B., Penney, J.B. (1986) Properties of D-2 dopamine receptor autoradiography: High percentage of high affinity agonist sites and increased nucleotide sensitivity in tissue sections. Brain Res. 383:121-128.

Richfield, E.K., DeBowey, D., Penney, J.B., and Young, A.B. (1987a) Basal ganglia and cerebral cortical distribution of dopamine D-1 and D-2 receptors in neonatal and adult cat brain. Neurosci. Lett 73:203-208.

Richfield, E.K., DeBowey, D., Penney, J.B., and Young, A.B. (1987b) Comparative distribution of dopamine D-1 and D-2 receptors in the basal ganglia of turtle, pigeon, rat, cat, and monkey. J. Comp. Neurol. (262:446-463).

Richfield, E.K., (1987c) Anatomical and pharmacological comparisons between dopamine D-1 and D-2 receptors in the central nervous system. Ph.D thesis, University of Michigan, Ann Arbor, Michigan.

Scatton, B., Rouquier, L., Javoy-Agid, F., and Agid, Y. (1982) Dopamine deficiency in the cerebral cortex in Parkinson disease. Neurology 32:1039-1040.

Seeman, P. (1980) Brain dopamine receptors. Pharm. Rev. 32:229-313.

Sibley, D.R., Delean, A., Creese, I. (1982) Anterior pituitary dopamine receptors: Demonstration of interconvertible high and low affinity states of the D-2 dopamine receptor. J. Biol. Chem. 257:6351-6361.

Stoff, J.C. and Kebabian, J.W. (1984) Two dopamine receptors: Biochemistry, physiology, and pharmacology. Life Sci. 35:2281-2296.

Swanson, L.W. (1982) The projections of the ventral tegmental area and adjacent regions: A combined fluorescent retrograde tracer and immunofluorescence study in the rat. Brain Res. Bull. 9:321-353.

Thierry, A.M., Stinus. L., Blanc, G., and Glowinsky J. (1973) Some evidence for the existence of dopaminergic neurons in the rat cortex. Brian Res. 50:230-234.

Ungerstedt, U. (1971) Stereotaxic mapping of the monoamine pathway in the rat brain. Acta Physiol. Scand. (suppl. 367) 82:1-48.

4

ARCHITECTURE OF CHOLINERGIC PRE- AND POSTSYNAPTIC MARKERS IN THE PRIMATE STRIATUM*

Deborah C. Mash

Departments of Neurology and Pharmacology
University of Miami School of Medicine

INTRODUCTION

The basal ganglia are the critical relay for the extrapyramidal system and the execution of motor function. Marsden has suggested that the basal ganglia are responsible for the automatic execution of learned motor plans (21). The basal ganglia may also play a role in higher cognitive functions. Injury to the basal ganglia is known to produce changes in a variety of behavioral functions in humans and animals including learning, language and personality (4, 32). The degree to which the basal ganglia are involved in cognitive processes is, at present, poorly understood. However, connectional and chemoarchitectural studies have demonstrated that the basal ganglia comprise highly heterogeneous and complex structures which are linked to different neural systems in brain.

Within the basal ganglia, the striatal complex exhibits a marked neurochemical compartmentalization at two different levels of organization. Staining for the degradative enzyme, acetylcholinesterase, has revealed that the caudate-putamen contains 300-600 micron zones of low enzyme activity termed striosomes (11,12) which correspond in the monkey to the cell islands and the surrounding matrix compartments (8). Graphical reconstruction of the histochemically-defined striosomes has shown that they form part of a complex three-dimensional labyrinth extending throughout the rostrocaudal extent of the striatum. The striosomes and the matrix in which they are embedded are thought to constitute biochemically distinct compartments which are related to the intrinsic structure of the striatal complex as well as to the organization of its afferent and efferent connections (10,14). Recent evidence suggests that the anatomical distribution of many neurotransmitter systems obeys this compartmental organization. Dopaminergic terminals, marked by catecholamine histofluorescence or tyrosine hydroxylase immunoreactivity, are concentrated in discrete patches called the "dopamine islands" (11). Opiate receptors are distributed in the rodent brain in a clustered pattern that is in register with the dopamine islands (16). Graybiel has suggested that the striosomal compartments may represent functionally distinct units in which groups of neurons can be modulated in a coordinated way (10).

The second level of organization is based on connectivity studies which have demonstrated that several neuromediator-specified fiber systems are not equally dispersed over the striatum but instead are segregrated into dorsolateral and ventromedial sectors. For example, somatostatinergic fibers exhibit a ventromedial to dorsolateral gradient of decreasing density (17). Evidence for a neurochemical compartmentalization along dorsolateral and ventromedial lines within the striatum has also been provided by in vitro receptor autoradiographic studies (19). In the rodent striatum, a sharp lateral to medial gradient of D2 dopamine receptors has been observed which corresponds to a similar gradient in markers for cholinergic nerve termi-

*Supported by a grant from the National Parkinson Foundation.

Table 1. Microdensitometric determinations of the densities of [^{3}H]-HC-3 binding, M1 and M2 receptor subtypes and putative neural nicotinic receptors in the primate striatum.

Area	Caudate		Putamen		N. Accumbens
	dorsolateral	ventromedial	dorsolateral	ventromedial	
[^{3}H]-HC-3	56.2 ± 3.2	26.1 ± 4.1	69.4 ± 7.0	23.7 ± 4.0	60.3 ± 4.6
M1 subtype	72.3 ± 5.0	98.3 ± 6.7	58.7 ± 7.6	87.4 ± 4.4	98.5 ± 10.4
M2 subtype	20.2 ± 3.0	25.2 ± 2.2	18.2 ± 5.0	22.2 ± 2.9	25.2 ± 1.8
[^{3}H]-nicotine	9.2 ± 1.1	3.9 ± 0.5	7.2 ± 0.6	1.2 ± 0.1	0.1 ± 0.05

Data are the mean $\pm$ S.E.M. of at least three densitometric determations. Values are given in fmol/mg tissue. Choline uptake site and nicotinic receptor autoradiograms were obtained by labelling sections with 10 nM [^{3}H]-HC-3 and 7.5 nM [^{3}H]-L-nicotine, respectively. M1 and M2 receptor subtypes were labelled according to the method of Mash and Potter (22).

nal densities (20). It is possible that the neurochemical gradients defined by the specific complement of postsynaptic receptor may be topographically related to functional subsets of extrapyramidal circuits. The distribution of cholinergic pre- and postsynaptic markers visualized by in vitro autoradiography as described in the primate striatum appears to reflect this latter organizational principle. Our studies demonstrate that cholinergic pre- and postsynaptic markers exhibit gradients in regional density throughout the caudate and putamen, and some local patchiness which may correspond to the striosomal compartments.

Distribution of cholinergic presynaptic markers in the primate

The cholinergic functions of the striatum are mediated by the actions of cholinergic interneurons. These cholinergic local circuit neurons are the rare "giant" cells scattered among the more prevalent medium-sized striatal- projection neurons (12). These perikarya express high levels of choline acetyltransferase (ChAT) and acetylcholinesterase (AChE), allowing their distribution in the striatum to be extensively studied by histochemical and immunocytochemical methods (27,36). In the striatal complex of the primate, there is a higher density of cholinergic cell bodies in the nucleus accumbens and the olfactory tubercle (ventral striatum) than in the caudate and putamen (dorsal striatum) (27). Using an improved protocol for ChAT fiber staining, Graybiel et al. (12) examined the distribution of ChAT-positive neuropil in the macaque striatum in relation to the regional variations in AChE histochemistry. The ChAT-immunostained neuropil included large-diameter dendrites, thin fibers and varicosities of many sizes, which were elevated in the extrastriosomal matrix compartment. Within the caudate and putamen, there was a strong dorsolateral gradient in staining for both cholinergic markers.

Hemicholinium-3 is a potent competitive inhibitor of high affinity choline uptake into presynaptic terminals. The distribution of high affinity choline uptake sites on cholinergic terminals has recently been visualized using [^{3}H]-hemicholinium-3 autoradiography (33). The autoradiographic distribution of choline uptake sites labelled with [^{3}H]-hemicholinium-3 has been shown to correlate with the immunocytochemical distribution of ChAT (33). In the primate striatum, [^{3}H]-hemicholinium-3 binding exhibited very dense patches in the caudate and in the lateral parts of the putamen. These dense patches of [^{3}H]-hemicholinium-3 binding were most prevalent in the dorsolateral sector of the striatum (Figure 1a). The lateral putamen exhibited a gradient in binding site density that was 3 times higher than the most medial portion of the putamen (Table 1). A few scattered large patches of dense choline uptake site labelling were also apparent ventrally within the nucleus accumbens. These observations demonstrate that in the primate striatum regional variations in choline uptake site labelling by [^{3}H]-HC-3 are largely coincident with the distribution of presynaptic cholinergic markers.

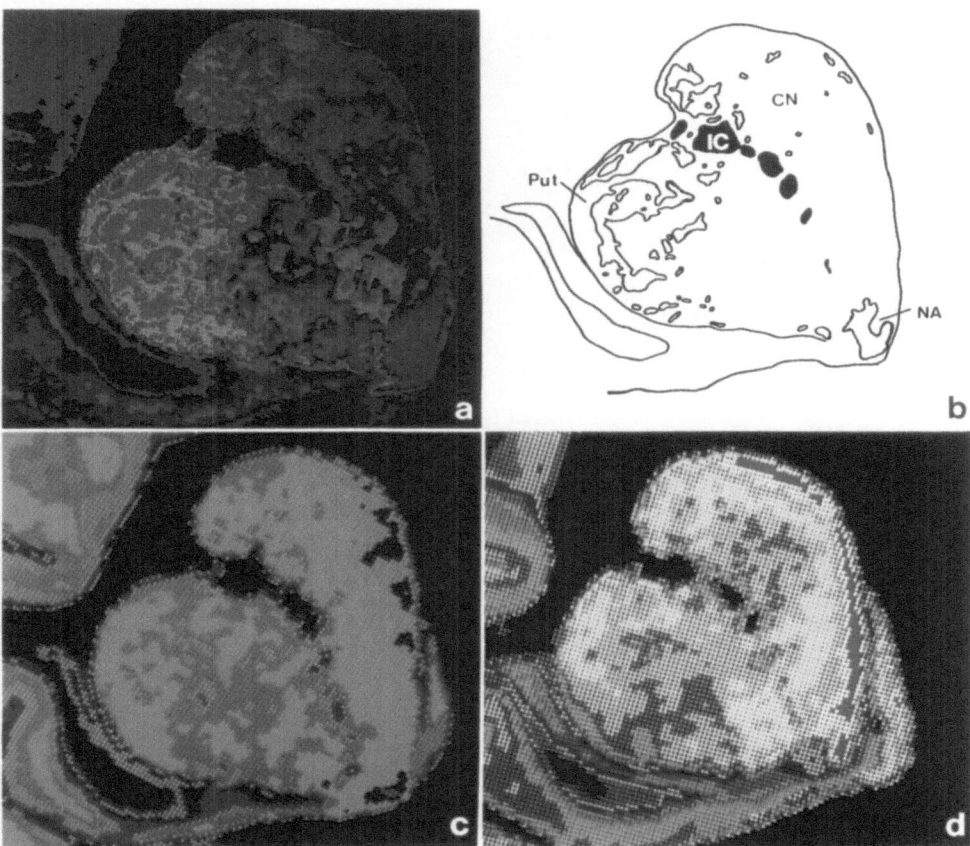

Figure 1. Pseudocolor density maps of cholinergic markers in the striatum of the monkey (Macaca mulatta). Red = highest densities; green = intermediate densities; gray = lowest densities (15 colors corresponding to a rainbow scale). **a.** The distribution of the presynaptic marker, [^{3}H]-hemicholinium-3 binding, was heaviest in the lateral sectors of the caudate and putamen. **b.** Schematic drawing of the areas of dense binding derived from the section shown in panel **a** outlining the patches. **c.** M2 receptor subtype labelling is homogeneous. **d.** M1 receptor subtype labelling is dense along the medial ventricular surface of the caudate extending into the nucleus accumbens.

Striatal distribution of subclasses of muscarinic receptors

The localization of receptors for acetylcholine has been studied in the striatum by in vitro autoradiographic techniques. These studies have demonstrated local patterning in the distribution of muscarinic receptors labelled with [3H]- propylbenzylcholine mustard and [3H]-quinuclidinyl benzilate. In the rat, cat, and human striatum, muscarinic receptor densities are elevated during early development in discrete patches that lie in register with high (rather than low) AChE staining and the fluorescent dopamine islands (28). In the mature striatum, the distribution of [3H]-propylbenzylcholine mustard binding was not uniform although the patches of elevated receptor labelling are only vaguely demarcated. This compartmentalization of muscarinic receptors suggets that there are different proportions of muscarinic receptors on neuronal elements found in and out of the striosomes and that the patch and nonpatch receptors mature at differing rates (28).

Recent observations on the distinct pharmacological and biochemical properties of muscarinic receptors in various tissues have led to the concept of at least two different recptor subclasses termed M1 and M2 (6,7,15,35). [3H]-Propyl-benzylcholine mustard and [3H]-quinuclidinyl benzilate do not distinguish among muscarinic receptor subtypes. [Putative M1 and M2 receptor subtypes have been distinguished by the M1-selective antagonist, pirenzepine, or the M2-selective agonist, carbachol (6,7,35).] The distribution of putative muscarinic receptor subtypes (M1 and M2) in the macaque brain have been investigated by in vitro receptor autoradiography (23,24). The autoradiographic visualization of muscarinic receptor subtypes in the primate cerebral cortex reveals a marked regional heterogeneity in M1 and M2 patterns which respects architectonic boundaries. The M1 receptor subtype is elevated in limbic and paralimbic cortical areas. In contrast, the M2 receptor subtype displayed peak densities in all primary sensory cortical areas.

Regional variations in putative subclasses of muscarinic receptors are also apparent in the primate striatum (Fig. 1c and d). Muscarinic receptor subtypes (M1 and M2) were labelled with [3H]-quinuclidinyl benzilate in the presence of sufficient cold pirenzepine or carbachol to selectively occlude binding of the radioactive ligand to M1 and M2 sites, respectively (22). The highest density of muscarinic receptors in the primate brain is found in the basal ganglia. The M1 receptor subtype is the predominate subclass of muscarinic receptor in the primate striatum (Table 1). Density gradients of the M1 receptor subtype spanned the rostrocaudal extent of the striatum. The nucleus accumbens had the highest density of the M1 receptor subtype. The M1 subtype labelling formed a dense strip extending along the medial edge of the caudate ventrally into the nucleus accumbens. A few large, very dense patches of M1 labelling were also apparent within the ventromedial sector of the putamen. The M1 receptor subtype pattern described here in the monkey differs from that reported previously in cat striatum (29). In the mature striatum of the cat, sites autoradiographically labelled under M1 conditions were in patches of particularly high density that matched the striosomes. In contrast, M2 binding appeared virtually homogenous. In the primate striatum, the distribution of the M2 receptor subtype exhibited a similar homogeneity in binding, although a few patches of elevated binding were apparent along the medial aspect of the caudate. These results suggest that muscarinic aspects of striatal function are subject to a regional and compartmental organization.

Mosaic architecture of neural nicotinic receptors

The use of a variety of probes for neural nicotinic acetylcholine receptors has indicated that the binding sites for the antagonist, [125I]-alpha bungarotoxin, are distinct from the high affinity sites labelled by L-[3-H]-nicotine (2). Putative neural nicotinic receptors have been tagged for autoradiographic studies by L-[3H]-nicotine (1,2). In the human cerebral cortex, L-[3H]-nicotine exhibits saturable binding to a single class of sites having nanomolar affinity for the ligand (5). We have examined the distribution of L-[3H]-nicotine binding in the primate striatum. Putative neural nicotinic receptor densities are markedly lower than the M1 receptor subtype throughout the striatal complex (Table 1). In contrast to the prevalence of the M1 receptor subtype within the ventromedial sectors of the putamen, high densities of nicotinic receptors were observed in the lateral sector of the putamen. The pattern of L-[3H]-nicotine binding in the putamen was similar to the topography of the corticostriate projections from the motor cortex. Radiotracer labelling of area 4 reveals an intense patch-like arrangement of fibers showing a greater density in the lateral sector of the putamen (18). A lateral to medial gradient of elevated L-[3H]-nicotine binding appeared as dense areas interdigitating with lighter

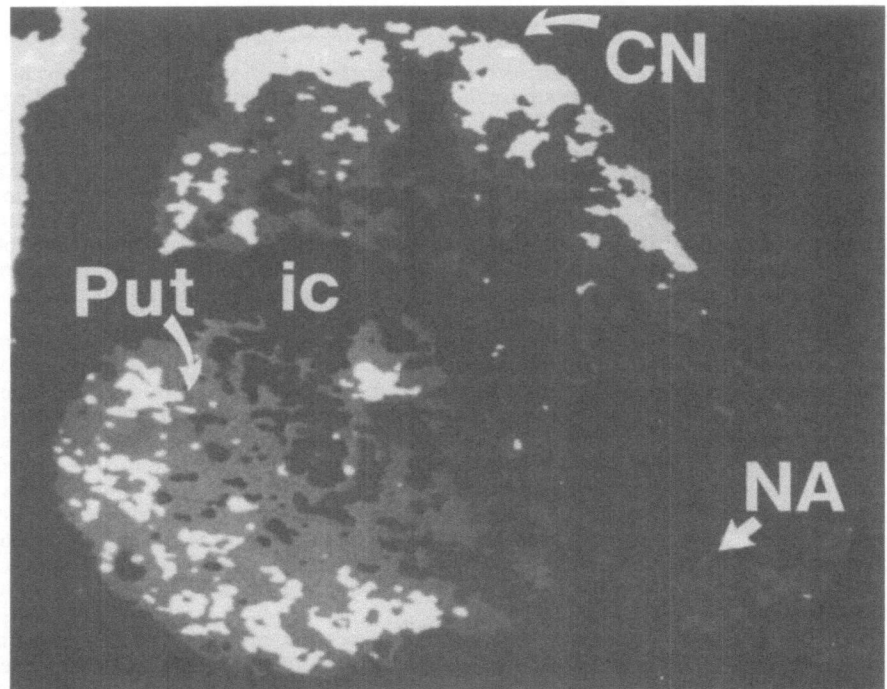

Figure 2. Autoradiogram of putative neural nicotinic receptors in the primate striatum. Patches of elevated nicotinic binding were highly represented in the caudate nucleus. In the putamen, dense labelling was apparent in the lateral sector. Nicotinic receptor labelling was virtually absent in the nucleus accumbens.

zones. Nicotinic receptor distributions in the caudate nucleus were less uniform than in the putamen. Small discrete patches of peak L-[3H] nicotine binding were observed throughout the rostrocaudal extent of the caudate nucleus (Fig. 2). Although there was not a precise fit, these small irregular patches of L-[3H]-nicotine binding appeared to align with acetylcholinesterase-poor zones visualized on adjacent sections (data not shown). This pattern suggests that the distribution of nicotinic sites may be related to the heterogeneity of dopaminergic input to the striatum.

Cholinergic modulation of striatal functions

The dorsal striatum has different behavorial specializations from the limbic striatum. With respect to the dorsal striatum it appears, based on connectivity studies, that the caudate nucleus may play a lesser role than the putamen in motor control (see Fig. 3). Although very little is known about the human limbic striatum, the nucleus accumbens has been suggested to be relevant for striatal channeling of drive and affect (10). Single unit studies in the rhesus monkey have demonstrated that putaminal neurons encode specific parameters of movement (3), while caudate neurons are activated during the performance of a delayed response task, a task associated according to Goldman-Rakic with the dorsolateral prefrontal cortex (9,34). The cortical afferents from the sensory-motor regions of the neocortex terminate primarily in the lateral sector of the putamen and extend into the ventrolateral part of the body of the caudate (18). In contrast, the orbitofrontal cortex, a paralimbic area, projects to the medial parts of the caudate, forming a wide band which extends from the dorsomedial ventricular surface to the ventromedial part of the precommisural putamen (34). On the output side, the caudate projects primarily to the substantia nigra, while the putamen projects primarily to the globus pallidus (30). These observations suggest that the putamen primarily subserves motor function while the caudate is preferentially involved in higher cognitive processing.

The arrangement of afferent and efferent fiber connections within the striatum is also related to the neurotransmitter content and complement of pre- and postsynaptic receptors. Within the striatum, acetylcholine can activate muscarinic (M1 and M2) and nicotinic receptors. The local patterning of cholinergic pre- and postsynaptic markers suggests that intrinsic cholinergic activity may gate particular subsets of striatal inputs and outputs. The presynaptic markers, ChAT and AChE, are most dense in the sensory-motor sectors of the striatum (12). The lateral to medial gradient of [3H]-hemicholinium-3 binding, further suggests an association between the density of cholinergic terminals in the dorsolateral sectors of the caudate and putamen, and the distribution of sensory-motor inputs. The cholinergic innervation of the primate cerebral cortex is heaviest in the limbic and paralimbic sectors (25,26). Based on these observations, Mesulam (26) has suggested that cortical cholinergic innervation may provide a mechanism for channeling motivationally relevant sensory information into and out of the limbic system. The distribution of M1 receptor subtypes in the primate cerebral cortex also displayed peak densities within limbic-paralimbic sectors (23,24). The enrichment of M1 receptors within the limbic parts of the striatum (ventromedial caudate and nucleus accumbens) fits with the areal trend of elevated M1 receptor labelling of limbic and paralimbic sectors of the cerebral cortex in the primate (23,24). Elevated M1 receptor labelling of the nucleus accumbens and ventromedial sector of the caudate, suggests that cholinergic activation of the M1 receptor subtype may be related to trans-striatal channeling of limbic and cognitive-affective processes. Nicotinic receptor labelling in the putamen appears to correspond to the terminal distribution of corticostriate projections, further suggesting a role for these sites in relaying the action of acetylcholine in the motor functions of the putamen. Selective cholinactive drugs may therefore have distinct effects on motor vs. executive functions of the striatum.

The cellular location of muscarinic receptor subtypes and neural nicotinic sites in the striatum remains unclear. A considerable proportion of the total muscarinic receptors in the striatum are assumed to be on intrinsic neurons. The dense patches of L-[3H]-nicotine binding in the caudate which correspond to AChE poor zones suggest that nicotinic receptors may be associated with nigrostriatal projections. The mosaic pattern of nicotinic receptors in the caudate fits with the larger representation of striosomes in the caudate nucleus than the putamen in higher primates (10). In the rat, there is evidence to suggest that nicotinic receptors have a presynaptic location on a subset of dopaminergic neurons in the pars compacta of the substantia nigra (1). A similar arrangement may exist in the monkey, since we have observed that L-[3H]-nicotine binding forms a dense band in the primate substantia nigra that appears to correspond to the densocellular zone of the pars compacta (data not shown). Understanding the role of muscarinic receptor subtypes, the relation of striatal nicotinic receptors, and the exact cellular locations of receptors for acetylcholine will further clarify dopaminergic-cholinergic interactions in the striatum.

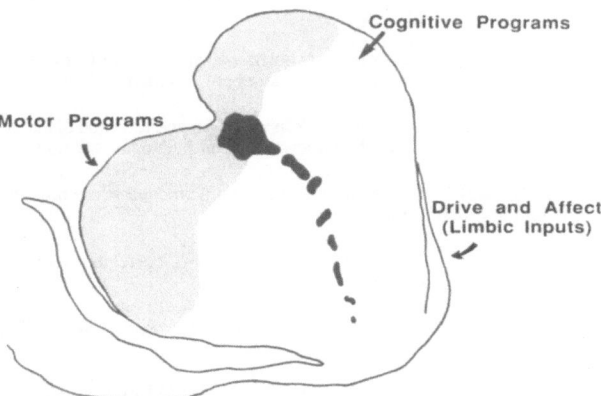

Figure 3. Drawing of a coronal section of the striatum illustrating proposed domains of motor, cognitive and limbic functions. Stippled area delineates the dorsolateral sector of the striatum as evidenced by the gradient in cholinergic presynaptic markers.

References

Clarke, P.B.S. and Pert, A. Autoradiographic evidence for nicotine receptors on nigrostratial and mesolimbic dopaminergic neurons. Brain Res. 348:355-358, 1985.

Clarke, P.B.S., R.D. Schwartz, S.M. Paul, C.B. Pert, and A. Pert (1985) Nicotine binding in rat brain: Autoradiographic comparison of [3H]-acetylcholine, [3H]-nicotine, and [125I]-bunagarotoxin. J. Neurosci. 5, 1307-1315.

Crutcher, M.D. and Delong, M.R. Single cell studies of the primate putamen. I. Functional organization. Exp. Brain Res. 53:244-258, 1983.

Damasio. A.R. Language and the basal ganglia. Trends Neurosci. 6:442-444, 1983.

Flynn, D.D., and D.C. Mash characterization if L-[3H]-nicotine binding in human cerebral cortex: Comparison between Alzheimer's disease and the normal. J. Neurochem. 47:1948-1954, 1986.

Flynn, D.D. and L.T. Potter Effect of solubilization on the distinct binding properties of muscarine receptors from rabbit hippocampus and brainstem. Mol. Pharm. 30:193-199, 1986.

Goldman-Rakic, P.S. Cytoarchitectonic heterogeneity of the primate neostriatum: Subdivision into island and matric cellular compartments. J. Comp. Neurol. 205:398-413, 1982.

Goldman, P.S., Roswold, H.E., Vest, B. and Galkin, T.W., Analysis of the delayed alternation deficit produced by dorsolateral prefrontal lesions in the monkey. J. Comp. Physiol. Psych. 77:262-280, 1971.

Graybiel, A.M. Dopamine-containing innervation of the striatum: Subsystems and their striated correspondents. In- Functions of the Basal Ganglia, Ciba Foundations Symposium 107: 114-144, 1984.

Graybiel, A.M. Correspondence between the dopamine islands and striosomes of the mammalian striatum. Neuroscience 13:1157-1187, 1984.

Graybiel, A.M., Baughman, R.W. and Eckenstein, F. Cholinergic neuropil of the striatum observes striosomal boundaries. Nature 323:625-627, 1986.

Graybiel, A.M. and Ragsdale, C.W. Jr., In- Emson, PC (ed) Chemical Neuronatomy. Raven Press, New York, 427-504.

Gerfen, C.R. The neostriatal mosaic: compartmentalization of corticostriatal input and strionigral output systems. Nature 311:461-464, 1984.

Haga, K. and Haga, T. Purification of the muscarinic acetylcholine receptor from porcine brain. J. Biol. Chem. 260:7927-7935, 1885.

Herkenham M, Moon-Edley, and S. Stuart, Cell clusters in the nucleus accumbens in the rat and the mosaic relationship of opiate receptors, acetylcholinesterase and subcortical afferent terminations. Neuroscience 11:561-593, 1984.

Johannsen, O. and Hokfelt, T., Thyrotropin releasing hormone, somatostatin, and enkephalin: distribution studies using immunohistochemical techniques. J. Histochem. Cytochem. 28:364-366, 1980.

Jones, E.G., Coulter, J.D., Burton, H., and Porter, H. Cells of origin and terminal distribution of corticostriatal fibers arising in the sensory-motor cortex of monkeys. J. Comp. Neurol. 173:53-80, 1977.

Joyce, J.N., Sapp, D.G. and Marshall, J.F. Human striatal dopamine receptors are organized in compartments. Proc. Natl. Acad. Sci. 83:8002-8006, 1986.

Joyce, J.N. and Marshall, J.F., Striatal topography of D-2 receptors correlates with indexes of cholinergic neuron localization. Neurosci. Lett. 53:127-131, 1985.

Marsden, C.D. The mysterious motor function of the basal ganglia: The Robert Wartenberg lecture. Neurology 32:514-539, 1982.

Mash, D.C. and Potter, L.T. Autoradiographic localization of M1 and M2 muscarine receptors in the rat brain. Neurosci. 19:551-564, 1986.

Mash, D.C. and Mesulam, M-M. Muscarine receptor distributions within architectonic subregions of the primate neocortex. Soc. Neurosci, Abstr. 12:809, 1986.

Mash, D.C., White, W.F. and Mesulam, M-M. Distribution of muscarinic receptor subtypes with architectonic sectors of the primate cerebral cortex. J. Comp. Neurol. -In Press.

Mesulam, M.-M., Mufson, E.J., Levey, A.I., and Wainer, B.H., Cholinergic innervation of cortex by the basal forebrain: Cytochemistry and cortical connections of the septal area, diagonal band nuclei, nucleus basalis (Substantia innominata), and hypothalamus in the rhesus monkey. J. Comp. Neurol. 214:170-197, 1983.

Mesulam, M.-M., Rosen, A.D. and Mufson, E.J. Regional variations in cortical cholinergic innervation: Chemoarchitectonics of acetylcholinesterase-containing fibers in the macaque brain. Brain Res. 311:245-258, 1984.

Mesulam, M.-M., Mufson, E.J., Levey, A.I. and Wainer, B.H. Atlas of cholinergic neurons in the forebrain and upper brainstem of the macaque based on monoclonal choline acetyltransferase immunohistochemistry and acetylcholinesterase histochemistry. Neuroscience 12: 669-686, 1984.

Nastuk, M.A. and Graybiel, A.M. Patterns of muscarinic cholinergic binding in the striatum and their relation to the dopamine islands and striosomes. J. Comp. Neurol. 237:176-194, 1985.

Nastuk, M.A. and Graybiel, A.M. Autoradiography of M1 and M2 muscarinic receptor binding in the striatum, In- Trends Pharmacol. Sci. (suppl) p. 92, Elsevier, Amsterdam, 1986.

Parent, A., Bouchard, C. and Smith, Y. The striopallidal and striatonigral projections: Two distinct fiber systems in the primate. Brain Res. 303:385-390, 1984.

Potter, L.T., Flynn, D.D. Hanchett, H.E., Kalinoski, D.L. Luber-Narod, J.L., and Mash, D.C., Independent M1 and M2 muscarine receptors: Ligands, autoradiography and functions. In- Trends Pharmacol. Sci (suppl) (eds. Hirschowitz and, B.I., Hammer, R., Giachetti, A., Keirns, J., and Levine, R.) pp. 22-31. Elsevier, Amsterdam, 1984.

Richfield, E.K., Twyman, R. Berent, S. Neurological syndrome following bilateral damage to the head of the caudate. Ann. Neurol. 22:768-771, 1987.

Rhodes, K.J., Joyce, J.N., Sapp, D. and Marshall J.F. [3H]-Hemicholinium-3 binding in rabbit striatum: correspondence with patchy acetylcholinesterase staining and a method for quantifying striatal compartments. Brain Res. 412:400-404, 1987.

Selemon, L.D. and Goldman-Rakic, P.S. Longitudinal topography and interdigitation of corticostriatal projections in the rhesus monkey. J. Neurosci. 5:776-794, 1985.

Watson, M., T.W. Vickroy, W.R. Roeske, and Yamamura, H.I., Subclassification of muscarinic receptors based upon the selective antagonist pirenzepine. In-Trends the Pharmacol. Sci. (eds. Hirschowitz and, B.I., Hammer, R., Giachetti, A., Keirns, J., and Levine, R.) pp. 9-11 Elsevier, Amsterdam, 1984.

Wolf, N.J. and Butcher LL., Cholinergic neurons in the caudate-putamen complex proper are intrinsically organized: a combined Evans Blue and acetylcholinesterase analysis. Brain Res. Bull., 7:487-507, 1981.

5

FUNCTIONAL ANATOMY OF DOPAMINE RECEPTORS

G. Frederick Wooten and Joel M. Trugman

Departments of Neurology and Neuroscience
University of Virginia Medical Center
Charlottesville, VA. 22908
(804) 924-8369

INTRODUCTION

The selective degeneration of pigmented dopaminergic neurons of the substantia nigra which results in striatal dopamine (DA) depletion is the critical pathologic process in Parkinson's disease (PD) (1). The administration of dopaminergic drugs that have access to the central nervous system results in amelioration of the clinical syndrome characteristic of PD (bradykinesia, rigidity and tremor) (2). Dopaminergic drugs reverse the symptoms of PD by interacting with DA receptors principally localized in the striatum. Biochemical studies of brain DA receptors suggest the existence of at least two types. The DA receptor is positively linked to adenylate cyclase while the D_2 DA receptor is negatively linked (3). Current evidence suggests that the anti-Parkinson efficacy of dopaminergic drugs is mediated primarily by the D_2 receptor (4).

Despite great progress in recent years in understanding both the neurochemical abnormalities in PD and the mechanism(s) of action of anti-Parkinson drugs, several fundamental questions remain. Data to be reviewed in this manuscript pertain to the question of which neural circuits are activated by D_1 and D_2 DA receptor stimulation in vivo. Selective lesions of the nigrostriatal pathway may be produced in rats by intracerebral microinjections of 6-hydroxydopamine (5). Rats with such lesions manifest a variety of motor abnormalities including brady- and hypokinesia (6) which are transiently improved by the administration of dopaminergic drugs (7). To identify which neural circuits are activated by DA receptor stimulation, we have employed the 2-deoxyglucose (2-DG) autoradiographic method of Sokoloff (8). Because glucose is the primary energy-producing substrate in the brain and the major expenditure of brain energy is for pumping ions across membranes, the autoradiographic mapping of regional cerebral glucose utilization (RCGU) affords a functional map of brain physiological activity (8,9). In the experiments summarized below, we have employed the general strategy of administering a variety of DA agonists (levodopa, specific D_1 and D_2 agonists, and several compounds currently in use for the treatment of PD) to rats with unilateral nigral lesions and then mapping RCGU by the 2-DG autoradiographic method.

METHODS

Animals and Lesions

Experiments were performed on male Sprague-Dawley rats weighing 275-325 grams. Following pretreatment with desipramine hydrochloride, 25 mg/kg, i.p., lesions were performed as previously described (10). Six-hydroxydopamine hydrobromide (6 μg/2 μl of 0.5% ascorbic acid in normal saline) was injected over 3 minutes into the left perinigral region. Rats were tested 10 days later with apomorphine hydrochloride, 0.5 mg/kg, i.p., to determine lesion efficacy. Only rats that turned a minimum of 5 rotations/min. at the peak action of apomorphine were used in subsequent 2-DG studies performed 3-6 weeks after lesion.

2-Deoxyglucose Autoradiography

^{14}C-2-DG autoradiography was performed according to the method of Sokoloff et al. (8). On the experimental day, after rats had been fasted overnight, a central venous catheter was inserted into the right external jugular vein under 1% halothane anesthesia. After 3 hours of recovery, rats were injected with the experimental drug followed by ^{14}C-2-DG 1 minute later. Forty-five minutes after the ^{14}C-2-DG injection, the rats were killed with a 50 mg/ml intravenous bolus of sodium pentobarbital, and the brains were rapidly removed and prepared for autoradiography. Film autoradiographs were analyzed with a Leitz variable aperture microdensitometer. Data for glucose utilization were expressed as a ratio of gray matter optical density (OD) to white matter OD.

RESULTS AND DISCUSSION

Effects of Levodopa (11)

Levodopa acts as a DA agonist via conversion in the brain to DA. As such, levodopa is a mixed agonist potentially activating both D_1 and D_2 receptors. Levodopa was administered in doses of 10, 25, or 50 mg/kg, s.c., 30 minutes after pretreatment with carbidopa, 25 mg/kg, i.p. Levodopa treatment markedly increased RCGU in the ipsilateral entopeduncular nucleus (EP) and substantia nigra pars reticulata (SNPR), cell groups that receive direct striatal input and function as major outflow pathways of corpus striatal activity (12). In contrast, levodopa did not alter RCGU in the globus pallidus (GP), supporting the thesis that DA has different effects on striatal outflow to the GP compared with outflow to both EP and SNPR (12). Moderate RCGU increases were observed in the ipsilateral subthalamic nucleus (STN), lateral midbrain reticular formation (LMRF), and deep layers of the superior colliculus (DLSC), all regions that receive direct projections from the EP, GP, or SNPR. Levodopa resulted in decreased RCGU in the ipsilateral lateral habenular nucleus (LHN) and increased RCGU in the contralateral LHN, changes that are likely mediated via altered neuronal activity in the striatum and EP (13). The results suggest that systemically-administered levodopa, after conversion to DA in the brain, interacts with supersensitive DA receptors in the DA-depleted striatum to selectively activate efferent pathways. Furthermore, the data suggest that the LMRF and DLSC are functionally activated during levodopa-induced turning in rats with unilateral nigral lesions and support the hypothesis that nigroreticular and nigrocollicular projections are of physiologic significance in the expression of striatal activity. Because levodopa is a mixed agonist, however, these results do not discriminate between the cerebral metabolic effects of selective activation of D_1 or D_2 DA receptors. Rather, they presumably reflect simultaneous coactivation of both types of DA receptors.

Effects of SKF-38393 (D_1 selective agonist) and LY-171555 (D_2 selective agonist) (14)

To determine the functional metabolic consequences of selective D_1 and D_2 DA receptor stimulation, we examined the effects of the selective D_1 agonist SKF-38393 and the D_2 agonist LY-171555 on RCGU in rats with unilateral nigral lesions (14). SKF-38393 (0.5-25 mg/kg) and LY-171555 (0.01-5.0 mg/kg) produced indistinguishable behavioral responses, including vigorous contralateral rotation. Treatment with each drug similarly increased glucose utilization, dose-dependently, in the parafascicular thalamus, STN, DLSC, and LMRF ipsilateral to the nigral lesion; RCGU was decreased in the ipsilateral LHN. In contrast, the D_1 and D_2 agonists differentially altered RCGU in the EP and SNPR (Figs. 1 and 2). SKF-38393, 5.0 and 25.0 mg/kg, increased glucose utilization 127 and 275%, respectively, in the SNPR ipsilateral to the lesion. LY-171555, 1.0 and 5.0 mg/kg, caused maximal contralateral turning yet did not alter glucose utilization in the ipsilateral SNPR. The glucose utilization response of the ipsilateral EP paralleled that of the SN, demonstrating large increases following the administration of SKF-38393 and minimal change following LY-171555. Thus, the selective D_1 agonist reproduces the marked glucose utilization increases (2-3-fold above control values) in the EP and

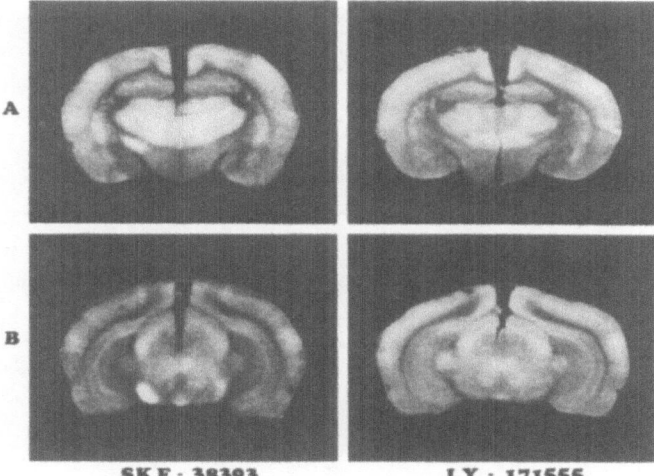

SKF - 38393   LY - 171555

Figure 1. 14C-2-DG reversed image autoradiography (printed directly from X-ray film) depicting RCGU patterns in the unilateral nigral-lesioned rat treated with SKF-38393, 25 mg/kg (D_1 agonist), and LY-171555, 1.0 mg/kg (D2 agonist). The substantia nigra pars compacta lesion is on the left side, as are the RCGU changes noted below. Light areas indicated high 2-DG uptake, while dark areas correspond to low 2-DG uptake. RCGU is markedly increased in the EP (A) and SNPR (B) ipsilateral to the nigral lesion following D_1 but not D_2 agonist treatment. (From reference 14, with permission).

SNPR that we previously observed with both levodopa (11) and apomorphine (15); in contrast, the D_2 agonist does not produce such changes. Both the EP and SNPR are selectively rich in D_1 receptors with $D_1:D_2$ receptor ratios of 20-30:1 (16). These results suggest that the large increases in RCGU caused by levodopa in the EP and SNPR may result, at least in part, from direct stimulation by DA of these receptors in the EP and SNPR, and do not reflect an exclusive drug effect on striatal DA receptors. Furthermore, given the magnitude of the RCGU response in EP and SNPR, the findings suggest that D_1 receptors in the EP and SNPR become functionally supersensitive following lesions of the substantia nigra.

Effects of Bromocriptine, Pergolide, and (+) PHNO

Following intravenous administration of bromocriptine (10 mg/kg), pergolide (0.04 mg/kg), or PHNO (0.04 and 0.4 mg/kg) to rats with unilateral nigral lesions, RCGU was unchanged in EP and SNPR ipsilateral to the lesion despite brisk contralateral rotation after each drug treatment. This pattern is characteristic of selective D_2 agonist stimulation and different from that produced by levodopa or apomorphine. Pergolide 0.4 mg/kg increased RCGU (about 2-fold) in the ipsilateral EP and SNPR, a metabolic effect indicative of selective D_1 DA agonist stimulation. Using quantitative in vitro autoradiographic methods, all three drugs competed for the D_2 DA receptor (using ^3H-spiperone to label striatal D_2 receptors) with IC_{50} values as follows: bromocriptine 8.5 nM, and PHNO 34 nM. The drugs were ineffective competitors for the D_1 site: bromocriptine 13 μM, pergolide 15 μM, and PHNO $\mu 100 > M$. Thus, bromocriptine and PHNO are selective D_2 DA agonists with markedly different in vivo potencies. In contrast, high dose pergolide has mixed D_1/D_2 agonist properties in vivo.

SUMMARY AND CONCLUSIONS

The administration of DA agonist drugs to rats with unilateral substantia nigra lesions produces essentially two different patterns of RCGU. Each pattern shared the following changes: increased RCGU in the ipsilateral parafascicular thalamus, STN, DLSC, and LMRF and decreased RCGU in the ipsilateral LHN. The two patterns differed with respect to the occurrence or absence of increases in RCGU in the ipsilateral EP and SNPR. Results are summarized in the table. Mixed agonists (i.e. with both D_1 and D_2 DA receptor agonist properties) such as apomorphine and levodopa produced large increases in RCGU in the ipsilateral EP and SNPR, a pattern characteristic of the selective D_1 DA receptor agonist SKF-38393. In contrast,

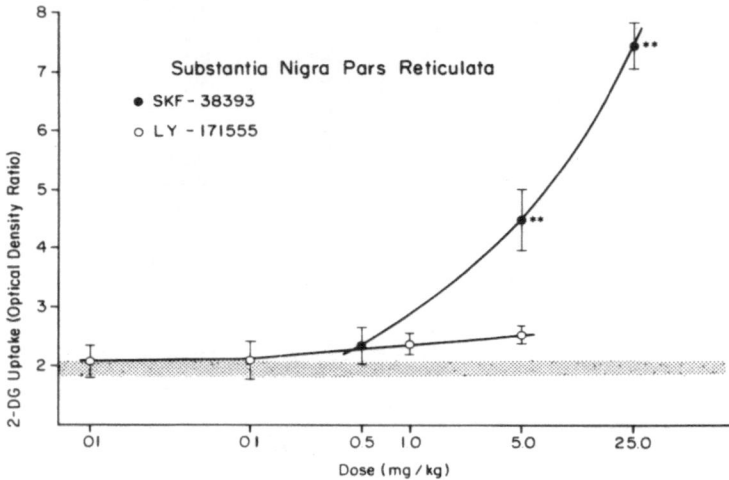

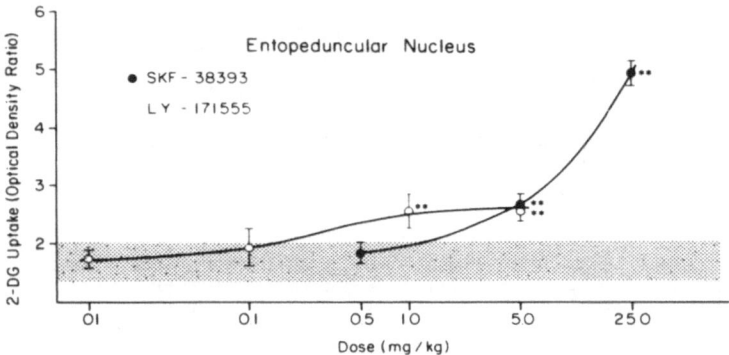

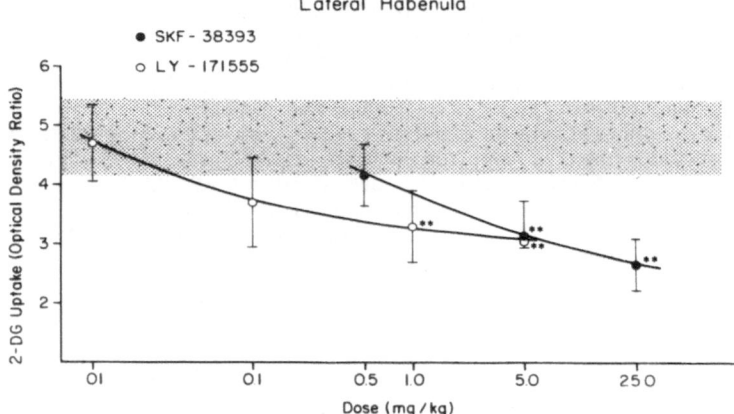

Figure 2. Log dose-response curves for 2-DG uptake in the SNPR, EP, and LHN ipsilateral to a nigral lesion. The shaded area indicates 2-DG uptake in the control group (lesioned rats given saline; n=6). For both the SNPR and EP, the slopes of the D_1 and D_2 dose response curves differ ($p<0.001$), indicating that the 2-DG uptake is differentially altered by these two drugs. The slopes of the dose-response curves in the LHN do not differ, indicating that 2-DG uptake in this nucleus is altered in a similar manner by both D1 and D2 agonists. Optical density ratios are significantly different from control values and are indicated as follows: *, $p<0.05$; **, $p<0.01$. (From reference 14, with permission).

the pattern produced by the D_2 specific agonist LY-171555 was shared by bromocriptine and (+) PHNO. Pergolide, however, appeared to be a specific D_2 agonist at low concentrations but also activated D_1 receptors at higher concentrations.

These results are significant in that they provide the first example of a selective, specific, and objective consequence of activation of D_1 DA receptors in the brain in vivo. Further studies are needed to determine if the increase in RCGU in the ipsilateral EP and SNPR is a consequence of D_1 receptor activation in the striatum, in the EP and SNPR directly, or simultaneously in both the striatum and EP-SNPR.

SUMMARY OF BEHAVIORAL, BIOCHEMICAL, AND CEREBRAL
METABOLIC PROPERTIES OF DOPAMINE AGONISTS

Drug (Dose mg/kg)	Stimulates Adenylate Cyclase in Vitro	%RCGU Increase in Ipsilateral SNPR	Contralateral Turning
Apomorphine 0.5	+	116	+ +
Levodopa 10 50	+	125 193	+ + + +
SKF-38393 5 25	+	127 275	+ + +
LY-171555 1 5	−	0 0	+ +
Bromocriptine 10	−	0	+
Pergolide 0.04 0.4	+	0 95	+ + + +
(+) PNHO 0.04 0.4	−	0 0	+ +

REFERENCES

1. Hornykiewicz, 0. (1982). Brain neurotransmitter changes in Parkinson's disease. In Movement Disorders, C.D. Marsden and S. Fahn, eds. Butterworths, London, pp. 41-58.

2. Cotzias, G.C., Papavasiliou, P.S., Fehling, C., et al. (1970). Similarities between neurologic effects of L-dopa and of apomorphine. New Engl. J. Med. 282:31-35.

3. Stoof, J.C. and Kebabian, J.W. (1984). Two dopamine receptors: biochemistry, physiology, and pharmacology. Life Sci. 35:2281-2296.

4. Schachter, M., Bedard, P., Debono, A.G., et al. (1980). The role of D1 and D2 receptors. Nature 286:157-159.

5. Ungerstedt, U. (1971). Stereotaxic mapping of monoamine pathways in the rat brain. Acta. Physiol. Scand. (Suppl) 367:1-48.

6. Marshall, J.F., Richardson, J.S., and Teitelbaum, P. (1974). Nigrostriatal bundle damage and the lateral hypothalamic syndrome. J. Comp. Physiol. Psychol. 87:808-830.

7. Marshall, J.F. (1978). Sensory inattention by 6-hydroxydopamine injections along the ascending dopaminergic fibers: spontaneous recovery and pharmacological control. Soc. Neurosci. Abstr. 4:46.

8. Sokoloff, L., Reivich, M., Kennedy, C., et al. (1977). The [14C] deoxyglucose method for the measurement of local cerebral glucose utilization: theory, procedure and normal values in the conscious and anesthetized albino rat. J. Neurochem. 28:879-916.

9. Mata, M., Fink, D.J., Gainer, H., et al. (1980). Activity-dependent energy metabolism in rat pituitary primarily reflects sodium pump activity. J. Neurochem. 34:213-215.

10. Wooten, G.F. and Collins, R.C. (1981). Metabolic effects of unilateral lesion of the substantia nigra. J. Neurosci. 1:285-291.

11. Trugman, J.M. and Wooten, G.F. (1986). The effects of L-DOPA on regional cerebral glucose utilization in rats with unilateral lesions of the substantia nigra. Brain Res. 379:264-274.

12. Carpenter, M.B. (1981). Anatomy of the corpus striatum and brain stem integrating mechanisms. In Handbook of Physiology - The Nervous System II, pp. 947-995.

13. Wooten, G.F. (1981). Dopamine neurons control metabolism in the lateral habenular nucleus indirectly via the entopeduncularis. Soc. Neurosci. Abstr. 7:851.

14. Trugman, J.M. and Wooten, G.F. (1987). Selective D1 and D2 dopamine agonists differentially alter basal ganglia glucose utilization in rats with unilateral 6-hydroxydopamine substantia nigra lesions. J. Neurosci. 7:2927-2935.

15. Wooten, G.F. and Collins, R.C. (1983). Effects of dopaminergic stimulation on functional brain metabolism in rats with unilateral substantia nigra lesions. Brain Res. 263:267-275.

16. Boyson, S.J., McGonigle, P., and Molinoff, I.P.B. (1986). Quantitative autoradiographic localization of the D1 and D2 subtypes of dopamine receptors in rat brain. J. Neurosci. 6:3177-3188.

6

NEW INSIGHTS INTO THE REGULATION OF DOPAMINE RECEPTOR SUBTYPES AND THEIR ROLES IN BEHAVIOR

Ian Creese

Rutgers State University of New Jersey
Newark, NJ 07102

Previous research has demonstrated the existence of two distinct dopamine receptor subtypes (Kebabian & Calne, 1979; Creese et al. 1983), possessing unique pharmacologic and biochemical properties. D_1 dopamine receptors stimulate adenylate cyclase activity, while D_2 dopamine receptors inhibit this enzyme and may well have other second messenger systems. However both receptor subtypes co-exist in many tissues making the determination of their respective physiological and behavioral roles difficult. All neuroleptics, commonly used drugs in the treatment of schizophrenia, have been shown to be either mixed D_1/D_2 receptor antagonists or selective D_2 receptor antagonists. These drugs, when administered to animals, induce profound motoric effects, termed catalepsy, which are characterized by the inability to initiate movement. This drug-induced behavioral syndrome is similar to some of the symptoms of Parkinson's disease. Indeed, when administered to man, a similar spectrum of "extrapyramidal motor effects" are induced – which are ameliorated by anti-parkinsonian drugs such as anticholinergics. Thus the behavioral effects of neuroleptics mimic some of the symptoms of Parkinson's disease which are probably mediated by identical underlying biochemical substrates. D_2 receptors have been implicated as the site mediating both the antipsychotic and the antidopaminergic, extrapyramidal motoric effects of neuroleptics (Creese et al., 1976; Seeman et al., 1976). By inference, D_2 receptors have been considered to be the mediators of dopaminergic agonists' behavioral effects. These behavioral effects are also motoric – agonists such as L-DOPA, apomorphine or amphetamine stimulating motor activity at low doses, while dyskinesias and stereotyped motor responses occur at higher doses in both animals and man. However, all of these agonists activate both D_2 and D_1 receptors.

Until now, the functional role of D_1 receptors could not be clearly defined due to the lack of selective, high affinity D_1 receptor agonists or antagonists. Recently, a D_1 selective antagonist, SCH 23390, was developed (Iorio et al., 1983) and has since been tritiated, making detailed D_1 receptor characterization possible. This novel benzazepine exhibits nanomolar potency in inhibiting dopamine stimulation of striatal adenylate cyclase activity and exhibits an apparent K_D value of approximately 0.5 nM for the striatal D_1 receptor. Hence, the advent of this first truly selective D_1 receptor antagonist has provided a tool for research into the functional correlates of D_1 receptor activation and antagonism. Neuroleptics, administered to rodents in vivo, besides inducing catalepsy, suppress conditioned avoidance responses, and antagonize the hyperactivity and stereotypy responses elicited by dopamine agonists. These behaviors are elicited by selective D_2 receptor antagonists or mixed D_1/D_2 receptor antagonists and are generally predictive of antipsychotic activity in man. Consequently, these behavioral effects have been used by the pharmaceutical industry as screens for potential neuroleptic drugs. Neuroleptics are also antiemetic and have been shown to induce hyperprolactinemia by blocking D_2 receptors in the anterior pituitary. Since this tissue lacks D_1 receptors it is clear that hyperprolactinemia is a response mediated solely by D_2 receptors. While SCH 23390 is a selective D_1 receptor antagonist, this benzazepine quinoline mimics many of the behavioral effects previously associated with D_2 receptor antagonists such as blocking agonist-induced stereotypy and hyperactivity, as well as inducing catalepsy (Iorio et al, 1983; Christensen et al., 1984; Mailman et al. 1984). However, SCH 23390 neither antagonizes emesis nor produces hyperprolactinemia – indicating that SCH 23390 administered in vivo is probably not acting as an antagonist at D_2 receptors. That SCH 23390 can induce these neuroleptic behavioral effects, without blockade of D_2 receptors, suggests D_1 receptors may have a role in the regulation of behaviors which had previously been associated with D_2 receptors alone. Thus D_1 receptor activation may provide a novel approach to the pharmacological management of Parkinson's disease which until now has depended on mixed D_1/D_2 receptor selective agonists or D_2 receptor selective agonists.

A major side effect of chronic neuroleptic administration used in treating schizophrenia is tardive dyskinesia, characterized by uncontrollable movements of the mouth, tongue and extremities. The extrapyramidal motor side effects are similar in some respects to the dyskinesias that can arise during the treatment of Parkinson's disease patients with L-DOPA. Chronic treatment of animals with neuroleptics does not usually result in such obvious dyskinetic movements but does produce enhanced sensitivity to the motoric effects of dopamine agonists (Tarsy and Baldessarini, 1974) with a concomitant increase in D_2 receptor number (Burt et al. 1977). It has been suggested that "pharmacological" denervation of dopamine receptors occurs during chronic antagonist treatment which results in the supersensitive response to subsequent agonist administration (Sibley and Creese, 1980). As Parkinson's disease patients also have up-regulated dopamine receptors because of the dopaminergic denervation characteristic of the disease, these observations suggest that dyskinesias are related in part to agonist stimulation of upregulated or "supersensitive" dopamine receptors. Since chronic treatment of rats with either D_2 receptor-selective or mixed D_1/D_2 receptor-selective antagonists produces identical behavioral supersensitivity but because D_2 receptor-selective antagonists cause a selective D_2 receptor upregulation, it has been concluded that an increase in D_2 receptors is both necessary and sufficient for the production of the enhanced behavioral sensitivity to dopamine agonists (Fleminger et al., 1983). It has therefore been suggested that the side effect of tardive dyskinesia may result from a selective D_2 receptor up-regulation. We reasoned that if chronic treatment with the D_1 receptor selective antagonist SCH 23390 does not increase D_2 receptors, it may lack the liability to produce tardive dyskinesia. Indeed, in a preliminary investigation of this hypothesis we found that after treating rats chronically with SCH 23390 (0.5 mg/kg/day, s.c. for 21 days), striatal D_1 receptors were significantly increased while no change was observed in the D_2 receptor population (Creese and Chen, 1985).

In order to investigate this hypothesis further, we examined the effect of chronic SCH 23390 treatment on dopamine receptor binding, and tested the rats for their locomotor activity and stereotypy responses to selective D_1 and D_2 receptor agonists (Hess et al., 1986). Since previous reports have suggested that tolerance develops to the cataleptogenic effects of D_2 selective neuroleptics (Ezrin-Waters and Seeman, 1977), we also investigated the cataleptogenic effects of both chronic SCH 23390 administration and chronic administration of the selective D_2 receptor antagoist spiperone (Hess et al., 1988). We expected that these studies would also provide information concerning the normal functioning of D_1 dopamine receptors.

Rats were injected daily (s.c.) for 21 days with 0.5 mg/kg SCH 23390, 0.2 mg/kg spiperone or 0.5 mg/kg SCH 23390 plus 0.2 mg/kg spiperone. An equal number of control rats were injected with the saline vehicle. Rats were tested on days 1,3,5,7,14 and 21 for the cataleptic response to their administered drug. Additionally, on day 22, half the group of chronic spiperone-treated animals were injected with 0.5 mg/kg SCH 23390 and half the group of chronic SCH 23390-treated animals were injected with 0.2 mg/kg spiperone and observed for their cataleptic response. To test for catalepsy, observations were made every 20 min for two hours following drug injection. Rats were individually placed in a box (20 x 20 x 30 cm) with a small bar diagonally placed across one corner, 10 cm above the floor. The rat's front paws were placed on the bar and time taken for the rat to remove both paws was measured, with a cutoff time of 120 sec (Sanberg et al., 1984) Two days following the 21st SCH 23390 injection the rats were tested for their behavioral response to either the selective D_1 receptor agonist, SKF 38393 (Setler et al., 1978) or the selective D_2 receptor agonist, quinpirole, (LY 171555)(Itoh et al., 1985). The rats were placed in individual wire mesh photocell cages (40 x 25 x 20 cm) which had 2 equally spaced horizontal infrared beams placed 2 cm high across the long axis of each cage. After 3 hours of habituation during which photocell beam interruptions were recorded every 10 min, each rat was removed and injected (s.c.) with either 3 mg/kg SKF 38393 (the selective D_1 receptor agonist) or 0.3 mg/kg quinpirole (the D_2 receptor agonist). Behavioral testing continued for 3 more hours. The rats were rated for stereotypy by 2 observers every 10 min for a 30 sec period. Rats were scored on the rating scale previously described by Cleese and Iversen (1975). Seventy-two hours following the last injection with SCH 23390 (the day following behavioral testing), the rats were killed by decapitation and their brains rapidly removed to chilled saline. The striata were dissected, frozen in liquid nitrogen and stored at $-70°C$ until biochemical assay for dopamine receptor characteristics.

Although catalepsy may be induced by a dose of SCH 23390 lower than 0.1 mg/kg (s.c.) because this study was examining the effect of chronic treatment with SCH 23390, 0.5 mg/kg SCH 23390 (s.c.) was used in an attempt to continually block of D_1 receptors. Significantly, in another experiment, treatment with SCH 23390 (0.5 mg/kg, s.c.) prior to peripheral administration of the drug N-ethoxycarbonyl-2-ethoxy-1,2-dithydroquinoline (EEDQ) which irreversibly blocks dopamine receptors, protected D_1 receptors from the irreversible blockade by EEDQ, but neither D_2 dopamine nor $5HT_2$ serotonin receptors were protected from EEDQ-induced modifications by SCH 23390 pretreatment. This indicates that SCH 23390, in all probability, interacts in vivo exclusively with D_1 receptors at the dose which was chronically administered to rats in this study. This refutes the suggestion that SCH 23390 administered in vivo may be converted to a D_2 receptor selective compound upon metabolism. These data are supported by Meller et al. (1985a) using a similar paradigm to examine the specificity of neuroleptics administered in vivo.

Consistent with previous reports by this laboratory (Creese and Chen 1985; Hess et al., 1986) chronic treatment with SCH 23390 resulted in a highly significant increase ($+16\%$) in D_1 receptor specific [^{3}H]SCH 23390 binding but no significant change in D_2 receptor specific [^{3}H]spiperone binding was observed. That no increase in D_2 receptors was observed indicated that SCH 23390 maintains its D_1 specificity in vivo. Chronic treatment with the D_2 receptor selective antagonist spiperone resulted in no significant change in [^{3}H]SCH 23390 binding while3 [$_3$spiperone binding was significantly increased by 24%. Chronic treatment with SCH 23390 plus spiperone resulted in a significant increase in both [^{3}H]SCH 23390 ($+19\%$) and [^{3}H]spiperone binding ($+32\%$). The increase in [^{3}H]spiperone binding in the animals treated with SCH 23390 plus spiperone was significantly greater than that increase observed in [^{3}H]spiperone binding in rats treated with spiperone alone. However, the increase observed in [^{3}H]SCH 23390 binding in rats treated with SCH 23390 plus spiperone did not differ significantly from that increase in [^{3}H]SCH 23390 binding in rats treated with SCH 23390 alone. The K_D of [^{3}H]SCH 23390 binding and the K_D of [^{3}H]spiperone binding was unchanged by chronic drug treatment in any experimental condition.

The overall catalepsy score collapsed over each 2 hour session of each daily treatment for rats treated with the D_2 receptor specific antagonist spiperone decreased over the course of the 21 day treatment. A significant reduction in the overall catalepsy score recoreded on day 1 of spiperone treatment occurred by day 5 and the response continued to progressively decrease over the 21 day period. However, complete tolerance to the cataleptogenic effects of spiperone was not observed: the overall catalepsy score on testing day 21 of rats treated chronically with spiperone was significantly greater than the day 21 catalepsy scores of rats treated chronically with saline.

In contrast to the results with spiperone and other classic neuroleptics where tolerance develops to their cataleptogenic effects after chronic treatment (Ezrin-Waters and Seeman, 1977), rats receiving chronic administration of the D_1 selective antagonist SCH 23390 for 21 days demonstrated no tolerance to its cataleptogenic action over the 21 day administration. Those rats receiving SCH 23390 plus spiperone demonstrated no tolerance to the cataleptic effect of this combination of a D_1 receptor antagonist plus D_2 receptor antagonist.

When rats were treated with SCH 23390 for 21 days and then administered an acute injection of spiperone, tolerance to the cataleptogenic effects of spiperone was observed. In fact, the catalepsy scores collapsed over the two hour session following acute challenge with spiperone after chronic 21 day SCH 23390 were not different from the scores of chronic spiperone treated rats on spiperone injection day 21. Conversely, rats challenged with an acute injection of SCH 23390 after 21 day spiperone treatment were as profoundly cataleptic as the rats which had received the 21 day chronic SCH 23390 treatment.

Unlike chronic neurolpetic treatment, where neither an increase nor decrease in spontaneous locomotor activity is observed (Tarsy and Baldessarini, 1974), rats chronically treated with SCH 23390 demonstrated a more than two fold higher level of locomotor activity than control animals during habituation to the locomotor cages two days after ending the chronic drug treatment. Similarly, the chronically SCH 23390 treated rats responded to the selective D_1 receptor agonist, SKF 38393, with potentiated locomotor behavior. Chronic SCH 23390 treated rats also showed a more intense stereotypy response to SKF 38393 than did the chronic saline treated rats. Unlike mixed D_1/D_2 agonists, SKF38393 induced stereotyped grooming in both groups of rats. However, grooming quickly diminished in the chronic saline-treated rats and they demonstrated locomotor stimulation. The chronic SCH 23390 treated rats, however, continued to groom and/or show stereotyped sniffing and rearing in one location over the first hour in response to SKF 38393.

Surprisingly, locomotor activity and stereotypy was potentiated in rats chronically treated with SCH 23390 after treatment with quinpirole, the selective D_2 receptor agonist. The time couse of the locomotor response, initially low, peaking and dropping away was explained by the stereotypy rating. Quinpirole initially induced intense stereotyped sniffing and rearing in one location in both groups. Between 50 and 80 min post injection, some of the chronic SCH 23390 treated rats were also showing stereotyped licking of the cage – although none of the control rats showed this intense response. The peak in crossovers at 140 min corresponded to the transition from stereotyped behavior in one location to locomotor activity in the chronically SCH 23390 treated rats.

That rats chronically treated with SCH 23390 demonstrated significantly higher locomotor activity than control animals during habituation to the locomotor cages suggests that D_1 receptor activation by endogenous dopamine is involved in regulating spontaneous activity. It also implies that the mechanisms by which SCH 23390 elicits its effects are different from and are not regulated in the same manner as those through which classic neurolpetics exert there effect. Other research supports this hypothesis. After treatment with either D_1 or D_2 receptor selective agonists, unilaterally 6-OHDA lesioned rats demonstrate similar contralateral circling behavior (reviewed in Arnt 1987). Furthermore, reserpinized or bilaterally 6-OHDA lesioned animals demonstrate similar hypermotility and oral stereotypy to either D_1 or D_2 receptor selective agonists. In these cases, behaviors induced by D_2 receptor agonists were blocked only by D_2 receptor antagonists and, similarly, D_1 receptor antagonists only blocked D_1 receptor agonist resonses. These data suggest that although D_1 and D_2 receptor-mediated resonses were overtly similar, they are independently modulated and can be distinguished from each other through these various pharmacologic treatments. Collectively, these data suggest that similar behaviors can be induced by the two distinct dopamine receptor systems.

In concert with these data, the chronically SCH 23390 treated rats responded to the selective D_1 receptor agonist, SCK 38393, with potentiated locomotor behavior. Surprisingly, a similar potentiation in locomotor activity and stereotypy was observed in chronically SCH 23390-treated rats after treatment with quinpirole, the selective D_2 receptor agonist. Quinpirole initially induced intense stereotyped sniffing and rearing in one location in both groups. Furthermore, between 50 and 80 min post injection, some of the chronic SCH 23390 treated rats were also showing stereotyped licking of the cage – although none of the control rats showed this response. This enhanced stereotypy was maintained in the chronically SCH 23390 treated rats. However, at the end of the recording session, locomotor activity was quickly attenutated in the chronically SC 23390 treated rats and each rat "instantaneously" went from "sniff/locomotion" to "freezing" in position. This frozen posture was maintained for 5-10 min before the rats closed their eyes and apparently went to sleep. Control rats did not show this freezing behavior. This response may potentially prove to be a useful model for the "on-off" phenomenon seen during dopamine agonist treatement of Parkinson's disease. One might speculate that this troublesome side effect may result from the upregulation of D_1 receptors in this disease.

In in vitro binding studies, quinpirole has extremely low affinity surprisingly ($K_D > 5$ x 10^{-6}M) for D_1 receptors, suggesting that it is unlikely that this drug might elicit these behaviors through D_1 receptors in vivo. Thus, the behavioral effects of a D_2 receptor agonist were potentiated after a selective D_1 receptor upregulation. Certainly this is an anomalous response give that SCH 23390 has no direct effect on D_2 receptors in vivo. Paralleling the results presented herein, it has been demonstrated that although SCH 23390 acts as a D_1 dopamine receptor antagonist, inducing catalepsy in rats, pretreatment with selective D_2 receptor agonists can prevent this response in a dose-dependent manner (Meller et al., 1985b). Likewise, the behavioral effects of the D_1 receptor agonist, SKF 38393, which include non-stereotyped sniffing, rearing and locomotor responses, could be partially reversed by the D_2 receptor antagonist metaclopramide (Molloy and Waddington, 1985). However, Breese and Mueller (1985) have demonstrated that SCH 23390 antagonism of locomotor activity induced by quinpirole may be prevented by 6-hydroxydopamine or reserpine treatment. That is, this D_1/D_2 receptor interaction is dependent on the integrity of catecholamine containing neurons. These behavioral data, then, suggest that D_1 and D_2 receptor-activated neuronal systems are not functionally isolated. Rather, the behavioral data presented herein suggest that although the networks through which D_1 and D_2 receptors mediate behavioral effects appear to be functionally independent (i.e., separate, parallel systems), these systems are interactive in their modulation of motor function. Thus, the enhanced behavioral effect of quinpirole stimulation of D_2 receptors seen in the present study may be the result of an enhanced action of endogenous dopamine at the upregulated D_1 receptors, with behavioral output based on the product of both D_1 and D_2 receptor stimulation. That the behavioral responses to quinpirole and SKF 38393 are not identical may reflect differences in the neuronal circuits activated by D_1 and D_2 receptors, differences which are accentuated after receptor upregulation.

Since many investigators have reported an interaction between D_1 and D_2 receptors in the regulation of behavior and electrophysiological responses (Carlson et al. 1987), to determine whether concomitant D_1 and D_2 receptor blockade could affect receptor upregulation of each receptor subtype, rats received simultaneous treatment with a D_1 receptor selective antagonist (SCH 23390) and a D_2 receptor selective antagonist (spiperone). Simultaneous D_1 and D_2 receptor antagonism resulted in an increase in the density of both D_1 and D_2 receptors with the drug doses employed in this study. Thus, simultaneous D_1 and D_2 receptor antagonism does not appear to reduce the abiltiy of D_1 receptors to upregulate. The increase in D_2 receptors observed after SCH 23390 plus spiperone treatment was significantly greater than the increase in D_2 receptors after spiperone treatment alone. Thus, chronic blockade of D_1 receptors, which alone does not result in a statistically significant D_2 receptor upregulation, potentiated the D_2 receptor upregulation resulting from chronic D_2 receptor blockade. Saller and Salama (1986) have recently demonstrated that SCH 23390 reduces the in vivo increase in dopamine metabolites elicited by D_2 receptor blockade. Thus, it is possible that the potentiation in D_2 receptor upregulation is due to a reduction in dopamine release onto D_2 receptors, caused by the D_1 receptor blockade by SCH 23390. The reduction in dopamine release would result in an effective increase in D_2 receptor blockade by spiperone.

The discovery of antagonist drug-induced receptor upregulation provided an attractive explanation for the behavioral tolerance to chronic antagonist administration (for review see Creese and Sibley, 1980). However, the results decribed herein demonstrate that this mechanism cannot explain the tolerance or lack of tolerance of the cataleptic response to the chronic administration of dopamine receptor antagonists. It is not yet clear why rats become tolerant to the cataleptogenic effects of D_2 receptor antagonists but do not become tolerant to the D_1 receptor antagonist SCH 23390. Tolerance to chronic spiperone administration occurs concomitantly with a D_2 receptor upregulation but does not occur to chronic SCH 23390 administration in spite of a similar concomitant D_1 receptor increase. Conversely, in spite of a normal compliment of D_2 receptors following chronic SCH 23390 treatment, rats are just as tolerant to spiperone as they are following chronic spiperone administration. It appears then that D_2 receptor upregulation is not necessary for the induction of behavioral tolerance. Although the D_2 receptor recognition site may not be upregulated, alterations in its second messenger system may have occurred. For example, in a previous study we demonstrated that chronic SCH 23390 treatment increased the ability of GTP to activate adenylate cyclase in rat striatal homogenate (Hess et al., 1986).

In conclusion, an increase in D_1 receptors induced by chronic SCH 23390 treatment may result in an overall increase in the tonic activity of the basa ganglia due to the enhanced action of endogenous dopamine at the upregulated D_1 receptors. This overall activation imparted by increased D_1 receptor activity may be great enough to impart what appears to be behavioral tolerance to a D_2 receptor antagonist by essentially "overriding" the D_2 antagonist signal. The results parallel our earlier findings where the behavioral effects of the D_2 receptor agonist quinpirole were potentiated after a selective D_1 receptor upregulation. After D_1 receptor upregulation, rats exhibited an increase in spontaneous locomotor activity while following D_2 receptor upregulation, however, neither an increase nor decrease in spontaneous activity has been observed (Tarsy and Baldessarini, 1974). Thus, enhanced D_1 receptor stimulation following D_1 receptor upregulation seems to promote an overall increase in activity levels which may result from the enhanced effectiveness of endogenous dopamine at the upregulated D_1 receptors. Conversely, D_2 receptor supersensitivity due to neurolepetic treatment is only apparent when the animal is challenged with a dopamine receptor agonist. These results support the hypothesis (Braun et al., 1986; Waddington, 1986; Hess et al., 1986) that D_1 receptor activation may provide a tonic background activation which may be equated with an overall level of arousal. In this model, additional D_2 receptor stimulation may be associated with the development of overt stereotyped behavior or dyskinesias which may be regulated by the level of D_1 receptor activation, with greater D_1 receptor activation yielding more intense agonist-induced D_2 receptor mediated responses.

These observations suggest that a selective D_1 receptor antagonist may be able to modulate the dyskinesias caused by L-DOPA treatment in Parkinson's disease. Alternatively, selective D_1 receptor agonists may activate without being dyskinetic.

Acknowledgements

I should like to thank Drs Ellen Hess and Andrew Norman for their contributions to the studies discussed in this chapter. This chapter is dedicated to Shirley and Alex Aidekman in recognition of their support of the Center for Molecular and Behavioral Neurosciences.

References

Arnt, J. Behavioural studies of dopamine receptors; evidence for regional selectivity and receptor multiplicity. In Dopamine Receptors, edited by I. Creese and C. Fraser, A.R. Liss, pp 199-232, 1987.

Braun, A.R., Barone, P. and Chase, T.N. Interaction of D_1 and D_2 receptors in the expression of dopamine agonist induced behaviors. In Neurobiology of Central D_1-Dopamine Receptors, I. Creese and G.R. Breese, eds., pp 151-166, Plenum Press, New York, (1986)

Breese, G.R. and Mueller, R.A. SCH-23390 Antagonism of a D-2 dopamine agonist depends upon catecholaminergic neurons. Eur. J. Pharmacol. 113:109-114, 1985.

Burt, D.R., Creese, I. and Snyder, S.H. Antischizophrenic drugs: Chronic treatment elevates dopamine receptor binding in brain. Science 196:326-328, (1977).

Carlson, J.H., Bergstrom, D.A. and Walters, J.R. Stimulation of both D_1 and D_2 dopamine receptors appears necessary for full expression of postsynaptic effects of dopamine agonists: A neurophysiological study. Brain Res. 400:205-216, 1987.

Christensen, A.V., Arnt, J. Hyttel, J., Larsen, J.J. and O. Pharmacological effects of a specific dopamine D-1 antagonist SCH 23390 in comparison with neuroleptics. Life Sci. 34:1529-1540, 1984.

Creese, I. and Chen, A. Selective D-1 dopamine receptor increase following chronic treatment with SCH 23390. Eur. J. Pharmacol. 109:127-128, 1985.

Creese, I. and Sibley, D.R. Receptor adaptations to centrally acting drugs. Ann. Rev. Pharmacol. Toxicol. 21:357-391(1980).

Creese, I., Burt, D.R. and Snyder, S.H. Dopamine receptor binding predicts clinical and pharmacological potencies of antischizophrenic drugs. Science, 192:481-483, 1976.

Creese, I. and Iversen, S.D. The pharmacological and anatomical substrates of the amphetamine response in the rat. Brain Res. 83:419-436, 1975.

Creese, I., Sibley, D.R., Hamblin, M.W. and Leff, S. The classification of dopamine receptors: relationship to radioligand binding. Ann. Rev. Neurosci. 6:43-71, 1983.

Ezrin-Waters, C. and Seeman, P. Tolerance to haloperidol catalepsy. Eur. J. Pharmacol. 41:321-327, 1977.

Fleminger, S., Rupniak N.M.J., Hall, M.D., Jenner, P. and Marsden, C.A. Changes in apomorphine-induced stereotypy as a result of subacute neuroleptic treatment correlates with increased D-2 receptors, but not with increases in D-1 receptors. Biochem. Pharmacol. 32:2921-2927, 1983.

Hess, E.J., Albers, L.J., Le, H. and Creese, I. Effects of chronic SCH 23390 treatment of the biochemical and behavioral properties of D_1 and D_2 dopamine receptors: Potentiated behavioral responses to a D_2 dopamine agonist after selective D_1 dopamine receptor upregulation. J. Pharmacol. Exp. Ther. 238:846-854 (1986).

Hess, E.J., Norman, A.B. and Creese, I. Chronic treatment with dopaine receptor antagonists: behavioral and pharmacologic effects of D_1 and D_2 dopamine receptors. J. Neuroscience, in press, 1988.

Iorio, L.C., Barnett, A., Leitz, F.H. Houser, V.P. and Korduba, C.A. SCH 23390, a potential benzazepine antipsychotic with unique interactions of dopaminergic systems. J. Pharm. Exp. Ther. 226:462-468, 1983.

Itoh, Y., Goldman, M.E. and Kebabian, J.W. TL333, a benzhydro[G]quinoline, stimulates both D-1 and D-2 dopamine receptors: Implications for the selectivity of LY 141865 towards the D-2 receptor. Eur. J. Pharmacol. 108:99-101, 1985.

Kebabian, J.W. and Calne, D.B. Multiple receptors for dopamine. Nature. 277:93-96, 1979.

Mailman, R.B., Schulz, D.W., Lewis, M.H., Staples, L., Rollema, H. and Dehaven, D.L. SCH 23390: A selective D_1 dopamine antagonist with potent D_2 behavioral actions. Eur. J. Pharmacol. 101:159-160, 1984.

Meller, E., Bohmaker, K. Goldstein, M. and Freidhoff, A.J. Inactivation of D_1 and D_2 dopamine receptors by N-ethoxycarbonyl-2-ethoxy-1,2-dihydroquinoline in vivo: Selective protection by neuroleptics. J. Pharmacol. and Exp. Ther. 233:656-6621, 1985a.

Meller, E., Kuga, S., Freidhoff, A.J. and Goldstein, M. Selective D_2 dopamine receptor agonists prevent catalepsy induced by SCH 23390, a selective D_1 antagonist. Life Sci. 36:1857-18641, 1985b.

7

DOPAMINE RECEPTORS AND SIGNAL TRANSDUCTION

R.G. MacKenzie and J.W. Kebabian

Abbott Laboratories
Abbott Park, IL 60064

INTRODUCTION

Dopamine receptors can be (and have been) identified on the basis of behavioral or physiological responses (such as turning behavior, supression of prolactin release or renal vasodilatation). However, by studying the signal transduction mechanisms used by the dopamine receptors, especially valuable insight into the identification and the classification of these receptors have been gathered. Thus, the dopamine-sensitive adenylate cyclase has proven to be a valuable model of the entity now known as the D-1 receptor (Kebabian et al., 1972). Furthermore, the now widely-accepted idea that there are two classes of dopamine receptor was based, in part, on the observation that certain drugs displayed inappropriate activity in the dopamine-sensitive adenylate cyclase assay (Kebabian and Calne, 1979).

The purpose of this chapter is to focus upon the different signal transduction mechanisms currently associated with either the D-1 or the D-2 dopamine receptor. It is our belief that from the understanding of the mechanisms used by dopamine receptors, it may be possible to better link the process of receptor stimulation (or blockade) to the final physiological event(s) occurring within the cells.

SIGNAL TRANSDUCTION AND D-1 RECEPTOR

The D_1 receptor has been identified with in vitro, ex vivo and in vivo assays. The in vitro assays are based on the ability of dopamine to stimulate either the production of cyclic AMP by intact cells or the activity of a dopamine-sensitive adenylate cyclase by cell-free homogenates of the tisseus (Brown and Makman, 1972; Dowling and Watling, 1981; Kebabian et al., 1972; Schultz et al., 1987). The in vivo and ex vivo assays are based on the measurement of a variety of tissue-specific responses to dopamine (Table 1). None of the in vivo assays offers the convenience and precision of the in vitro assay of enzyme activity (Table 1). Indeed, the 'proof' that any of these physiological events is the consequence of D_1 receptor stimulation requires the use of selective drugs whose pharmacological properties were characterized initially with the dopamine-sensitive adenylate cyclase assay (see Goldberg et al., 1984, Iorio et al., 1983). The D_1 receptor is, in many ways, similar to other receptors capable of activating adenylate cyclase. The available evidence suggests that the D_1 receptor consists of a ternary complex of receptor, stimulatory guanyl nucleotide binding protein (Gs) and adenylate cyclase.

The drug recognition site of the D_1 dopamine receptor can be directly identified in binding assays using a radiolabeled antagonist. Either [^{3}H]-SCH 23390 or its iodinated congener [^{125}I]-SCH 23892 interact with a binding site possessing drug recognition properties similar to those of the receptor regulating the dopamine-sensitive adenylate cyclase (Billard et al., 1984; Sidhu et al., 1986). The specific binding sites are found at high concentration in regions of the brain known to contain the dopamine-sensitive adenylate cyclase activity. The binding sites are lacking from non-dopaminergic brain regions or tissues rich in D_2 receptors. The D_1 receptor

Table 1
Models of the D_1 receptor

Tissue	Effect of dopamine	Parameter measured
Kidney	vasodilitation	renal blood flow
Parathyroid gland (bovine) release	enchanced exocytosis	parathyroid hormone
Brain	increased neuronal activity	increased glucose utilization
	rotary behavior cAMP production	number of turns adenylate cyclase activation

References:

Kidney: Goldberg et al., 1984; Parathyroid gland: Brown and Aurbach, 1980; Brain: Trugman and Wooten, 1987; Arnt, 1985; Kebabian et al., 1972.

has not yet been solubilized, isolated and puried in a manner similar to the D_2 receptor (see Caron, this volume).

GTP is essential for the demonstration of the activation by dopamine of adenylate cyclase. Sufficient GTP exists in a homogenate of brain tissue to support the activation by dopamine of enzyme activity (Clement-Cormier et al., 1975). Thus, when the soluble fractions of brain homogenate are replaced with buffer, the stimulatory effect of dopamine upon adenylate cyclase activity is lost. Readdition of GTP restores the effect of dopamine (Chen et al., 1980). In the bovine parathyroid, the presence of GTP is also essential for the demonstration of the maximal increase in cAMP production (Attie et al., 1980). These effects of GTP have been taken as evidence that Gs, the stimulatory guanyl nucleotide binding protein, is interposed between the D_1 receptor and the enzyme adenylate cyclase (Rodbell, 1980). However, it should be noted that in no case has such a 'G protein' been shown to be necessary and sufficient for the activation of the enzyme. Such a demonstration will require the isolation of the receptor and its reconstitution into a well-defined assay system.

The existence of D_1 receptors can be demonstrated in many tissues and organs. However, for each receptor, there is a gap between the dopamine-stimulated increase in cAMP production and the final physiological response of the tissue. The origin of this gap is the lack of understanding about the linkage between the biochemistry and the physiology of the individual tissues.

How does cAMP promote hormone release from the parathyroid gland or relaxation of vascular smooth muscle? We simply don't know. Within the central nervous system, the issues are even more complex. Within the dopaminergic regions of the brain, which cells possess the D_1 receptors? What is the physiological response to the stimulation of the D_1 receptor? How is a change in neuronal activity 'translated' into the characteristic turning behavior? The answer to each of these questions awaits further experimental investigation.

DOPAMINE D-2 RECEPTORS AND SIGNAL TRANSDUCTION

What are the signal transduction pathways linked to dopamine D-2 receptors?

Inhibition of Adenylate Cyclase

Agonist binding to the dopamine D-2 receptor has been associated with the activation of multiple transmembrane signalling pathways. Of these, the best characterized is the inhibition of adenylate cyclase and consequent inhibition of the accumulation or efflux of cAMP (Enjalbert and Bockaert, 1983; Schettini et al., 1983; Stoof and Kebabian, 1982). These effects have been observed in a variety of known dopaminergic targets including the lactotrophs of the anterior pituitary (Swennen and Denef, 1982), the melanotrophs of the intermediate lobe (Cote et al., 1981), and the striatum (Onali et al., 1985). Not surprisingly, dopamine D_2 inhibition of cyclase is best observed in tissues, such as the intermediate lobe, where a single or predominant cell type expresses the D_2 receptor negatively coupled to adenlyate cyclase. Thus, dopamine D_2 inhibition of adenylate cyclase is a much more robust phenomenon in the intermediate lobe than in the brain (compare Battaglia, et al., 1985; Weiss et al., 1985 to Cote, et al., 1981). In this regard, the recent observation of dopamine D_2 inhibition of adenylate cyclase in a rat anterior pituitary clonal cell line should prove most useful (Judd et al., 1987). Dopamine D_2 agonists have been shown to inhibit basal cyclase activity as well as to reverse cyclase stimulation induced by agonists acting at receptors positively coupled to cyclase (Onali et al., 1981; Cote et al., 1981) or by agents such as cholera toxin (Cronin and Thorner, 1982) and forskolin (Miyazaki et al., 1984) which act at post-receptor membrane proteins.

The dopamine D_2 receptor appears to be linked to adenylate cyclase via a member (or members) of the family of guanine-nucleotide binding proteins (G-proteins) (Birnbaumer et al., 1987) as shown by the guanine-nucleotide requirement of dopamine D_2 inhibition of the enzyme (Cote et al., 1982; Cooper et al., 1986). Moreover, pre-treatment with pertussis toxin blocks dopamine D_2 inhibition of cyclase and results in the ADP-ribosylation of a protein within the molecular weight range of proteins thought to mediate agonist inhibition of adenylate cyclase (Cronin et al., 1983; Enjalbert et al., 1986). The purified dopamine D_2 receptor of rat anterior pituitary co-elutes from an affinity chromatography column with a unique G-protein but it is not known whether this G-protein participates in dopamine D_2 inhibition of cyclase (see Caron this volume).

Effects on Ion Channels

Electrophysiological studies on dopamine D_2 effects on ion channel have been performed primarily in pituitary cells and neurons of the substantia nigra. A consistent effect of dopamine on lactotrophs, melanotrophs and neurons of the substantia nigra is to decrease the rate of cell firing; either spontaneous cell firing or that evoked by injection of depolarizing current (Ingram et al., 1986; Douglas and Taraskevich, 1985; Grace and Bunney, 1983). Inhibition of cell firing has been linked to a dopamine D_2 hyperpolarization due to increased membrane potassium conductance (Lacey et al., 1987) although other suggestions of a direct dopamine D_2 inhibition on positive inward currents have been put forth (Grace and Bunney, 1985). The coupling between the dopamine D_2 receptor and the potassium channel is not known. The observation that pertussis toxin pretreatment abolishes dopamine inhibition of cell firing (Innis and Aghajanian, 1987) points to a G-protein mediation of the response but the effector active at the potassium channel itself remains to be determined. In one study, using an enriched population of lactotrophs and a voltage sensitive dye to measure membrane potential, it was shown that dopamine D_2 hyperpolarization, although blocked by pertussis toxin treatment, was dissociable from dopamine D_2 inhibition of adenylate cyclase (Malgaroli et al., 1987).

Dopamine has also been shown to inhibit calcium currents (measured by calcium-sensitive dyes) via the D_2 receptor in anterior pituitary cells (Magaroli et al., 1987; Anderson et al., 1987) and striatal slices (Fujiwara et al., 1987). In pituitary cells, dopamine D_2 activation inhibits the rise in intracellular calcium to stimulatory agents such as thyrotropin releasing hormone (TRH), angiotensin-II (AII) or elevated extracellular potassium (Malgroli et al., 1987; Anderson et al., 1987) and in striatal slices, D_2 activation inhibits calcium mobilization following high external potassium and electrical field depolarization (Fujiwara et al., 1987). In anterior pituitary cells, TRH and AII elicit biphasic increases in intracellular calcium characterized by a brief initial peak lasting less than a minute followed by a plateau phase of elevated intracellular calcium lasting for many minutes (Malgaroli et al., 1987; Anderson et al., 1987). The peak has been attributed to calcium mobilization from intracellular stores while the plateau is thought to reflect an influx of calcium from outside the cell (Malgaroli et al., 1987). Moreover, the biphasic calcium response has been proposed to reflect agonist stimulation of phospholipase C resulting in the production of two second messengers; inositol phosphates for the early peak phase and diacyglcerol for the stimulation of protein kinase C which has been proposed to mediate the late plateau phase (Rasmussen et al., 1986). Both phases of TRH-induced calcium mobilization have been shown to be inhibited by dopamine D_2 activation (Malgaroli et al., 1987).

The coupling of dopamine D_2 receptor occupancy to inhibition of calcium currents appears to be G-protein mediated because pre-exposure to pertussis toxin blocks the effect (Malgaroli et al. 1987). It remains to be determined whether dopamine D_2 inhibition of calcium currents is a direct membrane effect of D_2 occupancy or whether it is consequent to other D_2 actions. For example, dopamine D_2 elevations in potassium currents and subsequent hyperpolarization might lead to a decrease in calcium currents through voltage-sensitive calcium channels as recently demonstrated for somatostatin-induced hyperpolarization and subsequent inhibition of calcium influx in GH_4C_1 pituitary cells (Koch and Schonbrunn, 1988). The D_2 inhibition of adenylate cyclase is dissociable from D_2 inhibition of the longer lasting plateau phase of stimulated increases in intracellular calcium but D_2 inhibition of the early peak phase, reflecting calcium mobilization from internal stores, does appear to be mediated by the lowering of intracellular cAMP. Agents which reverse D_2 effects on cyclase also reverse D_2 inhibition of the TRH-induced calcium peak but have no effect on D_2 inhibition of the TRH-induced calcium plateau (Malgaroli et al., 1987). As mentioned above, TRH is thought to influence calcium fluxes via stimulation of phospholipase C and therefore inhibition of this enzyme has become a candidate mode of action for activated dopamine D_2 receptors (see section C below).

The complexity of dopamine D_2 effects on ion channels is further confounded by a recent report on subtypes of rat anterior pituitary cells with differential internal calcium responses to dopamine. In cells with elevated intracellular calcium, dopamine D_2 activation decreased the intracellular levels of the cation, but in many cells exhibiting normal resting calcium levels there was a D_2-induced increase in intracellular calcium (Winiger et al., 1987).

Effects on Phosphatidyl Inositol Hydrolysis

Receptor-mediated enhancement of membrane phosphatidyl inositol hydrolysis via the stimulation of phospholipase C and resultant pleiotropic production of inositol phosphates and diacyglycerol are well-characterized transmembrane signalling events (Berridge, 1984; Nishizuka 1984). Only dopamine, however, acting through the D_2 receptor, has been implicated in receptor-mediated inhibition of phospholipase C. The initial observation consistent with such an action of dopamine involved dopamine D_2 inhibition of [^{32}Pi] incorporation into phosphatidyl inositol of rat hemipituitaries (Canonico et al., 1982). This finding led several investigators to obtain more direct evidence for dopamine D_2 inhibition of phosphatidyl inositol hydrolysis using the production of labelled inositol phosphates from cells pre-labelled with [^{3}H]myo-inositol as a measure of phospholipase C activity. Stimulation of the D_2 receptor has been shown to inhibit basal phospholipase C activity in striatal slices (Pizzi et al., 1987) as well as activity stimulated by AII and TRH in dissociated anterior pituitary cells (Simmonds and Strange, 1985; Enjalbert et al., 1986; Journot et al. 1987).

Dopamine D_2 inhibition of phospholipase C appears to be G-protein mediated since the effect is abolished following pretreatment with pertussis toxin (Journot et al., 1987). Moreover, a claim has been made that the D_2 effect on phospholipase C is independent of dopamine D_2 effects on either adenylate cyclase or calcium currents but data in support of this remain unpublished (Journot et al., 1987). It should be noted that the group responsible for the initial observation on [^{32}Pi] incorporation have been consistently unable to reproduce reports by others on [^{3}H]inositol phosphate production and have interpreted their original finding as a dopamine D_2 inhibition of phosphatidyl inositol synthesis (Canonico et al., 1986). Receptor-mediated inhibition of phospholipase C is a potentially important action of dopamine and more work must be done to fully characterize this phenomenon.

What are the physiological consequences of dopaminergic D-2 signal transduction in model systems?

Inhibition of Prolactin Release From the Anterior Pituitary Lactotroph

Prolactin secretion by the anterior pituitary lactotroph is under the inhibitory control of the hypothalamus via the release of dopamine from tuberoinfundibular neurons. These neurons arise in the arcuate and periarcuate nuclei of the hypothalamus and project to the median eminence where released dopamine is taken up into the portal blood and transported to the anterior pituitary for binding to dopamine D_2 receptors of the lactotrophs. Dopamine D_2 activation has a profound inhibitory effect on prolactin synthesis (Maurer, 1980) and degradation (Dannies and Rudnick, 1980) and might actually influence the growth and maintenance of the lactotrophs themselves (Miyazaki et al., 1985). However, the acute dopamine D_2 inhibition of secretion to a variety of secretagogues has been the most closely examined dopaminergic action on these cells and serves as a model of the physiological consequences of dopamine D_2 signal transduction.

Vasoactive intestinal polypeptide (VIP) elevates intracellular cAMP levels in anterior pituitary cells and stimulates prolactin release from the lactotrophs. Since both VIP effects are reversed by dopamine D_2 activation and because elevated cAMP levels induced by other agents also stimulate prolactin release, it has been proposed that dopamine inhibits VIP-stimulated prolactin secretion via inhibition of adenylate cyclase (Onali et al., 1981). However, it is apparent that dopamine can inhibit prolactin secretion via cAMP-independent mechanisms since secretion evoked by the addition of cAMP analogs is also reversed by dopamine D_2 activation (Delbeke and Dannies, 1985). Moreover, dopamine inhibits AII-stimulated prolactin release and AII, like dopamine, inhibits AII-stimulated prolactin release and AII, like dopamine, inhibits adenylate cyclase in anterior pituitary homogenates. AII is thought to have this action on lactotroph membranes since the effect is not additive with dopamine inhibition (Enjalbert et al., 1986) and most of the specific AII binding is to lactotrophs (Aguilera et al., 1982). AII does stimulate anterior pituitary phospholipase C and it has been proposed that D_2 inhibition of AII stimulation of this enzyme might mediate AII stimulation of prolactin (Enjalbert et al., 1986).

The pattern of TRH-stimulated prolactin release has been studied and shown to be similar to the pattern of TRH-stimulated calcium flux (see above, section B); namely a brief, early peak followed by a sustained plateau phase (Delbeke and Dannies, 1985). The early peak is mimicked by the calcium ionophore A23187, whereas the plateau is mimicked by a phorbol ester which activates protein kinase C; the biphasic TRH response being reproduced by the combination of A23187 and phorbol ester (Delbeke and Dannies, 1985). TRH stimulates anterior pituitary phospholipase C and it is thought that the peak and plateau responses reflect the bifurcation of this transmembrane signalling pathway into production of inositol phosphates and diacyglycerol respectively (Rasmussen et al., 1986). Not surprisingly, D_2 inhibition of phospholipase C has been proposed to mediate D_2 inhibition of TRH-stimulated prolactin release (Journot et al., 1987). However, dopamine D_2 activation not only lowers the peak and plateau prolactin responses to TRH but also blocks the prolactin responses elicited by A23187 and phorbol ester and these agents act distal to phospholipase C (Delbeke and Dannies, 1985). Interestingly, D_2 inhibition of these secretagogues is blocked by cAMP-elevating agents (Delbeke and Dannies, 1985) suggesting that the cAMP-lowering effect of D_2 activation plays some inhibitory role. It is known that calmodulin stimulates anterior pituitary adrenylate cyclase (Schettini et al., 1983) and it is possible that cAMP is the final common pathway for secretagogues acting via cyclase or phospholipase C. Indeed, forskolin, which increases intracellular cAMP levels, has been shown to act synergistically with TRH, A23187 and phorbol ester on prolactin secretion from GH_4C_1 cells (Delbeke et al., 1984).

In summary, dopamine, by binding to the lactrotroph D_2 receptor, inhibits prolactin release through a lowering of intracellular cAMP and through one or more cAMP-independent mechanisms. It remains to be seen whether D_2 effects on ion channels and phospholipase C contribute to dopaminergic inhibition of prolactin secretion.

REFERENCES

Aguilera, G., Hyde, C.L. and Catt, K.J. Angiotensin II receptors and prolactin release in pituitary lactotrophs. Endocrinology 111:1045-1050 (1982).

Anderson, J.M., Yasumoto, T. and Cronin, M.J. Intracellular free calcium in rat anterior pituitary cells monitored by fura-2. Life Sci. 41:519-526 (1987).

Arnt, J. (1985) Behavioural stimulation is induced by separate dopamine D_1 and D_2 receptor sites in reserpine-pretreated but not in normal rats. Eur. J. Pharmacol. 113:79-88.

Attie, M.F., Brown, E.M., Gardner, D.G., Spiegel, A.M. and Aurbach, G.D. (1980) Characterization of the dopamine-responsive adenylate cyclase of bovine parathyroid cells and its relationship to parathyroid secretion. Endocrinology 107:1776-1781.

Battaglia, G., Norman, A.B., Hess, E.J. and Creese I. D_2 dopamine receptor-mediated inhibition of forskolin-stimulated adenylate cyclase activity in rat striatum. Neurosci. Lett. 59:177-182 (1985).

Berridge, M.J. Inositol trisphosphate and diacyglycerol as second messengers. Biochem. J. 220:345-360 (1984).

Billard, W., Ruperto, V., Crosby, G., Iorio, L.C. and Barnett, A. (1984) Characterization of the binding of 3H-SCH 23390, a selective D_1 receptor antagonist ligand, in rat striatum. Life Sci. 35:1885-1893.

Birnbaumer, L., Codina, J., Mattera, R., Yatani A., Scherer, N., Toro, M.-J. and Brown, A.M. Signal transduction by G proteins. Kidney Internatl. 32:S14-S37 (1987).

Brown, E.M. and Aurbach, G.D. (1980) Role of cyclic nucleotides in secretory mechanisms and actions of parathyroid hormone and calcitonin. Vitam. Horm. 38:205-256.

Brown, J.H. and Makman, M.H. (1972) Stimulation by dopamine of adenylate cyclase in retinal homogenates and of adenosine-3':5'-cyclic monophosphate formation in intact retina. Proc Natl Acad Sci USA 69:539-543.

Canonico, P.L., Valdenegro, C.A. and MacLeod, R.M. Dopamine inhibits $^{32}P_i$ incorporation into phos phatidylinositol in the anterior pituitary gland of the rat. Endocrinology 111:347-349 (1982).

Canonico, P.L., Jarvis, W.D., Judd, A.M. and MacLeod, R.M. Dopamine does not attenuate phosphoinositide hydrolysis in rat anterior pituitary cells. J. Endocrinol. 110:389-393 (1986).

Chen, T.C., Cote, T.E. and Kebabian, J.W. (1980) Endogenous components of the striatum confer dopamine-sensitivity upon adenylate cyclase activity: the role of endogenous guanyl nucleotides. Brain Res 181:139-149.

Clement-Cormier Y.C., Parrish, R.G., Petzold, G.L., Kebabian, J.W., and Greengard, P. (1975) Characterization of a dopamine-sensitive adenylate cyclase in the rat caudate nucleus. J. Neurochem. 25:143-149.

Cooper, D.M.F., Bier-Laning, C.M., Halford, M.K., Alijanian, M.K. and Zahniser, N.R. Dopamine, acting through D_2 receptors, inhibits rat striatal adenylate cyclase by a GTP-dependent process. Mol. Pharmacol. 29:113-119 (1986).

Cote, T.E., Grewe, C.W., Tsuruta, K., Stoof, J.C., Eskay, R.L. and Kebabian, J.W. D_2 dopamine receptor-mediated inhibition of adenylate cyclase activity in the intermediate lobe of the rat pituitary gland requires guanosine 5'-triphosphate. Endocrinology 110:812-819 (1982).

Cote, T.E., Grewe, C.W. and Kebabian, J.W. Stimulation of a D_2 dopamine receptor in the intermediate lobe of the rat pituitary gland decreases the responsiveness of the beta-adrenoceptor: Biochemical mechanism. Endocrinology 108:420-426 (1981).

Cronin, M.J. and Thorner, M.D. Dopamine and bromocriptine inhibit cyclic AMP accumulation in the anterior pituitary: The effect of cholera toxin. J. Cyc. Nuc. Res. 8:267-275 (1982).

Cronin, M.J., Myers, G.A., MacLeod, R.M. and Hewlett, E.L. Pertussis toxin uncouples dopamine agonist inhibition of prolactin releaes. Am. J. Physiol. 244:E499-E504 (1983).

Dannies, P.S. and Rudnick, M.S. 2-Bromo-alpha-ergocryptine causes degradation of prolactin in primary cultures of rat pituitary cells after chronic treatment. J. Biol. Chem. 255:2776-2781 (1980).

Delbeke, D. and Dannies, P.S. Stimulation of the adenosine 3',5'-monophosphate and the Ca^{2+} messenger systems together reverse dopaminergic inhibition of prolactin release. Endocrinology 117:439-446 (1985).

Delbeke, D., Kojima, I., Dannies, P.S. and Rasmussen, H. Synergistic stimulation of prolactin release by Phorbol ester, A23187 and forskolin. Biochem. Biophys. Res. Comm. 123:735-741 (1984).

Douglas, W.W. and Taraskevich, P.S. The elctrophysiology of adenohypophyseal cells. In: A.M. Poisner and J.M. Trifaro (eds.) The Electrophysiology of the Secretory Cell: The Secretory Process, Vol II, Elsevier Sciene Publishers, New York, 63-92 (1985).

Dowling, J.E. and Watling, K.J. (1981) Dopaminergic mechanisms in the teleost retina. II. Factors affecting the accumulation of cyclic AMP in pieces of intact carp retina. J. Neurochem. 36: 569-579.

Enjalbert, A., Sladeczek, F., Guillon, G., Bertrand, P., Shu, C., Epelbaum, J., Garcia-Sainz, A., Jard, S., Lombard, C., Kordon, C. and Bockaert, J. Angiotensin II and dopamine modulate both cAMP and inositol phosphate productions in anterior pituitary cells J. Biol. Chem. 262:4071-4075 (1986).

Enjalbert, A. and Bockaert, J. Pharmacological characterization of the D_2 dopamine receptor negatively coupled with adenylate cyclase in rat anterior pituitary. Mol. Pharmacol. 23:576-584 (1983).

Fujiwara, H., Kato, N., Shuntoh, H. and Tanaka, C. D_2-dopamine receptor-mediated inhibition of intracellular Ca^{2+} mobilization and release of acetlcholine from guinea-pig neostriatal slices. Br. J. Pharmac. 91:287-297 (1987).

Goldberg, L.I., Glock, D., Kohli, J.D. and Barnett, A. (1984) Separation of peripheral dopamine receptors by a selective DA1 antagonist, SCH 23390. Hypertension 6(2 Pt 2): 25-30.

Grace, A.A. and Bunney, B.S. Intracellular and extracellular electrophysiology of nigral dopaminergic neurons – 1. Identification and characterization. Neuroscience 10:310-315 (1983).

Grace, A.A. and Bunney, B.S. Low doses of apomorphine elicit two opposing influences on dopamine cell electrophysiology. Brain Res. 333:285-298 (1985).

Ingram, C.D., Bicknell, R.J. and Mason, W.T. Intracellular recordings from bovine anterior pituitary cells: Modulation of spontaneous activity by regulators of prolactin secretion. Endocrinology 119:2508-2518 (1986).

Innis, R.B. and Aghajanian, G.K. Pertussis toxin blocks autoreceptor-mediated inhibition of dopaminergic neurons in rat substantia nigra. Brain Res. 411:139-143 (1987).

Iorio, L.C., Barnett, A., Leitz, F.H., Houser V.P., Korduba, C.A. (1983) SCH 23390, a potential benzazepine antipsychotic with unique interactions on dopaminergic systems. J. Pharmacol. Exp. Ther. 226:462-468.

Journot, L., Homburger, V., Pantaloni, C., Priam, M., Bockaert, J. and Enjalbert, A. An islet activation protein-sensitive G protein is involved in dopamine inhibition of angiotensin and thyrotropin-releasing hormone-stimulated inositol phosphate production in anterior pituitary cells. J. Biol. Chem. 262:15106-15110 (1987).

Judd, A.M., Login, I.S. and MacLeod, R.M. Dopamine inhibits prolactin release and cAMP generation in the MMQ cell, a homogeneous prolactin-secreting cell line. Society for Neuroscience Abstract 13:192 (1987).

Kebabian, J.W. and Calne, D.B. (1979) Multiple receptors for dopamine. Nature 277:93-96.

Kebabian, J.W., Petzold, G.L., and Greengard P. (1972) Dopamine-sensitive adenylate cyclase in caudate nucleus of rat brain, and its similarity to the "dopamine receptor". Proc. Natl. Acad. Sci. USA 69:2145-2149.

Koch, B.D. and Schonbrunn, A. Characterization of the cyclic AMP-independent actions of somatostatin in GH cells. J. Biol. Chem. 263:226-234 (1988).

Lacey, M.G., Mercuri, N.B and North R.A. Dopamine acts on D_2 receptors to increase potassium conductance in neurones of the rat substantia nigra zona compacta. J. Physiol. 392:397-416 (1987).

Malgaroli, A., Vallar, L., Elahi, F.R., Pozzan, T., Spada, A. and Meldolesi, J. Dopamine inhibits cytosolic Ca^{2+} increase in rat lactotroph cells. J. Biol. Chem. 262:13920-13927 (1987).

Maurer, R.A. Dopaminergic inhibition of prolactin synthesis and prolactin messenger RNA accumulation in cultured pituitary cells. J. Biol. Chem. 255:8092-8097 (1980).

Miyazaki, K., Dambrosia, J.M. Kebabian, J.W. Dopaminergic modulation of the diethylstilbestrol-induced proliferation of the anterior pituitary gland of the Fisher 344 rat. Neuroendocrinology 41:405-408 (1985).

Miyazaki, K., Goldman, M.E. and Kebabian, J.W. Forskolin stimulates adenylate cyclase activity, adenosine, 3',5'-monophosphate production and peptide release from the intermediate lobe of the rat pituitary gland. Endocrinology 114:761-766 (1984).

Nishizuka, Y. Turnover of inositol phospholipids and signal transduction. Science 225:1365-1370 (1984).

Onali, P., Schwartz, J.P. and Costa, E. Dopaminergic modulation of adenylate cyclase stimulation by vasoactive intestinal peptide in anterior pituitary. Proc. Natl. Acad. Sci. USA 78:6531-6534 (1981).

Onali, P., Olianas M.C. and Gessa G.L. Characterization of dopamine receptors mediating inhibition of adenylate cyclase activity in rat striatum. Mol. Pharmacol. 28:138-145 (1985).

Pizzi, M., D'agostini, F., DaPreda, M., Spano, P. F. and Haefely, W.E. Dopamine D_2 receptor stimulation decreases the inositol trisphosphate level of rat striatal slices. Eur. J. Pharmacol. 136:263-264 (1987).

Rasmussen, H., Apfeldorf, W., Barrett, P., Takuwa, N., Zawalich, W., Kreutter, D., Park, S. and Takuwa, Y. Inositol Lipids: Integration of cellular signalling systems. In: J.W. Putney (ed). Phosphoinositides and Receptor Mechanisms. Receptor Biochemistry and Methodology Series: Vol. 7, 109-147 (1986).

Rodbell, M. (1980) The role of hormone receptors and GTP-regulatory proteins in membrane transduction. Nature 284:17-22.

Schettini, G., Cronin, M.J. and MacLeon R.M. Adenosine 3',5'-monophosphate (cAMP) and calcium-calmodulin interrelation in the control of prolactin secretion: Evidence of dopamine inhibition of cAMP accumulation in prolactin release after calcium mobilization. Endocrinology 112:1801-1807 (1983).

Schultz, P.J., Sedor, J.R. and Abboud, H.E. (1987) Dopaminergic stimulation of cAMP accumulation on cultured rat mesangial cells. Am. J. Physiol 253 (Heart Circ. Physiol 22) H358H-H364.

Sidhu, A., van Oene, J.C., Danridge P., Kaiser, C. and Kebabian, J.W. (1986) [^{125}I]SCH 23982: the ligand of choice for identifying the D_1 dopamine receptor. Eur. J. Pharmacol. 128:213-220.

Simmonds, S.H., Strange, P.G. Inhibition of inositol phospholipid breakdown by D_2 dopamine receptors in dissociated bovine anterior pituitary cells. Neurosci. Lett. 60:267-272 (1985).

Stoof, J.C. and Kebabian J.W. Independent in vitro regulation by the D_2 dopamine receptor of dopamine-stimulated efflux of cyclic AMP and K^+-stimulated release of acetylcholine from rat neostriatum. Brain Res. 250:263-270 (1982).

Swennen, L. and Denef, C. Physiological concentrations of dopamine decrease adenosine $3',5'$-monophosphate levels in cultured rat anterior pituitary cells and enriched populations of lactotrophs: Evidence for a causal relationship to inhibition of prolactin release. Endocrinology 111:398-405 (1982).

Trugman, J.M. and Wooten, G.F. (1987) Selective D_1 and D_2 dopamine agonists differentially alter basal ganglia glucose utilization in rats with unilaterial 6-hydroxydopamine substantia nigra lesions. J. Neurosci. 7:2927-2935.

Weiss, S., Sebben, M., Garcia-Sainz J.A., and Bockaert J. D_2-dopamine receptor-mediated inhibition of cyclic AMP formation in striatal neurons in primary culture. Mol. Pharacol. 27:595-599 (1985).

Winiger, B.P, Wuarin, F., Zahnd, G.R., Wollheim, C.B. and Schlegel, W. Single cell monitoring of cytosolic calcium reveals subtypes of rat lactotrophs with distinct responses to dopamine and thyrotropin-releasing hormone. Endocrinology 121:2222-22228 (1987).

8

THE THERAPEUTIC POTENTIAL OF D_1 AND D_2 DOPAMINE AGONISTS IN PARKINSON'S DISEASE

Menek Goldstein[1] and Ariel Y. Deutch[2]

[1]New York University Medical Center,
Neurochemistry Research Laboratories
New York, NY 10016 and [2]Yale University, Departments of Psychiatry
and Pharmacology, New Haven, CT 06510

ABSTRACT

The neurological and biochemical deficits in monkeys with unilateral ventromedial tegmental lesions of the brain stem were compared with those treated with MPTP. The effects of D_1 an D_2 dopamine agonists on parkinsonian-like symptoms were investigated in both non-human primate models. The stimulation of D_2 dopamine receptors elicits a complete relief of tremor in monkeys with unilateral ventromedial tegmental lesions of the brain stem and a relief of parkinsonian-like symptomatology in MPTP-treated monkeys. The combined administration of bromocriptine with L-dopa produces an increased duration of the relief of tremor in monkeys with unilateral ventromedial tegmental lesions of the brain stem when compared with each drug alone. It is postulated that the presence of synaptic dopamine enhances the dopamine agonist activity of bromocriptine at the D_2 dopamine receptors. Dopamine agonists which effectively stimulate dopamine autoreceptors and postsynaptic supersensitive, but not normosensitive dopamine receptors, effectively relieve tremor and do not produce severe abnormal involuntary movements in monkeys with unilateral ventromedial tegmental lesions of the brain stem. The therapeutic potential of selective dopamine agonists in treatment of Parkinson's disease has to be further evaluated.

INTRODUCTION

The concept of substituting the missing striatal dopamine (DA) in Parkinson's disease with administered levodopa or with DA agonists has gained wide use in treatment of Parkinson's disease. During the last decades substantial progress was made in the understanding of the physiology and biochemistry of the dopaminergic neurotransmission and its relevance to the pathology of Parkinson's disease. Animal models which replicate certain neurological and biochemical deficits of Parkinson's disease were useful in studies on the neurobiology of Parkinson's disease and in evaluation of the antiparkinsonian activity of putative drugs. These models are based on the disruption of central dopaminergic transmission in non-human primates by stereotaxical lesions of the nigrostriatal dopaminergic neuronal systems or by treatment with the neurotoxin 1-methyl-4-phenyl-1,2,3,6-tetrahydropyridine (MPTP). An understanding of the usefulness of these animal models could be obtained by comparing the biochemical and neurological deficits in monkeys with unilateral surgical lesions of the ventromedial tegmental (VMT) area of the brain stem and those treated with MPTP, and in those which occur in Parkinson's disease. In this presentation we describe both animal models and their utility for evaluating the antiparkinsonian activity of various types of DA agonists.

Comparison of neurological and biochemical deficits in monkeys with unilateral VMT lesions of the brain stem with those treated with MPTP

Monkeys with unilateral lesions in the VMT area of the brain exhibit tremor and hypokinesia similar to that observed in Parkinson's disease (Poirier and Sourkes, 1965). Mon-

Table 1

A comparison of deficits in monkeys with unilateral VMT lesions of the brain stem
with those monkeys treated with MPTP

Monkey with VMT lesions	MPTP-treated monkeys
Degeneration of the nigrostriatal and mesolimbic DA neurons.	Degeneration of the nigrostriatal and partial degeneration of the mesolimbic DA neurons. Loss of TH-positive neurons in the locus coeruleus and subcoeruleus aged monkeys.
Contralateral hypokinesia and tremor	Transient and immediate effects: stiffening, slow movements. Permanent effects: akinesia- lack of spontaneous movements, rigidity.
Supersensitivity of the post-synaptic DA receptors	Supersensitivity of the post-synaptic DA receptors.

keys treated with MPTP first exhibit transient neurological deficits such as slow movements and stiffening of the limbs, and subsequent irreversible deficits (e.g. akinesia, rigidity, etc.). A decrease in DA and tyrosine hydroxylase (TH) activity occurs on the lesioned side of the striatum in monkeys with unilateral VMT lesions of the brain stem (Goldstein et al., 1973) and on both sides of the striatum and, to a lesser extent, in the mesolimbic areas in monkeys treated with MPTP (Elsworth et al., 1987). In addition, MPTP treatment produces an extensive loss of TH-positive norepinephrine (NE) neurons in the locus coeruleus of an aged monkey, while no

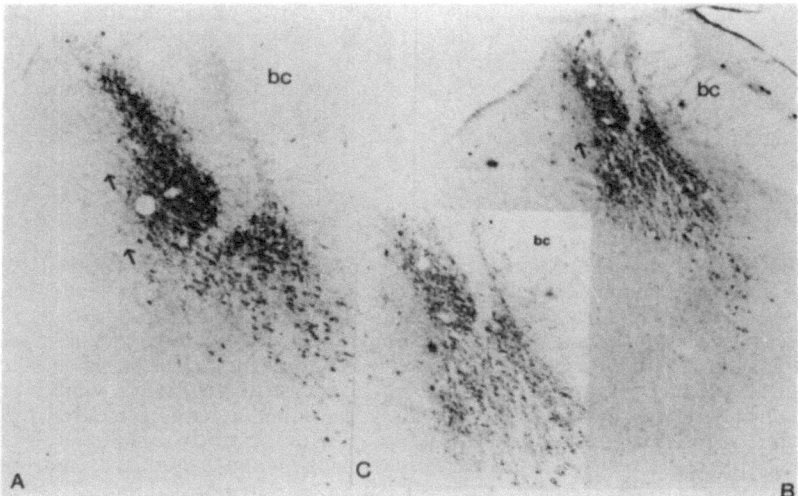

Figure 1. The noradrenergic neurons of the locus coeruleus (LC) region of MPTP-treated monkeys, revealed by TH immunohistochemistry.

A. The noradrenergic neurons of the LC in an MPTP-treated monkey. The striatum of this animal was virtually devoid of TH-positive fibers. The neurons of the LC are clearly visible, as are the (NE) cells of the subcoeruleus region. A dense dendritic arborization of the NE neurons, extending into the laterodorsal tegmental nucleus, is present (arrows). bc, brachium conjunctivum.

B. The LC of an aged African green monkey treated with MPTP. In contrast to the LC from a young animal, shown in panel A, there is an extensive loss of TH-positive neurons in the LC and subcoeruleus; gliosis is present.

C. A higher magnification of the LC region from the monkey shown in panel B. The extensive loss of both the TH-positive neurons and the severe pruning of the dendrites is apparent.

Table 2

Comparative properties of DA agonists

Drug	Effect on DA receptors		Efficacy	
	D_1	D_2	Antitremor[a]	Antiparkinsonian[b]
Dopamine	Stimulation	Stimulation	Very effective	Very effective
Apomorphine	Stimulation	Stimulation	Very effective	Very effective
Bromocriptine	Blockade[c]	Stimulation	Effective	Effective
Lergotrile	Blockade[c]	Stimulation	Effective	Effective
Lisuride	Blockade[c]	Stimulation	Effective	Effective
Pergolide	Stimulation Blockade[c]	Stimulation	Very effective	Very effective (in some patients)
LY 141865	?	Stimulation	Effective	?
SKF 38393	Stimulation	?	?	?

[a]Antitremor efficacy evaluated in monkeys with VMT lesions of the brain stem.
[b]Antiparkinsonian efficacy evaluated in parkinsonian patients.
[c]Blockade at higher concentrations of the drug.

real loss is observed in the younger animals (A.Y. Deutch, K. Fuxe and M. Goldstein, unpublished data) (Table 1, Fig. 1). The DA denervation following the surgical lesion or MPTP treatment produces D_2 DA receptor supersensitivity in the striatum while changes in D_1 DA receptors have not yet been determined (Table 1).

The effects of D_1 and D_2 DA agonists on the amelioration of parkinson-like symptoms

The question whether the amelioration of parkinson-like symptoms by DA agonists is associated with stimulation of D_1 or D_2 or both D_1 and D_2 DA receptors was investigated. It is evident from the results presented in Table 2 that DA agonists which stimulate both D_1 and D_2 DA receptors effectively relieve tremor in monkeys with unilateral VMT lesions of the brain stem and are effective as antiparkinsonian agents in man. The selective D_2 DA agonist LY 141865 (partial ergoline) is very effective in relieving tremor in the monkeys with VMT lesions of the brain stem, but clinical studies with this agent have not yet been performed. The selective D_1 DA agonist SKF 38393 in the tested dose (3 mg/kg; i.m.) was not effective as an antitremor agent in this animal model and no clinical trials are available. The results presented in Table 3 show the effects of DA agonists on parkinsonian symptoms in monkeys treated with MPTP. It is evident that some recently described potent and selective D_2 DA agonists, (+)-4-

Table 3

Comparative properties of DA agonists in monkeys
treated with MPTP

Drugs (dose, mg/kg)	Stimulation of DA receptors	Effect on akinesia (duration)
(+) PHNO (0.01-0.04, i.m.)	D_2	Effective (4-5 hrs)
LY 163502 (0.01-0.02, i.m.)	D_2	Effective (4-5 hrs)
Bromocriptine (6.0, i.m.)	D_2	Slightly effective (3-4 hrs)
L-dopa (25, i.m.) + MK 486 (10.0, i.m.)	$D_1 + D_2$	Effective (30-60 min)
CY 208-24 (0.5-2.0 per os)	D_1	Not effective

Table 4

Effect of αMPT on DA agonist-induced rotations in 6-OH-DA lesioned rats and computer derived parameters of dissociation constants for DA agonists

DA agonists	Ratio[a] αMPT/No αMPT	K_L/K_H[b]
Dopa (DA)	1.0	80
Apomorphine	1.0	55
SKF 38393	0.85	—
Pergolide	0.65	33
CQ 32804	0.65	11
LY 141865	0.33	Approx 1
RU 24213	0.25	"
Bromocriptine	0.17	"

[a]The data was published (Gershanik et al., 1983).
[b]The data is reported in the literature.

propyl-9-hydroxynaphthoxazine (+)PHNO (Martin et al., 1984) or LY 163502, effectively relieve parkinsonian symptoms for several hours. Bromocriptine, which is a less potent D_2 DA agonist, is less effective, and L-dopa in combination with the peripheral L-aromatic amino acid decarboxylase (AAAD) inhibitor MK 486 is effective, but only for a short period of time. The recently described selective D_1 DA agonist CY 208-24 was not effective in the tested doses. Thus, it appears that stimulation of D_2 DA receptors is associated with the antiparkinsonian activity of DA agonists in both animal models.

D_1 and D_2 DA agonist synergism

There are a number of behavioral studies which indicate a synergistic effect of stimulation of D_1 and D_2 DA receptors by DA agonists (Arnt et al., 1987). Evidence was presented that D_1 DA receptor activation is required for postsynaptic expression of D_2 agonist effects (Walters et al., 1987). We have observed that treatment with bromocriptine and levodopa prolongs the antitremor effectiveness of levodopa in monkeys with unilateral VMT lesions of the brain stem. The question was therefore addressed whether the effects of combined treatment with D_1 and D_2 DA agonists are produced by simultaneous stimulation of D_1 and D_2 DA receptors or whether D_1 agonists enhance the intrinsic activity of D_2 DA agonists by promoting coupling of the DA receptor with the guanine nucleotide regulatory protein (N) of the adenylate cyclase system (Goldstein et al., 1985).

In our earlier studies we have observed that the antitremor activity of the DA agonist trivastal and that of bromocriptine is diminished by prior administration of the DA synthesis inhibitor α-methyl-p-tyrosine (αMPT) (Goldstein et al., 1973). It was also reported that D_2 agonist-induced contralateral rotation in rats with unilateral 6-OH-DA lesions of the nigrostriatal DA neurons is greatly diminished by treatment with αMPT (Gershanik et al., 1983). These findings led to the idea that DA agonists which differentiate between high and low affinity states of the receptor ($K_L/K_H > 1$) (Table 4) are not affected by pretreatment with MPT, while those which do not differentiate ($K_L/K_H = 1$) are affected by pretreatment with the DA synthesis inhibitors (Goldstein et al., 1985). This might imply that DA agonists with low intrinsic efficacy ($K_L/K_H = 1$) require DA or another DA agonist with high intrinsic efficacy to induce a conformational change at the receptor which leads to conversion of the D_2 DA receptor from a low to a high affinity state (Goldstein et al., 1985). In order to test this hypothesis the antitremor activity of a long-acting bromocriptine preparation (Parlodel L.A.) alone and in combination with L-dopa was evaluated.

The administration of a low dose of Parlodel L.A. by itself (5 mg/kg, i.m.) has no effect on tremor and does not produce abnormal involuntary movements (AIM's), while the administration of a low dose of L-dopa (25 mg/kg, i.m.) relieves tremor for approximately 1 hr and 45 min and produces AIM's for approximately 1 hr. Figure 2 shows that when L-dopa is given to a monkey which was pretreated 2 hrs before with Parlodel L.A., the antitremor action persists for a longer period of time than when L-dopa is given alone. The duration of AIM's is not significantly increased. The administration of L-dopa two, six and seven days after pretreatment with Parlodel L.A. also produces a longer antitremor effect than L-dopa alone. However, when L-

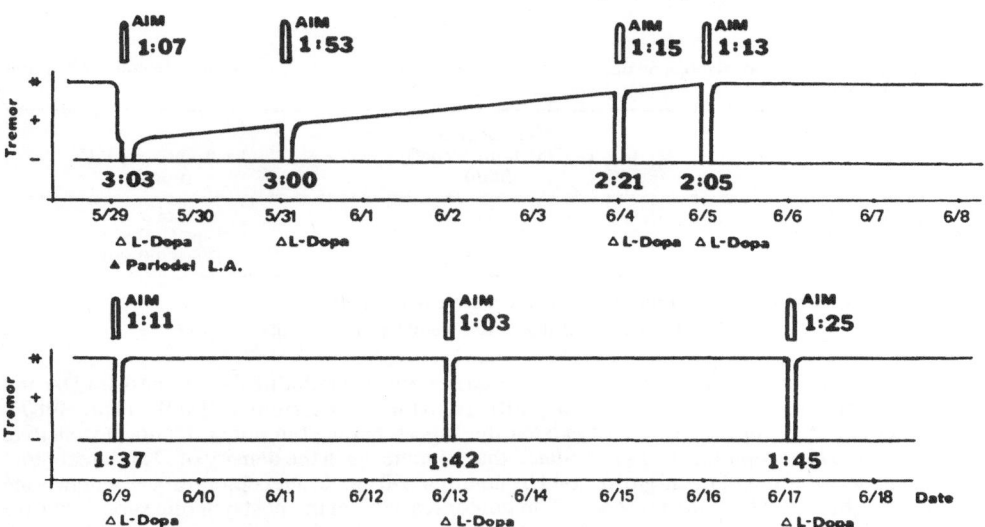

Figure 2. The effect of the administration of Parlodel, L.A. and L-dopa on tremor and occurrence of AIM's in monkeys with unilateral VMT lesions of the brain stem.

A single dose of Parlodel, L.A. (5 mg/kg; i.m.) was given. L-dopa (25 mg/kg; i.m.) was given two hrs and at different days after administration of Parlodel, L.A. (the dates when the drugs were administered are registered above).

dopa is given 10-11 days after pretreatment with Parlodel, the prolonged antitremor effect disappears.

These results show that the co-administration of bromocriptine with L-dopa prolongs the antitremor effects of L-dopa without significantly affecting the duration of AIM's. The increased duration of the antitremor action of DA in presence of bromocriptine supports the previously proposed mechanism that DA induces a conformational change of the D_2 DA receptor which leads to the formation of a ternary complex between the transmitter, receptor protein and the nucleotide regulatory protein (Goldstein et al., 1985). Bromocriptine does not differentiate between the high and low affinity states of the receptor ($K_L/K_H = 1$), but it might replace DA from the ternary complex and, by virtue of its higher affinity, enhance the duration of the agonist activity. Based on this hypothesis, we predicted that bromocriptine and other DA agonists with low intrinsic activity will be therapeutically more effective when combined with low doses of L-dopa. Additional clinical studies are required to test this hypothesis and to evaluate the clinical utility of the combined treatment with low doses of both L-dopa and bromocriptine.

The antiparkinsonian activity of partial DA agonists

Recently a number of DA agonists with selective activity for the DA autoreceptors and for supersensitive, but not for normosensitive DA receptors, were developed. Evidence from several different models indicates that compounds such as 3-PPP (Hjorth et al., 1981), EMD 23,448 (Goldstein et al., 1987), and ciladopa (Koller et al., 1986) selectively stimulate central DA autoreceptors and supersensitive postsynaptic DA receptors. It is evident from the data summarized in Table 5 that EMD 23,448 and ciladopa effectively relieve tremor in monkeys with unilateral VMT lesions of the brain stem. The relief of tremor is concomitant with the occurrence of Type I and not Type II AIM's (Table 5). Thus, in contrast to the classical DA agonists such as DA, apomorphine, etc., these partial DA agonists produce only the less severe Type I AIM's. Based on these studies one might predict that selective DA agonists are clinically advantageous over classical DA agonists since their antiparkinsonian activity is associated with the occurrence of less severe AIM's.

The mechanism which might be responsible for the selective stimulation of DA autoreceptors and supersensitive postsynaptic DA receptors was investigated. Evidence was ob-

Table 5

Effect of EMD 23,448 and ciladopa on tremor and occurrence of AIM's in monkeys with unilateral VMT lesions of the brain stem

Drug (mg/kg; i.m.)	Effect of tremor (Duration of relief) (min)	Occurrence of AIM's (mins)
EMD 23,448 (1.0)	30	none
EMD 23,448 (2.0)	120 - 160	Type I (60-90)
Ciladopa (5.0)	120 - 180	Type I (60-90)

Classification of AIM's: AIM I - Restlessnes, slight stereotype movements
AIM II - Chorea-like and intensive stereotype movements

tained that a receptor reserve exists at the DA autoreceptors regulating the synthesis of DA and at those modulating the release of the transmitter from the nerve terminals (Meller et al., 1987). A similar DA receptor reserve might exist at the supersensitive but not at the normosensitive postsynaptic DA receptors. It was postulated that an increase in the density of D_2 DA receptors following denervation of the nigrostriatal DA neurons will create a receptor reserve where none had existed before. Thus, the presence of "receptor reserve" at the postsynaptic supersensitive DA receptors may explain, in part, why DA agonists with low intrinsic activity, ineffective at normosensitive receptors, will be effective at supersensitive receptors (Meller et al., 1987).

REFERENCES

Arnt, J., Hyttel, J. and Perregaard, J. Dopamine D-1 receptor agonists combined with the selective D-2 agonist quinpirole facilitate the expression of oral stereotyped behaviour in rats. Eur. J. Pharmacol. 133:137-145, 1987.

Elsworth, J.D., Deutch, A.Y., Redmond, D.E., Jr., Sladek, J.R., Jr. and Roth, R.H. Effects of 1-methyl-4-phenyl-1,2,3,6-tetrahydropyridine (MPTP) on catecholamines and metabolites in primate brain and CSF. Brain Res. 415:293299, 1987.

Gershanik, O., Heikkila, R.E. and Duvoisin, R.C. Effects of dopamine depletion on rotational behavior to dopamine agonists. Brain Res. 261:358-360, 1983.

Goldstein, M., Battista, A.F., Ohmoto, T., Anagnoste, B. and Fuxe, K. Tremor and involuntary movements in monkeys: Effect of L-Dopa and of a dopamine receptor stimulating agent. Science 179:816-817, 1973.

Goldstein, M., Lieberman, A. and Meller, E. A possible molecular mechanism for the antiparkinsonian action of bromocriptine in combination with levodopa. Trends Pharmacol. Sci. 6:436-437, 1985.

Goldstein, M., Fuxe, K., Meller, E., Seyfried, C.A., Agnati, L. and Mascagni, F.M. The characterization of the dopaminergic profile of EMD 23,448, an indolyl-3-butylamine: selective actions on presynaptic and supersensitive postsynaptic DA receptor populations. J. Neural Transm. 70:193-215, 1987.

Hjorth, S., Carlsson, A., Wikstrom, H., Lindberg, P., Sanchez, D., Hacksell, U., Arvidsson, L.E., Svensson, U. and Nilsson, J.L.G. 3-PPP, a new centrally acting DA receptor agonists with selectivity for autoreceptors. Life Sci. 28:1225-1238, 1981.

Koller, W.C., Fields, J.Z., Gordon, J.H. and Perlow, M.J. Evaluation of ciladopa hydrochloride as a potential anti-parkinson drug. Neuropharmacol. 25:973979, 1986.

Martin, G.E., Williams, M., Pettibone, D.J., Yarbrough, G.G., Clineschmidt, B.V. and Jones, J.H. Pharmacologic profile of a novel potent direct-acting dopamine agonist, (+)-4-propyl-9-hydroxynaphthoxazine [(+)-PHNO]. J. Pharmacol. Exp. Ther. 230:569-576, 1984.

Meller, E., Bohmaker, K., Namba, Y., Friedhoff, A.J. and Goldstein, M. Relationship between receptor occupancy and response at striatal dopamine receptors. Mol. Pharmacol. 31:592-5898, 1987.

Poirier, L.J. and Sourkes, T.L. Influence of the substantia nigra on the catecholamine content of the striatum. Brain 88:181, 1965.

Walters, J.R. Bergatrom, D.A., Carlson, J.H., Chase, T.N. and Braun, A.R. D_1 dopamine receptor activation required for postsynaptic expression of D_2 agonist effects. Science 236:719-722, 1987.

9

ISOLATION AND BIOCHEMICAL CHARACTERIZATION OF THE D$_1$ AND D$_2$ DOPAMINE RECEPTORS

Jay A. Gingrich, Susan E. Senogles, Nourdine Amilaiky, Wei K. Chang*, Joel G. Berger*, and Marc G. Caron

Departments of Physiology Medicine, and Howard Hughes Medical Institute Laboratories, Duke University Medical Center, Durham, NC 27710 and *Research Division, Schering-Plough Corp., Bloomfield, NJ 07003

INTRODUCTION

The catecholamine, dopamine, exerts physiologic effects in both the central nervous system and the periphery. Dopamine systems in the central nervous system (CNS) have been implicated in several neurologic and psychiatric disorders such as Parkinsonism, schizophrenia, Huntington's chorea, Tourette's syndrome, and Lesch-Nyhan syndrome. In the periphery, dopamine plays a role in processes controlling renal vascular tone and release of the hormones, prolactin from the anterior pituitary gland and parathyroid hormone from the parathyroid gland. The symptoms of some pathological processes are ameliorated through the use of drugs targeted toward the receptors for dopamine. For example, dopamine agonists have been successfully used to treat Parkinson's disease, hyperprolactinemia (prolactinomas), and to prevent renal ischemia associated with cardiovascular collapse (shock), and dopamine antagonists have proven useful in the treatment of schizophrenia, and Tourette's syndrome.

An important advance in the understanding of dopamine physiology was the recognition that multiple receptor subtypes are involved in mediating the effects of dopamine. At least two major receptor subtypes have been identified on the basis of pharmacologic and biochemical criteria and are referred to as D$_1$ and D$_2$ dopamine receptors (for review see, Kebabian and Calne, 1979; Stoof and Kebabian, 1984). Biochemically, the D$_1$ receptor mediates the stimulatory effect of dopamine on the enzyme adenylyl cyclase, while the D$_2$ receptor is coupled to the inhibition of adenylyl cyclase. Recent evidence, however, has suggested that the D$_2$ receptor may also be involved in the control of K$^+$ fluxes and phosphotidyl inositol turnover (Sasaki and Sato, 1987; Enjalbert et al., 1987; Margaroli et al., 1987).

Attempts have been made to assign the various effects of dopamine to the activation or inhibition of one or the other receptor subtype. Until recently, almost all the effects of dopamine in the CNS were thought to be mediated by the D$_2$ receptor. These included: antipsychotic effects, antistereotypic effects and avoidance inhibition effects of dopaminergic antagonists (for review see, Seeman, 1980; Arnt, 1987). However, studies which examined the behavioral effects of the first D$_1$ selective antagonist, SCH 23390, found this compound to elicit many of the same effects expected of a D$_2$ selective antagonist (Iorio et al., 1983; Christensen et al., 1984; Waddington, 1986). Subsequent studies have suggested that despite their apparent opposing effects on cAMP production D$_1$ and D$_2$ receptors in the CNS exert complementary or synergistic effects on behavior and neuronal firing (Barone et al., 1986; Robertson & Robertson, 1986; Walters et al., 1987). In dopamine depleted animals, D$_1$ receptor activation appears to be required for expression of D$_2$ mediated effects, suggesting that the D$_1$ receptor may act in a "permissive" or "enabling" capacity for D$_2$ activation under these circumstances (Gershanik et al., 1983; Walters et al., 1987; for review see, Clark and White, 1987). Indeed, the availability of selective pharmacologic tools for both the D$_1$ and D$_2$ receptors has yielded many unexpected findings which have suggested a previously unappreciated degree of interaction between these receptor subtypes in the CNS.

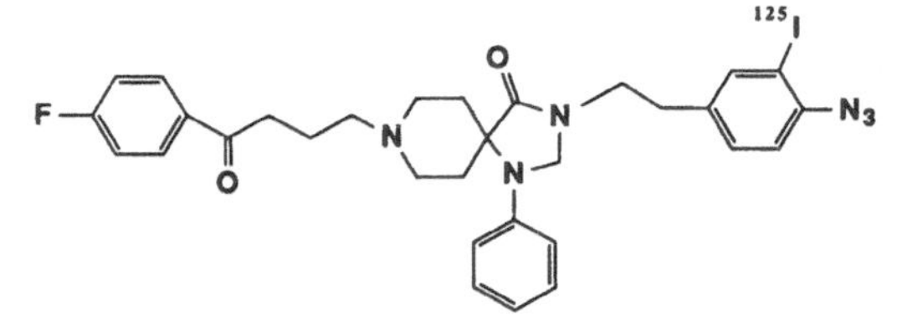

N-(p-azido-m-[^{125}I]-iodophenethyl)-spiperone ([^{125}I]N$_3$ NAPS)

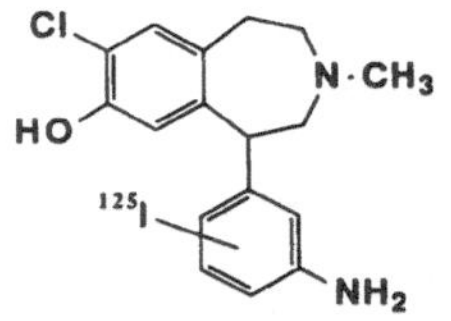

[^{125}I]SCH 38548 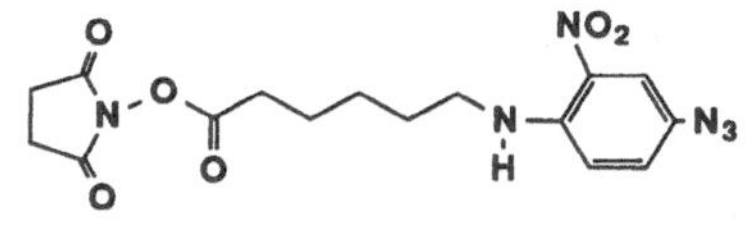   N-succinimidyl-6-(4'-azido-2'-nitrophenyl-amino)-hexanoate
(SANPAH)

Figure 1. Structures of [^{125}I]N$_3$-NAPS, [^{125}I]SCH 38548, and SANPAH.

The focus of work in this laboratory has been the purification and biochemical characterization of the receptors for dopamine and the study of their molecular mechanisms of signal
transduction. These studies should ultimately provide a clearer understanding of the effects of
chronic drug therapy on these receptors and yield some insights into the paradoxical interactions these receptors exert in the CNS. The purpose of this chapter is to review some of the
progress our laboratory has made towards the molecular characterization of the two receptor
subtypes for dopamine. In particular, the following sections will describe results obtained describing the identification of the ligand binding subunits of each receptor subtype using photoaffinity labelling techniques, and the progress we have made towards the isolation of each receptor subtype. Considerable progress has been made in the isolation and characterization of
the D$_2$ dopamine receptor subtype. Indeed, the D$_2$ receptor from bovine anterior pituitary
gland has recently been purified to apparent homogeneity (Senogles et al., 1988). This work
will be reviewed briefly. More recently, work directed towards the isolation of the D$_1$ receptor
has been initiated in our laboratory, and the work describing the successful solubilization of
and application of an affinity chromatography procedure to the purification of this receptor will
be discussed.

Identification of the Ligand Binding Subunits of the D$_1$ and D$_2$ Receptors By Photoaffinity Labelling

As a first step in the molecular characterization of the dopamine receptor subtypes, the
ligand binding subunits of these receptors have been identified by covalent labelling with
selective radioiodinated ligands. Fig. 1 shows the structure of the two probes designed for this
purpose. The compound [^{125}I]SCH 38548 is a close structural analog to the well characterized
D$_1$ selective antagonist, SCH 23390 – the amine moiety in the 3' position allows both the
radioiodination of this compound as well as providing a reactive group that can be used in concert with the photo-activatable, heterobifunctional crosslinking reagent, N-succinimidyl-6-
(4'-azido-2'-nitrophenyl-amino)-hexanoate (SANPAH) to label the receptor. Using
[^{125}I]SCH 38548 and SANPAH, a protein of Mr 72,000 can be covalently labelled and visualized by autoradiography after sodium dodecylsulfate polyacrylamide gel electrophoresis
(SDS-PAGE) from rat striatum (see Fig. 2). The incorporation of [^{125}I]SCH 38548 can be
blocked in a stereoselective fashion (SCH 23390>SCH 23388, cis flupenthixol) > trans
flupenthixol) and by agonists with the rank order of potency expected of the D$_1$ receptor
(SKF 38393)> apomorphine> dopamine), thus demonstrating the appropriate characteristics
of the ligand binding subunit of the D$_1$ receptor (Amlaiky et al., 1987).
The ligand binding subunit of the D$_2$ receptor subtype has also been identified using a
photosensitive derivative of spiperone, [^{125}I]N-(p-azidophenethyl) spiperone ([^{125}I]N$_3$-
NAPS) – the structure of which is shown in Fig. 1. In contrast to [^{125}I]SCH 38548, [^{125}I]N$_3$-
NAPS covalently incorporates into a protein of Mr~94,000-120,000 and displays a phar-

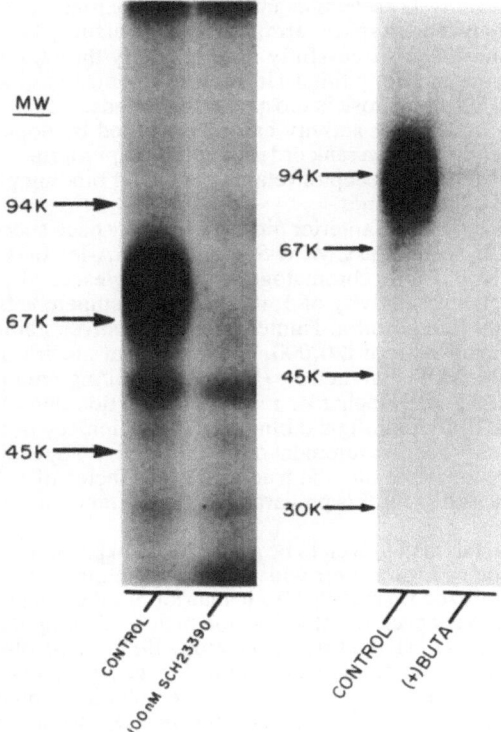

Figure 2. Identification of the Ligand Binding Subunit of the D_1 and D_2 Dopamine Receptors from Rat Striatum by Photolabelling. Rat striatal membranes were incubated with [^{125}I]SCH 38548 (left) or [^{125}I]N$_3$-NAPS (right) as described in Amlaiky et al., 1987 & Amlaiky and Caron, 1985. Shown are the autoradiograms of SDS-PAGE. Molecular weight standards are indicated on the sides.

macologic specificity consistent with that of the D_2 receptor (N-propylnorapomorphine (NPA)>apomorphine>dopamine; (+) butaclamol>> (−)butaclamol) (Amlaiky and Caron, 1985, 1986). That the D_1 and D_2 receptors exist on separate peptides is clearly seen in Fig. 2 which shows the relative positions of the ligand binding subunits of these two receptors on SDS-PAGE after covalent photolabelling and subsequent autoradiography.

It should be noted that in most tissues where D_2 receptors have been labelled by the photoaffinity probe [^{125}I]N$_3$-NAPS a major peptide of Mr 94,000 has been characterized (Amlaiky and Caron, 1985). However in some tissues, such as the neurointermediate lobe of the rat pituitary, and bovine anterior pituitary gland, a species of higher Mr (120,000) can be found (Amlaiky and Caron, 1986).

Progress Towards the Isolation and Biochemical Characterization of the D_1 and D_2 Dopamine Receptors

D_2 Dopamine Receptor

The D_2 receptor has been solubilized from several species and tissues sources with a variety of detergents and chaotropic agents (for review see Caron et al., 1983). Using the detergent digitonin, we have been able to solubilize D_2 receptor binding activity for bovine and porcine anterior pituitary gland (Kilpatrick and Caron, 1983, 1984) with complete retention of ligand binding pharmacology. The D_2 dopamine receptor from these digitonin solubilized preparations has been substantially enriched by biospecific affinity chromatography on CMOS-Sepharose (Senogles et al., 1986). CMOS (carboxymethyleneoximinospiperone) is a derivative of the high affinity D_2 selective antagonist, spiperone, and when immobilized on Sepharose 4B through an extended spacer arm, retains high affinity for the D_2 receptor. The resulting affinity matrix has been successfully used to purify the D_2 dopamine receptor more than 1000-fold from bovine anterior pituitary in a single chromatographic step. The interaction between the D_2 and CMOS-Sepharose is biospecific since adsorption of receptor activity can be blocked and elution of receptor activity can be promoted by dopamine agonists and antagonists with stereoselectivity and a rank order of potency appropriate for the D_2 receptor. The affinity-purified reconstituted D_2 receptor retains the ligand binding pharmacology expected of the D_2 receptor found in membranes.

The D_2 receptor from bovine anterior pituitary has now been successfully purified to apparent homogeneity using sequential CMOS-Sepharose, DSA-lectin (Datura stramonium lectin-agarose), and hydroxylapatite chromatography (Senogles et al., 1988). The resulting purified material has a specific activity of 5,400 pmol [^{3}H]spiperone binding/mg protein and migrates as a single band (after Bolton-Hunter iodination, silver staining or Coomassie blue staining) on SDS-PAGE with a Mr of 120,000. The theoretical specific activity of the D_2 receptor (based on a Mr of 120,000) is predicted to be 8,300 pmol/mg protein. However, the D_2 receptor after elution from CMOS-Sepharose requires reinsertion into phospholipid vesicles in order to measure high affinity radioligand binding. The efficiency of this reconstitution process is ~60-70%, meaning that measurement of specific activity by radioligand binding in this system will necessarily underestimate the true value by a factor of ~1.4-1.7 fold. Thus, the final specific activity measured for this preparation is consistent with the predicted theoretical specific activity.

CMOS-Sepharose has also proven to be a useful tool in probing the nature of the interaction of the anterior pituitary D_2 receptor with a specific guanine nucleotide binding protein. Early studies of the solubilized D_2 receptor from anterior pituitary gland found that occupancy of the receptor with an agonist prior to solubilization increased the apparent size of the receptor on molecular exclusion HPLC (Kilpatrick and Caron, 1983). This observed increase in size was reversible by treatment with guanine nucleotides, suggesting that the apparent increased size of the agonist occupied receptor was due to the co-solubilization of a receptor-G protein complex. Furthermore, reinsertion of the affinity-purified D_2 receptor back into phospholipid vesicles yielded biphasic, high affinity agonist binding in this preparation. This high affinity agonist binding was shifted to low affinity by guanine nucleotides or pertussis toxin treatment of the reconstituted material suggesting that a guanine nucleotide binding protein was co-purifying with the D_2 receptor (Senogles et al., 1986, 1987). Using selective antibodies and peptide mapping, the α subunit of the G protein which co-purified with the D_2 receptor was characterized and found to have a Mr of 40000 and to bear similarities to members of the G_i and G_o class of G proteins (Senogles et al., 1987). Although the exact identity of the G-protein remains to be determined, this work represents the first demonstration of a functional association of a particular G protein with a dopamine receptor and should provide a unique approach for probing the details of the signal transduction pathway of the D_2 receptor.

The D_1 Dopamine Receptor

Although the first description of dopamine mediated stimulation of adenylyl cyclase was reported over 17 years ago (Kebabian and Greengard, 1971), the receptor which mediates this effect of dopamine had until recently resisted biochemical characterization. The lack of progress in the isolation and characterization of this receptor subtype was largely due to the lack of selective pharmacologic tools necessary for biochemical studies. In 1981, SCH 23390, the first compound with the properites of a D_1 selective antagonist was described. SCH 23390 has high affinity and selectivity for the D_1 receptor in ligand binding studies (Billard et al. 1984), making this compound an ideal candidate for the development of photoaffinity labelling probes and ligands for affinity chromatography of the D_1 receptor. Modification of the 1-phenyl ring of SCH 23390 by the addition of an amino group to either the 3'(SCH 38548) or 4' positions (SCH 39111) produces ligands which retain high affinity for the D_1 receptor (Amlaiky et al., 1987, and unpublished observations). The amine moiety of SCH 39111 has allowed this compound to be immobilized to Sepharose 6B via an extended spacer arm to create a matrix for affinity chromatography of the D_1 receptor (see Fig. 5 inset).

As a first step towards the isolation and purification of the D_1 receptor, conditions were sought which would allow the solubilization of the receptor from the membrane with retention of its ligand binding properties. Using homogenates of rat corpus striatum, the detergent digitonin solubilized the D_1 receptor in good recovery and resulted in preparations with relatively

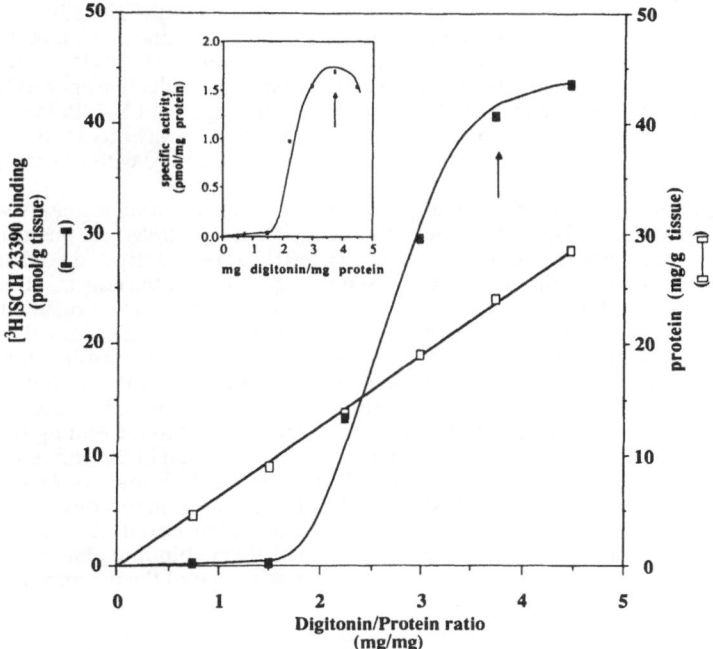

Figure 3. Solubilization of SCH 23390 Binding Activity from Rat Striatal Homogenates with Digitonin. Membranes (100 μl) from rat striatal tissue was resuspended in increasing volumes of solubilization buffer containing 1% digitonin in order to determine the optimal digitonin/protein ratio for solubilization of receptor activity. The membranes were incubated with the solubilization buffer for 1 hr on ice and then particulate material was pelleted by centrifugation for 90 min at 45,000 x g. The activity remaining in the supernatant was then assayed for both [³H]SCH 23390 binding activity and protein. The solid squares indicate receptor binding activity in pmol/g striatal tissue and the open symbols indicate protein solubilized in mg/g original striatal tissue weight. INSET: the inset figure plots the specific activity of [³H]SCH 23390 binding in pmol/mg protein over the same range of digitonin/protein ratios. The arrow in each figure indicates the digitonin/ratio protein ratio routinely used for solubilization. This ratio represents a volume of 25 ml of 1% digitonin solution per g of striatal tissue. The results shown are representative of 4 experiments.

high specific activity. As seen in Fig. 3, low ratios of detergent to protein (eg. 0-1.5 mg/mg) solubilized little or no receptor activity, but higher ratios (2.5-4.5 mg/mg) solubilized up to 43 pmol of [³H]SCH 23390 binding sites and 24 mg of protein per g of striatal tissue. Ratios higher than this, although yielding slightly more receptor, resulted in preparations of lower specific activity (see inset Fig. 3).

The solubilized receptor displayed the expected pharmacologic properties of a D₁ dopamine receptor in ligand binding experiments. As shown in Fig. 4a, [³H]SCH 23390 retained high affinity for solubilized receptor in direct saturation experiments of (K_D of 1.0 ± / − 0.2 nM). Fig. 4b demonstrates the competition of [³H]SCH 23390 binding to solubilized receptors by various antagonists. The solubilized receptor exhibits highest affinity for SCH 23390 as well as for the thiothixene derivatives piflutixol and cis flupenthixol (see Fig. 4b). Intermediate affinity is observed for perphenazine. The low affinity observed for promethazine, is also consistent with data from membrane binding (Leff et al., 1985). The solubilized receptor displays the stereoselectivity expected of the D₁ dopamine receptor: SCH 23390>SCH 23388, cis flupenthixol>>trans flupenthixol and (+)butaclamol >>(−)butaclamol (data not shown). The [³H]SCH 23390 binding activity in the digitonin-solubilized preparation was clearly not a result of cross reactivity with another receptor type. Excluding the D₁ receptor, the serotonin S_2 and S_{1c} receptor subtypes have the highest known affinity for SCH 23390 (Christensen et al., 1984), however, the high affinity S_2 (the predominant sertonin subtype found in striatum) antagonist, ketanserin, had only micromolar affinity for competing with SCH 23390 (data not shown). As expected antagonists selective for other receptor types – sulpiride (D_2 dopaminergic), propranolol (β-adrenergic), prazosin ($α_1$-andrenergic) and yohimbine ($α_2$-adrenergic) – exhibited low affinity for competing with [³H]SCH 23390 for binding to these solubilized preparations (data not shown).

Fig. 4c shows the ability of several agonists to displace [³H]SCH 23390 from the solubilized receptor. The rank order of potency for the agonists examined was found to be: SKF
38393>apomorphine>dopamine> (−)epinephrine>>serotonin>LY 171555, and is clearly
characteristic of a dopamine receptor. The high affinity for the D_1 selective agonist SKF 38393
and the very low affinity found for the D_2 dopamine selctive agonist LY 171555 confirms that
this dopamine receptor is of the D_1 and not the D_2 subtype. The low affinity found for serotonin
is consistent with the lack of cross reactivity exhibited by SCH 23390 with serotonin receptors
in this preparation.

Once the D_1 receptor was successfully solubilized, affinity chromatography was utilized
to enrich the preparation in D_1 receptor activity (Gingrich et al., 1988). As mentioned above,
an affinity matrix was synthesized by coupling SCH 39111 to epoxy activated Sepharose 6B via
an extended spacer arm (see Fig. 5 inset for the putative structure of the matrix). Batchwise incubation of digitonin-solubilized membranes with SCH 39111-Sepharose resulted in the adsorption of 75-85% of the [³H]SCH 23390 binding activity to the matrix while the bulk of the
protein (~90%) was not retained. Fig. 5 shows a typical elution profile obtained using
SCH 39111-Sepharose. As shown in Fig. 5, washing the gel with a variety of buffers removed
most of the protein while eluting little receptor (typically less than 15%). Biospecific elution of
the receptor was achieved by addition of 100 μM (+)butaclamol to the eluting buffer. Using
this procedure, 35-55% of the adsorbed receptor could be recovered in the eluate with a resulting 150-240 fold purification (360 pmol/mg protein) over the digitonin-solubilized preparation. The resulting specific activity indicates that the affinity-purified receptor is only about 40
fold short of theoretical specific activity (13,889 pmol/mg protein based on a molecular weight
of 72,000) in the peak fractions or about 2-3% pure. The ligand binding pharmacology of the
affinity purified D_1 receptor was found to be consistent with that of the receptor in solubilized
preparations.

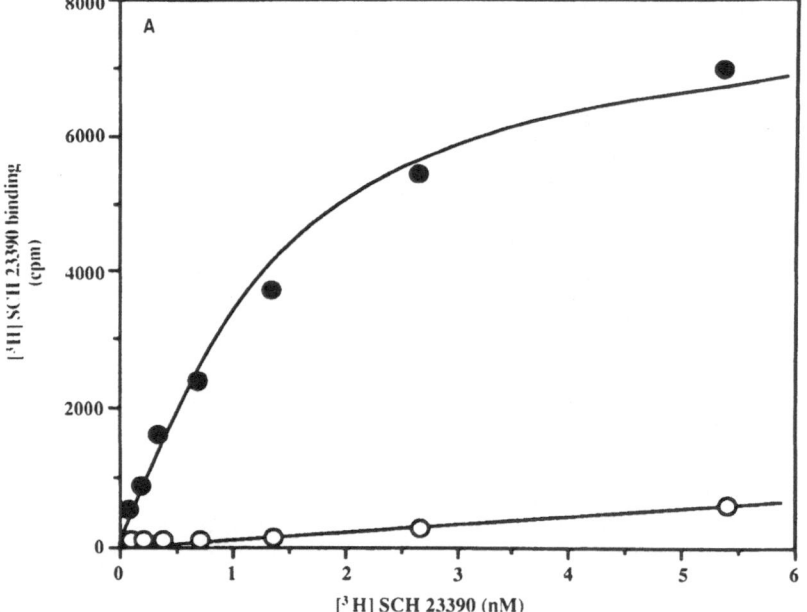

Figure 4. Ligand Binding Pharmacology of Solubilized Striatal Preparations.
A: Solubilized receptor was assayed using [³H]SCH 23390 (Amersham) 78 Ci/mmol in a volume of 0.25 ml containing 0.10 ml of soluble, and 0.15 ml of 0.1% digitonin, 50 mM HEPES (pH 7.2), 100 mM NaCl, 5 mM EDTA,
and various concentrations of [³H]SCH 23390 (filled circles). Nonspecific binding was determined in the presence of 4 μM SCH 23390 (open circles). Samples were incubated for either 2 hr at 22° C or 16 hr at 4° C. Each
conditon yielded equivalent results. Bound and free radioligand were separated at 4°C by gel filtration (Sephadex
G-50, fine) on 0.6 x 13.5 cm columns equilibrated with 0.1% cholate, 50 mM TRIS (pH 7.0, 22° C), 120 mM
NaCl, and 2.5 mM EDTA. Nonspecific binding was typically less than 0.25% of the total added radioactivity and
less than 5% of that which was bound.

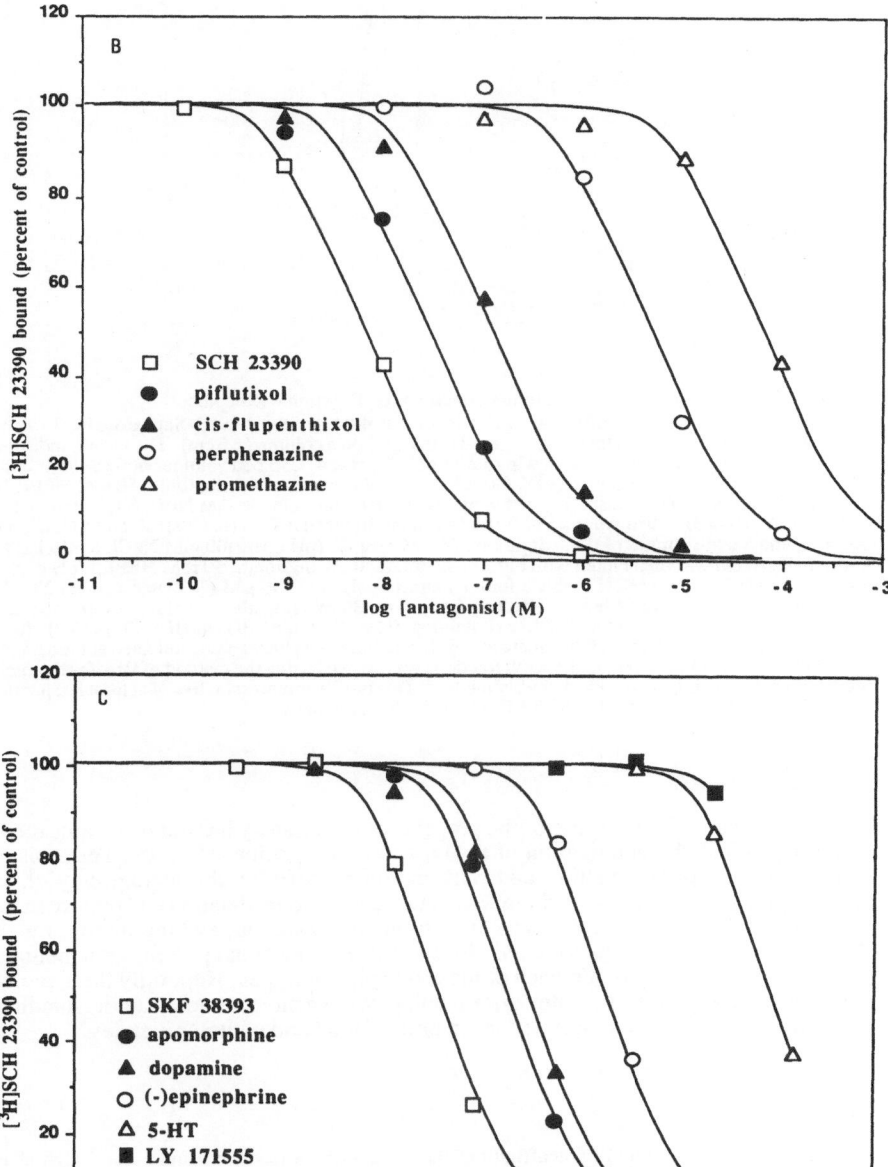

B AND C: 0.10 ml of solubilized receptor was incubated with 4-5 nM [³H]SCH 23390 and the indicated concentration of competing ligand in a total volume of 0.25 ml for 2 hr at 22° C. Data were analyzed by computer modelling methods previously described (De Lean et al., 1982).

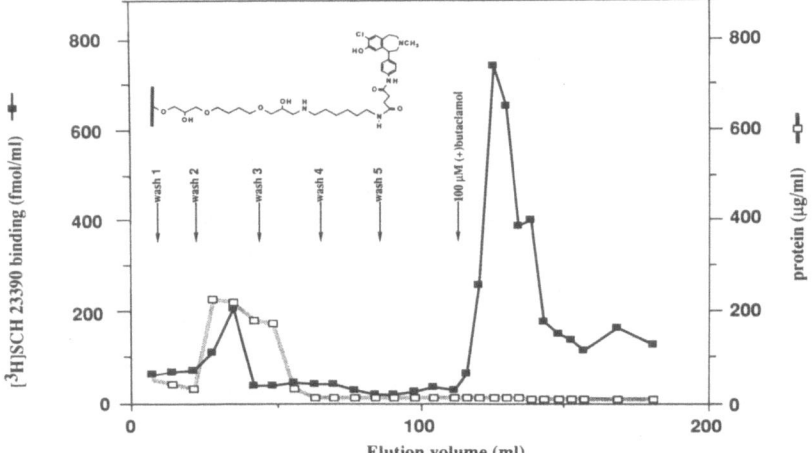

Figure 5. Affinity Chromatography of D_1 Dopamine Receptor.
Digitonin-solubilized membranes (45 ml) were batch adsorbed to 4.5 ml of SCH 39111-Sepharose by slowly mixing at 4° C for 20 hr and then the gel slurry was washed (1ml/min) on a column (1.5 cm). The arrows indicate the beginning of each washing step using the following series of buffers at 4° C: 5 bed volumes of 0.1% digitonin, 50 mM HEPES (pH 7.2), 100 mM NaCl, 5 mM EDTA (wash 1); 5 bed volumes of 1% digitonin, 50 mM HEPES (pH 7.2), 250 mM NaCl, 5 mM EDTA (wash 2); 5 bed volumes of 0.1% digitonin, 50 mM HEPES (pH 7.2), 500 mM NaCl, 5 mM EDTA (wash 3); 5 bed volumes of 0.1% digitonin, 50 mM HEPES (pH 7.2), 100 mM NaCl, 5 mM EDTA (wash 4); and 5 bed volumes of 0.1% digitonin, 50 nM HEPES (pH 6.0), 100 mM NaCl, 5 mM EDTA at 22° C (wash 5). The resin was then eluted with buffer containing: 0.1% digitonin, 50 mM HEPES (pH 6.0), 100 mM NaCl, 5 mM EDTA, 0.1% MeOH (vehicle for (+)butaclamol), and 100 μM (+)butaclamol at 22° C at a flow rate of 5 ml/hr over a period of 6 hours collecting fractions at 30 min interals. The eluates were collected on ice in tubes containing an equal volume of buffer containing: 0.1% digitonin, 50 mM HEPES (pH 7.2), 100mM NaCl, 5 mM EDTA to readjust the pH of the eluate to ~6.8. The final two eluates were collected at the same flow rate but with 3 hrs per fraction. Protein in the wash fractions was assayed using the method of Bradford, protein in the eluates was measured using the Amido-Schwarz method. This profile is representative of at least 8 separate experiments: INSET: putative structure of SCH 39111 Sepharose affinity matrix.

Summary and Perspective

This chapter summarizes some of the progress our laboratory has made towards the isolation and biochemical characterization of the D_1 and D_2 dopamine receptors. The tools now exist to allow for the photolabelling and purification by affinity chromatography of each dopamine receptor subtype. These techniques should allow a more detailed and precise inquiry into the molecular details of receptor structure, signal transduction, and regulation. Furthermore, these purification techniques should also facilitate the eventual protein sequencing and molecular cloning of the gene(s) for each of these receptor subtypes. Hopefully these molecular approaches to the dopamine receptor system will provide a more complete understanding of how this important receptor system functions in both normal and pathologic states.

References

Amlaiky, N. & Caron,, M.G. (1985) Photoaffinity of the D2-dopamine receptor using a novel high affinity radio-iodinated probe. J. Biol. Chem. 260: 1983-1986.

Amilaiky. N. Berger, J.G., Chang, W., McQuade, R.J., & Caron, M.G. (1987) Identification of the binding subunit of the D1-dopamine receptor by photoaffinity crosslinking. Mol. Pharmacol. 31: 129-134.

Amlaiky, N. and Caron, M.G., (1986) Identification of the D2-dopamine receptor binding subunit in several mammalian tissues and species by photoaffinity labelling. J. Neurochem. 47: 196-204.

Arnt, J. (1987)Behavioral studies of dopamine receptors: evidence for regional selectivity and receptor multiplicity. in Dopamine Receptors, (Creese, I., & Fraiser, C.M., eds.) pp. 199-231, Alan R. Liss, New York.

Barone, P., Davis, T.A., Braun, A.R., & Chase, T.N. (1986) Dopaminergic mechanisms and motor function: characterization of D-1 and D-2 dopamine receptor interactions. Eur. J. Pharmacol. 123: 109-114.

Billard, W., Ruperto,V., Crosby,G., Iorio, L. C., & Barnett, A. (1984) Characterization of the binding of [³H]SCH 23390, a selective D-1 receptor antagonist ligand in rat striatum. Life Sci. 35: 1885-1893.

Caron, M.G., Kilpatrick, B.F., and De Lean, A. (1983) The dopamine receptor of the anterior pituitary gland: Ligand binding and solubilization studies, in Dopamine Receptors (C. Kaiser and J.W. Kebabian, eds). ACS Symposium Series 224, pp. 73-92.

Christensen, A.V., Arnt, J., Hyttel, J., Larsen, J., & Svendsen, O. (1984) Pharmacological effects of specific dopamine D-1 antagonist SCH 23390 in comparison with neuroleptics. Life Sci. 34: 1529-1540.

Clark, D., and White, F.J., (1987) Review: D1 dopamine receptor-The search for a function: A critical evaluation of the D1/D2 dopamine receptor classification and its functional implications. Synapse. 1:347-388.

DeLean, A.U., Kilpatrick B.R., & Caron, M.G. (19820 Dopamine receptor of the porcine anterior pituitary gland: Evidence for two affinity states discriminated by both agonists and antagonists. Mol. Pharmacol. 22:290-297.

Enjalbert, A., Sladeczek, F., Guillon, G., Bertrand, P., Shu, C., Epelbaum, J., Garcia-Sainz, A., Jard, S., Lombard, C., Kordon, C., & Bockaert, J., (1986) Angiotensin II and dopamine modulate both cAMP and inositol phosphate productions in anterior pituitary cells. J. Biol. Chem. 261:4071-4075.

Gershanik, O., Heikkila, R.E., & Duvoisin, R.C. (1983) Behavioral correlations of dopamine receptor activation. Neurology. 33:1489-1492.

Gingrich, J.A., Amlaiky, N., Senogles, S.E., Chang, W.K., McQuade, R.D., Berger, J.G., Caron, M.G., (1988) Affinity chromatography of the D1 dopamine receptor from rat corpus striatum, FASEB abstract.

Iorio, L.C., Barnett, A., Leitz, F.H., Houser, V.P., & Korduba, C.A. (1983) SCH 23390, a potential benzazepine antipsychotic with unique interactions on dopaminergic systems. J. Pharmacol. and Exp. Therap. 226:63-468.

Kebabian, J.W. & Calne, D.B. (1979) Multiple receptors for dopamine. Nature 277:93-96.

Kebabian, J.W. & Greengard, P. (1971) Dopamine-sensitive adenyl cyclase: possible role in synaptic transmission. Science 174:1346-1349.

Kilpatrick, B.F. and Caron, M.G. (1983) Agonist binding promotes a guanine nucleotide reversible increase in the apparent size of the bovine anterior pituitary dopamine receptor. J. Biol. chem 258:13528-13534.

Kilpatrick, B.F. and Caron, I. (1984) Dopamine receptor of the porcine anterior pituitary band: Solubilization and characterization. Biochem. Parmacol. 33:1981-1988.

Leff, S.E., & Creese, I. (1985b) Interactions of dopaminergic agonists and antagonists with dopaminergic D3 binding sites in rat striatum. Mol. Pharmacol. 27:184-192

Margaroli, A., Vallai, L., Elaki, f.R., Pozzan, T., Spada, A., and Meldolesi, J. (1987) Dopamine inhibits cytosolic Ca+ + increases in rat lactotroph cells: Evidence of a dual mechanism of action. J. Bil. Chem. 262:1390-13927.

Robertson, G.S. & Robertson, H.A. (1986) Synergistic effects of D1 and D2 dopamine agonists on turning behavior in rats. Eur. J. Pharmacol. 384:387-390.

Sasaki, K. & Soto, M. (1987) A single GTP-binding protein regulated K+ -channels coupled with dopamine, histamine and acetylcholine receptors. Nature. 325:259-262.

Seeman, P. (1980) Brian dopamine receptors. Pharm. Rev. 32:229-313.

Senogles, S.E., Amlaiky, N., Johnson, A.L., & Caron, M.G. (1986) Affinity chromotography of the anterior pituitary D2-dopamine receptor. Biochemistry 25:749-753.

Senogles, S.E., Benovic, J.L., Amlaiky, N., Unson, C., Milligan, G., Vinitksy, R., Spiegel, A., and Caron, M.G., (1987) The D2 dopamine receptor of anterior pituitary is functionally associated with a pertussis toxin-sensitive guanine nucleotide binding protein. J. Biol. Chem. 262:3106-3113.

Senogles, S.E. Amlaiky, N., & Caron, M.G., (1988) Purification of the D2 dopamine receptor from anterior pituitary. FASEB abstract.

Stoff, J.C. & Kebabian, J.W. (1984) Two domapine receptors: biochemistry and pharmacology. Life Sci, 35:2281-2296.

Waddington, J.L. (1986) Behavioral correlates of the action of selective D-1 dopamine receptor antagonists. Biochem. Pharmacol. 35:3661-3667.

Walters, J.R., Bergstrom, D.A., Carlson, J.H., Chase, T., & Braun, A.L. (1987) D1 Dopamine receptor activation required for postsynaptic expression of D2 agonist effects Science 236:719-722.

10

MPTP: TWENTY QUESTIONS

J. William Langston

Institute for Medical Research
San Jose, California 95128

INTRODUCTION

The last five years have witnessed a remarkable outpouring of scientific literature on 1-methyl-4-phenyl-1,2,3,6-tetrahydropyridine or MPTP, a newly described parkinsonogenic neurotoxin. Perhaps the only thing that has surpassed this meteoric rise has been the speed with which concepts regarding its mechanism of action have left the realm of active investigation and gained status as conventional wisdom. From a scientific standpoint, one could argue that this is unfortunate. It will be the thesis of this article that it is time to step back and critically look at what we actually know and don't know about this compound.

To do this, I have prepared the following list of 20 questions about MPTP. The list is not meant to be authoritative or balanced, but rather provocative. Nor is it meant to be all-inclusive. It is suggested that each point will be given a fair hearing with the hope that it might provide a nugget of truth worthy of further exploration. If one or more of these questions provokes a fresh insight or new approach to the study of this most interesting parkinsonogenic agent, the major goal of this chapter will have been achieved.

THE QUESTIONS

Question 1. <u>How good a model for human Parkinson's disease is MPTP-induced parkinsonism in the primate?</u>

The answer to this question is that the model is remarkably good - particularly as a behavioral model of the disease. Virtually all species of non-human primates given MPTP exhibit most, if not all, of the motor features of Parkinson's disease. Even subtle features, such as loss of facial expression and diminished blinking, have been noted in some species. Not content with motor similarities, some laboratories are now investigating neuropsychologic changes in these animals as well. At first it appeared that the only deficient motor feature of the model was the absence of a clear-cut resting tremor, but recent observations in the African green monkey suggest that this is also a feature of MPTP-induced parkinsonism in non-human primates.

At the neuropathological level, the model is surprisingly good, though not yet perfect. Degeneration of the substantia nigra and locus ceruleus, both prominent features of Parkinson's disease, are seen in non-human primates lesioned with MPTP. Some investigators have reported degenerative changes in the ventral tegmental area. Of critical importance to those anxious to reproduce the entire neuropathological picture of Parkinson's disease, however, is the Lewy body, an eosinophilic intraneuronal inclusion which is highly characteristic of, though not specific for, Parkinson's disease. We have identified eosinophilic intraneuronal inclusions in aged primates given MPTP which have the same distribution as Lewy bodies in Parkinson's disease. However, further work needs to be done to further characterize these structures and define their exact relationship to human Lewy bodies (see Question 2). Even if no relationship is found, it will be an interesting project to determine how MPTP is inducing these intraneuronal inclusions.

Finally, there is the issue of distribution of lesions. In Parkinson's disease, other areas of the brain are affected in addition to the substantia nigra and locus ceruleus, including other pigmented nuclei in the brainstem and the nucleus basalis of Meynert. A major question remains as to whether or not it will be possible to induce lesions in these areas as well. This might be done by altering the route of administration, dosage schedule, age, or even species of animal. Obviously, this type of endeavor is extremely important as the closer the model comes to the idiopathic disease, the more interesting the inducing agent becomes as a possible etiological candidate.

Question 2: <u>What is the significance of inclusion in primates?</u>

Electron microscopic studies will go a long way towards answering this question. A preliminary ultrastructural study involving one monkey has revealed what may have been the early stages of Lewy body formation, but structures with the classical appearance of Lewy bodies were not encountered. On the other hand, it may prove that MPTP simply does not produce true Lewy bodies, or that new world primates (we use the squirrel monkey in our laboratory) do not generate structures which are identical to human Lewy bodies. Recently, we have found filamentous structures within these inclusions, an observation which could be quite important, as disturbances in cytoarchitecture are thought to be fundamental to a number of neurodegenerative diseases of aging, including Parkinson's disease. These filaments are somewhat larger in diameter than those seen in human Lewy bodies, raising the possibility that they might represent microtubules. Further work using electron microscopy and immunohistochemistry may answer the many questions surrounding these MPTP-induced inclusions.

Question 3: <u>Will analogs of MPTP produce an even more characteristic neuropathological picture of Parkinson's disease in primates?</u>

After the discovery of MPTP, it seemed likely that other active analogs would be found, a prediction which has since proved to be the case. Most would agree that it would be a remarkable coincidence if a compound accidentally synthesized and sold to drug addicts proved to be the precise etiological agent in Parkinson's disease. It would not be surprising, therefore, if one of these other analogs proves to be more important in the study of Parkinson's disease. One way to pursue this question is to administer these analogs to primates to see if the neuropathological features of the disease can be even more precisely reproduced. Unfortunately, this work is time consuming and expensive, but in this author's opinion, it is very important to do.

Question 4: <u>Is the mouse a good model for the primate?</u>

For reasons of accessibility and expense, much of the work on MPTP is being done in the mouse. The majority of this work is carried out using striatal dopamine depletion as a marker for toxicity (few studies have focused on cell loss which, in our hands at least, is more difficult to induce in the mouse unless aged animals are used). Much of this work rests on the assumption that discoveries in the mouse can be generalized to both non-human and human primates. Is this a valid assumption? In many instances it probably is, but there is a real danger in assuming that such is always the case. A few exceptions are worth noting:

1) One of the major discoveries regarding the mechanism of action of MPTP is the fact that its dopamine-depleting effects in the mouse striatum can be prevented by dopamine uptake blockers (see Question 15). To date, we have not been able to protect monkeys against the toxicity of MPTP using dopamine uptake blockers. While this may simply be a technical problem because of differences in the half-life of 1-methyl-4-phenylpyridine (NPP +) in the mouse and the monkey (see below), this important question has yet to be satisfactorily answered.

2) Analogs may act quite differently between these two species. We have identified one analog[1,2,3,6-tetrahydro-1-methyl-4-(methylpyrrol-2-yl)pyridine(TMMP)] which is more potent than MPTP in the rodent, but less potent in the primate. Thus, there is a reversal of relative effectiveness in these two species.

3) It has been known for some time that there is a striking difference in the half-life of the putative toxic metabolite of MPTP, I-methyl-4-phenylpyridinium, when the primate is compared to the rodent (see Questions 5 and 6).

The reasons for these species differences need to be pursued and kept in mind when using rodent data to make generalizations about the effects and mechanism of action of MPTP. This is particularly true if the goal is to use the mouse to study Parkinson's disease, a disease which is unique to humans, and for which primates probably continue to be the best model.

Question 5: <u>What are the reasons for species differencs?</u>

While there is a sense that the pieces of this puzzle are starting to come together, it is by no means certain that the final answers are in. Why are primates so much more sensitive to MPTP than lower species such as the rat and the mouse? There are a number of interesting observations on this subject which represent partial solutions. A number of investigators have raised the possibility that neuromelanin plays a key role in MPTP toxicity. Since higher species such as the primate have the most melanin, they would therefore be the most susceptible. Yet, the role of neuromelanin in MPTP toxicity has been challenged repeatedly and is far from being finally determined (see Question 11). One group of investigators (see Harik et al, this symposium) have found much higher concentrations of MAO B in the capillary endothelia of the cerebral vasculature in the rat. Thus, MPTP may be rapidly converted to MPP+ within the blood-brain barrier and unable to reach the nervous system in this species. Another explanation may relate to the fact that rodents are able to eliminate MPP+ from the nervous system much more rapidly than primates (see Question 6).

Once the relative importance of each of these elements has been clarified, it may be possible to manipulate them in order to make rodents more sensitive to MPTP. Thus it might be possible to develop an even more useful model in this abundant and inexpensive species, perhaps approaching that which we now have in the primate.

Question 6: <u>Is there an active transport system for MPP+ in rodents which is absent in the primate?</u>

As mentioned above, one of the most striking primate/rodent differences relates to the rapid disappearance of MPP+ in the rodent nervous system. We have proposed the existence of an active transport system in the rodent. Presumably this system is deficient or absent in the primate. If this hypothesis is correct, it should be possible to find agents to block the transport system, and thereby increase the half-life of MPP+ in the central nervous system (CNS) of the rodent. In fact, preliminary evidence in our laboratory suggests that this may be one of the actions of diethyldithiocarbamate (DDC). If so, one would predict that DDC would not exacerbate toxicity in primates as there is no transport system to block. If evidence can be accumulated in support of this hypothesis, it would be particularly interesting as it might suggest a mechanism by which higher species become more susceptible to the neurodegenerative effects of certain toxins. This may represent a true parallel with diseases such as Parkinson's disease because the latter does not appear to occur in any other species besides the primate (i.e., humans).

Question 7: <u>Can nigrostriatal nerve terminal damage be disasscociated from neuronal death?</u>

As noted above, many of the studies of MPTP use dopamine depletion in the striatum as the primary marker of toxicity. However, this depletion could be either a pharmacological effect or it could be due to terminal damage. In either case, the following question arises: Is damage to the nerve terminal sufficient to cause the entire neuron to die? There are several reasons to suspect that this is not the case. For example, complete infarction of the basal ganglia in humans, where the terminals are totally and abruptly destroyed, induces little in the way of overt cell loss in the substantia nigra in the majority of cases. In regard to MPTP, studies in younger mice have shown nearly complete recovery of striatal dopamine depletion over time. While there are a number of possible explanations for this recovery, including nerve terminal regeneration and compensatory changes in remaining terminals, it is by no means clear that damage to terminals necessarily is accompanied by actual neuronal degeneration. There even remains that possibility that the mechanism of damage to terminals is different in some fundamental way from that which causes death of neurons. A corollary question which should also be raised at this point is whether or not loss of tyrosine hydroxylase immunoreactivity can be employed as a marker of cell loss; there is increasing evidence that it cannot.

Question 8: Where is the cellular site of action of MPTP/MPP+?

This question is proving to be particularly controversial. The conventional wisdom is that after arriving in the CNS, MPTP is converted to MPP+ in the extraneuronal compartment, probably in glia, by monoamine oxidase B. It then gains entrance to dopaminergic terminals via their reuptake system. However, it is quite unclear what happens next. As discussed in the previous question, terminal damage alone may not account for death of the neuron. Is MPP+ transported in a retrograde manner to cell bodies, where the final act of killing the cell occurs? This does seems possible, but such a process has yet to be proven. Taking an opposing view, one could even argue that damage to nerve terminals might be a protective mechanism, in essence "cauterizing" them, thereby preventing further uptake of MPP+.

Alternatively, there is the possibility that nerve cells are directly damaged. The dendrites of nigral dopaminergic neurons are thought to have catecholaminergic reuptake sites. Thus, it might be possible for MPP+ to gain direct access to nerve cell bodies. Once in them, the mechanism of cell death could be quite similar to that which damages terminals or, conceivably, quite different. There is considerable controversy at present as to whether or not MPP+ accumulates within the substantia nigra or indeed can even be found in this region. It should be pointed out that most studies done to date do not differentiate between MPP+ in cell bodies, surrounding glia, other nondopaminergic nerve cells or terminals, or even the ground substance. Perhaps immunohistochemical techniques or electron microscopic autoradiography might help to answer this question.

Question 9: Is the free radical hypothesis of MPP+ toxicity dead?

The studies to date are mixed as to whether or not antioxidants protect against MPP+ neurotoxicity. At least one report has suggested that they do not prevent MPTP-induced parkinsonism in primates. Further, there seems to be a gathering consensus that the redox cycling of MPP+ is not a likely biologic phenomenon and that its mechanism of action probably differs from that of the structurally similar toxin, paraquat. However, it does seem possible that some type of free radical generation still may play a role in generating the neurotoxic effects of MPTP.

Question 10: Is the "reactive-intermediate" hypothesis dead?

One of the earliest and most attractive theories of the mechanism of action of MPTP was that an active intermediate generated during the biotransformation of MPTP to MPP+ might damage neurons. With the gathering evidence that the biotransformation of MPTP to MPP+ takes place outside neurons, enthusiasm for this theory has waned. However, it may be premature to abandon it entirely. Perhaps one or more active intermediates are actually taken up into the neurons as well.

Question 11: What is the role of neuromelanin?

The presence of neuromelanin within neurons has provided the basis of an attractive hypothesis which might explain both species (see Question 5) and anatomical selectivity. There is abundant evidence that MPTP and MPP+ bind to neuromelanin, suggesting that cells containing this substance might be prone to accumulate either or both. Further, data have been published indicating that chloroquine partially protects against MPTP toxicity in primates, presumably by binding to neuromelanin, thus blocking the binding of MPP+ to this substance.

However, there have always been a number of arguments against a key role for neuromelanin in regard to MPTP neurotoxicity. Not all areas which contain neuromelanin are affected by MPTP. Secondly, it is now quite clear that mice, particularly aged mice, do have cell loss after exposure to MPTP, yet these animals are not thought to have neuromelanin. For the moment, the role of neuromelanin remains uncertain, just as it does as an etiopathogenic agent in Parkinson's disease itself.

Question 12: Is MPP+ the metabolite that kills nigral neurons?

This may seem an outlandish question given the widespread acceptance of MPP+ as the toxic metabolite. Further, there is no doubt that MPP+ is toxic. However, it should be pointed out that to date a well-designed study showing a concentration-effect relationship between MPP+ and the actual number of degenerating nigral neurons has yet to be carried out. Until this is done, the possibility remains that some other intermediate or metabolite might be sequestered in the nigra, causing cell death. Although many would argue this is unlikely, the suggestion here is that such a possibility should not be dismissed entirely until such a direct relationship has been shown.

Question 13: <u>Does the effect of MPP+ on mitochondria explain its selective neurodegenerative effects?</u>

At the present time, perhaps the most actively investigated theory regarding the mechanism of action of MPTP/MPP+ relates to the effects of MPP+ on mitochondrial oxidation. The evidence is increasingly convincing that this is a major mechanism of action of MPP+. However, it appears that mitochondria from any tissue are sensitive to this effect of MPP+, and that damage to any neuronal population (or even hepatocytes for that matter) is induced in a fairly non-selective way by MPP+ (after all, the compound was developed at one time as a possible herbicide).

How do we get from this non-selective effect to selective neurotoxicity? Many would argue that the answer is simply quantitative or kinetic, with a selective uptake of MPP+ providing high enough intraneuronal concentrations in certain cell groups to cause their demise. However, this may not be an easy proposition to prove and most of the studies to date on the effects of MPP+ on mitochondria must be said to be related to non-selective toxicity. For example, studies on striatal slices are probably not evaluating nigrostriatal toxicity, but rather the toxic effects of MPP+ on the intrinsic neuronal population in the striatum, an area which is not affected either by MPTP or in Parkinson's disease.

Might there still be an undiscovered feature of nigrostriatal dopaminergic nigral neurons that makes them much more sensitive to MPP+ (or conceivably even some other metabolite or intermediate)? Unfortunately, a pure dopaminergic nigral cell tissue culture preparation, which would be of extraordinary value in addressing this question experimentally, is not yet available.

Question 14: Will the effects of MPP+ on mitochondria usher in the new area of Parkinson's disease study?

For years investigators have speculated that the generation of free radicals from the enzymatic and non-enzymatic oxidation of dopamine might play a critical role in causing degeneration of nigrostriatal neurons in Parkinson's disease. In fact, one could argue that we are currently in the epoch of the "free radical hypothesis of Parkinson's disease." This era was largely ushered in by the work of Doyle Graham and Gerald Cohen (see Cohen, this symposium). We may have reached a critical juncture for this hypothesis, as it is now being tested as part of a national, multi-institutional study entitled the Deprenyl and Tocopherol Antioxidative Therapy of Parkinsonism (DATATOP). As part of this study, antioxidants are being tested to see if they can slow or halt the progression of Parkinson's disease. Should this prove to be the case, it could confirm the free radical hypothesis of Parkinson's disease. On the other hand, if (1) antioxidants fail to show a protective effect and (2) the actions of MPP+ on mitochondria are found to be responsible for many or all of the effects of MPTP/MPP+ on the nervous system, perhaps the next epoch of investigation into the cause of cell death in Parkinson's disease will focus on mitochondria. Parkinson's disease is generally not thought of as a mitochondrial disease, but it is interesting to speculate that we might be heading in that direction, particularly if we are to accept MPTP as our guide.

Question 15: <u>Does catecholamine uptake alone explain selectivity?</u>

As mentioned previously, it is now a conventional wisdom that uptake of MPP+ by the dopamine uptake system may be the key to selective toxicity (see Question 3). However, it still seems possible that other factors might be involved, such as regional differences in numbers of MAO B containing glia, or as yet unidentified factors within nigral neurons which enhance their susceptibility to the toxin. Once again, a key question is whether or not selectivity can be explained on a kinetic basis (i.e., greater delivery of the toxin to its target). Interestingly, little attention has been paid to the mechanism of transport of MPP+ into and out of glia, a question which might also be relevant to the anatomic selectivity of MPTP as well. Finally, dopamine uptake as the sole determinate of selectivity is also troubled by increasing evidence that MPTP damages non-dopaminergic areas of the brain (see for example, Questions 16 and 17).

Question 16: How does MPTP damage the locus ceruleus?

If MPP+ does indeed gain entrance to neurons via the dopamine uptake system, accumulating an adequate amount to kill nerve cells, how then do we explain locus ceruleus damage? Perhaps it is a matter of the noradrenergic uptake system being essentially the same as the dopaminergic system. It has been suggested by one group of investigators that the uptake system of dopaminergic nerve terminals surrounding the neurons of the locus ceruleus may in fact provide a protective mechanism. However, if these "protective" catecholaminergic terminals were damaged by repeated doses of MPTP, this protective mechanism might be lost. On the other hand, it may prove that MPTP is a less selective toxin than previously suspected and that other mechanisms are involved in its ability to induce neuronal degeneration.

Question 17: How do we explain the effects of MPTP in the periamygdaloid region?

Virtually all of the current hypotheses on the mechanism of action of MPTP require dopaminergic uptake sites, yet one of the most striking areas in which inclusion bodies have been observed is in the periamygdaloid cortex, an area which is not dopaminergic. Interestingly, this region of the brain is also a predilection site for cytoarchitectural changes in both Parkinson's and Alzheimer's disease. Perhaps in older animals an entirely different mechanism is at work, or we are seeing some form of secondary degeneration. These are important and potentially very exciting questions if we are to use "degenerative neurotoxins" to study neurodegenerative diseases of aging.

Question 18: Are MPTP-like compounds produced endogenously?

To date, there is no evidence that the brain, through its normal metabolic processes, produces an MPTP-like compound. However, such compounds might exist in very small quantities which could be quite difficult to identify. Perhaps the strongest argument against an intrinsic build-up of an MPTP-like substance in the brain as a cause of Parkinson's disease relates to heredity. Because the accumulation of such an endogenous compound would presumably be dependent on inherited enzymatic processes, Parkinson's disease itself might then be likely to show a hereditary tendency. To date, twin studies have not shown this to be the case.

Question 19: Do MPTP-like compounds exist in the environment?

This continues to be, at least in this author's opinion, the "missing link" in the MPTP story. As the model becomes more and more authentic in its reduplication of the features of Parkinson's disease, this issue becomes central. To date, no one has identified MPTP or any of its active analogs in the environment. However, finding this compound could be extraordinarily difficult and the search is clearly in its infancy.

There are some hints that such compounds could exist in the environment. For example, there is the entity of yellow star thistle toxicity. Horses eating enough of this plant, which is endogenous to the Sacramento Valley in California, develop a form of nigropallidal degeneration. A search has been going on for years to determine what compound in the plant produces this selective damage to nigral neurons. Might an MPTP analog be present? This question has yet to be answered.

Question 20: The final, and perhaps the major question, that which has driven many of the MPTP studies is as follows; What is the relationship of MPTP to Parkinson's disease?

Of course, the answer to this question is not yet known. Is it just a remarkable coincidence that MPTP can produce so many of the features of Parkinson's disease, or might an MPTP-like substance play a direct role in the causation of the disease? Of all of the questions listed in this treatise, perhaps this is the most important to answer. On the other hand, even if MPTP or one of its analogs proves not to play a role in Parkinson's disease, there are so many parallels between the two that suspicion must remain high that this compound is telling us something of the processes underlying the idiopathic disease. For example, might there be a commonality of factors between those which make certain neuronal systems vulnerable to MPTP and the process which underlies the pathogenesis of Parkinson's disease?

CONCLUSION

In the author's opinion, these are all important questions, and yet this list remains only partial. Questions ranging from "Why do the effects of MPTP increase with age?" to "What is the mechanism of the uptake system for MPP+ in mitochondria?" have not been addressed at all. Further, each time one of these questions is answered, several new ones are likely to take its place. What I have tried to point out here is that we may be reaching an important juncture in the history of MPTP research. Re-examining what have already become "old" concepts and conclusions, and questioning our basic assumptions, may be both timely and productive at this point, and in fact may be mandatory if we are to continue to progress.

New directions are also beckoning. As good as the MPTP model has become, a strong case can be made that studies of the macroenvironment (neuroepidemiology) should be pursued with full vigor. Perhaps the time has come to aggressively employ this less than exact science in our search for toxins in the environment, for it is conceivable we might be closing in on the cause for this all too common neurodegenerative disease of aging.

ACKNOWLEDGEMENTS

This work was supported in part by the United Parkinson Foundation, the Parkinson's Disease Foundation, and NIEHS Grant 1 R01 ES0-3697-03. We would also like to thank ZoAnn McBride and David Rosner for assistance with manuscript preparation.

11

SEARCH FOR ENVIRONMENTAL OR ENDOGENOUS NEUROTOXINS RELATED TO MPTP

S.P. Markey, H.Ikeda, S.-C. Yang, C.J. Markey,
A.M. Marini*, and J.N. Johannessen

Laboratory of Clinical Science, National Institute of Mental Health and
* Clinical Neuroscience Branch, National Institute of Neurological and
Communicative Disorders and Stroke, Bethesda, MD

INTRODUCTION

The cause of Parkinson's disease remains unknown despite clinical investigations of infectious, genetic, or other contributory etiologic factors (1). Neurotoxins have been considered as causative agents in past clinical investigations due particularly to the observations of parkinsonian symptomatology caused by manganese, 6-hydroxydopamine, neuroleptics, carbon monoxide, and carbon disulfide (2). The clinical responses to each of these neurotoxins differed distinctly from Parkinson's disease, and thus when MPTP-induced neurotoxicity was first observed (3, 4), it generated considerable interest because it closely paralleled the idiopathic disease condition in nearly every aspect. MPTP-induced parkinsonism in non-human primates differs from the idiopathic disease in humans only in its rapid course, the apparent lack of progression following the toxic insult, the absence of the neuropathological hallmark Lewy bodies, and the limited involvement of neurons other than the nigro-striatal in MPTP-affected animals (5). These observations have reawakened interest in a neurotoxic etiology of Parkinson's disease and have enabled us to define new stategies and tools for these investigations.

One testable hypothesis is that there is a compound, either environmentally or endogenously generated, which is a structural analog of MPTP or its oxidized metabolite MPP +. Our studies have demonstrated that MPP + is retained in primate striatal tissue for weeks (10-20 days half-life) after exposure. Thus a neurotoxic analogue of MPP + may be accumulated and appear in post-mortem brain tissue extracts. We have raised antibodies to MPP +, as well as antibodies to MPTP, by sensitizing rabbits to bovine serum albumin diazo-linked amino substituted analogues of MPP + or MPTP (6, 7). The antibodies recognize structural analogues to MPP + or MPTP, albeit with different affinities, and have been used to define a quantitative ELISA assay. Thus, a screening methodology for compounds structurally related to a known parkinsonian syndrome producing neurotoxin has been developed.

A second testable hypothesis is that there is a neurotoxin structurally unrelated to MPTP or MPP + which shares its mechanism of selective neurotoxicity. To screen for such a compound, a functional assay using a cell culture sensitive to MPP + has been developed. Several cell lines have been cultured by investigators studying MPTP toxicity, and it is clear that MPTP oxidation to MPP + is required for neurotoxicity expression. A cell type which responds directly to MPP + -like compounds, but not to MPTP, is cerebellar granule cells from rat. These cells may be kept in culture for up to two weeks, but die over several days when exposed to 50-100 μM MPP +. Measurement of MPP + levels in affected primate brain suggest that nigral neurons have accumulated levels of 20-50 μM when exposed to toxic doses of MPTP (8). Consequently, cultured cells which have similar responsivity are a useful basis for an in vitro screening assay.

A limitation of a functional assay for neurotoxins is the requirement for larger amounts of suspected neurotoxins than in a chemical assay. If one assumes that the unknown neurotoxin might be present in brain at one-tenth the concentration of MPP+, then tissue might contain 1-5 μmole/kg. A one gram aliquot of dissected tissue would contain only 1-5 μmole of material. However, if that were able to be redissolved and tested in a total cell culture volume of 1 ml, then cell viability could be tested at 1-5 μM concentrations. Consequently, our functional assay has been scaled to provide a system compatible with the tissue extracts being prepared.

Methods

Immunoassay - The antibodies described previously (6,7) have been employed in an amplified ELISA system adapted from Carr et al. (9). Microtiter plates (Costar) were coated first with keyhole limpet haemocyanin conjugated with diazotized m-amino MPP+ (1 μg/ml) in coating buffer at 4°C overnight. The plates were rinsed three times with phosphate buffered saline (PBS) containing 0.05% Tween-20 and then 70 μl rabbit antibody added (1/27,000 dilution) along with 70 μl of MPP+ standard or an extract to be assayed. Each sample was assayed in six wells simultaneously. After 30 minutes, the incubation was terminated by washing the wells three times with PBS-Tween-20, and then sheep anti-rabbit IgG bound to alkaline phosphatase (Sigma) added at a dilution of 1/1000 and incubated for 1.5 hr. The plates were emptied and washed four times with PBS-Tween and then primary substrate (NADP in diethanolamine buffer) added and kept in the dark at room temperature for 15 min as described by Carr et al. (9). The secondary substrate (alcohol dehydrogenase, diaphorase, and iodonitrotetrazolium violet) was added to provide enzyme amplification through NADH-NAD cycling to reduce the indicator to a purple formazan which absorbs at 490 nm. The reaction was quenched after 15 min by the addition of 25 μl of 0.4 M HCl to each well and read using a Titertek Multiskan reader.

Cell Culture - Primary cultures of rat cerebellar granule cells were prepared from 8 day old pups according to Gallo et al. (10). Briefly, 8 day old Sprague-Dawley rat pups are decapitated, their cerebella isolated and the meninges removed by mechanical dissection. The cells are then prepared by trypsinization followed by mechanical dissociation. After 5 days incubation in vitro various concentrations of MPTP, MPP+, and their analogs were added to culture dishes and incubated under standard conditions to maintain the cells at a final volume of 2 ml. Cell death was scored only when a significant decrease in cell number was observed.

After 48 hours incubation with MPP+ and its analogues, culture dishes were washed twice with 1 ml prewarmed (37°C) buffer solution containing 154 mM NaCl; 5.6 mM KCl; 1 mM MgCl2; 2.3 mM CaCl2; 5.6 mM glucose; 8.6 mM HEPES; pH 7.4. Dishes were incubated with 1 ml fluorescein diacetate solution diluted in Locke solution for 5 min at 37°C. Only viable cells take up the dye and retain it. A 5 mg/ml solution of fluorescein diacetate in acetone was diluted 1/1000 in buffer solution for fluorescence staining (11).

Brain Tissue Extraction and Fractionation: Frozen human brain tissues were received from the Brain Tissue Resource Center at Harvard (Dr. E.D. Bird) and the National Neurological Research Bank (Dr. W.W. Toutellote) in Los Angeles. Weighed, dissected regions were homogenized in water (1/4; wt/vol) and 1 ml aliquots containing 200 mg tissue used for extraction. Homogenates were extracted first with 15 ml and then 2 more times with 10 ml volumes of ethanol, the ethanol concentrated to dryness, and the residue redissolved in 2 ml water. The aqueous solution was then basified with NH_4OH, and extracted four times with 1 ml benzene to separate MPTP-like amines from the more polar quaternary substances. Benzene extracts have been acidified, concentrated to dryness and stored for subsequent analysis. Aqueous solutions were filtered and either concentrated to dryness for cell culture assay, or applied to a weak cation resin (Amberlite CG-50, 200-400 mesh). The resin columns were 5 mm i.d. x 20 mm long. The sample application was followed with 12 x 1 ml water washes and then 12 x 1 ml 2 M ammonium acetate. MPP+ standards eluted in fractions 3-5 of the NH_4OAc buffer. All water and ammonium acetate fractions were analyzed individually in ELISA wells without further concentration.

Results

Immunoassay of brain extracts: Our prior efforts to analyze brain extracts using an ELISA procedure similar to that described by Niewola for the quantification of paraquat (12) led to uninterpretable results when applied to human brain extracts to which small amounts of MPP+ were added. Non-specific interferences in concentrated brain extracts required two major changes in procedure - the enhancement of assay sensitivity and the preparation of a protein-free brain extract likely to contain MPP+-like substances. The first goal was attained by

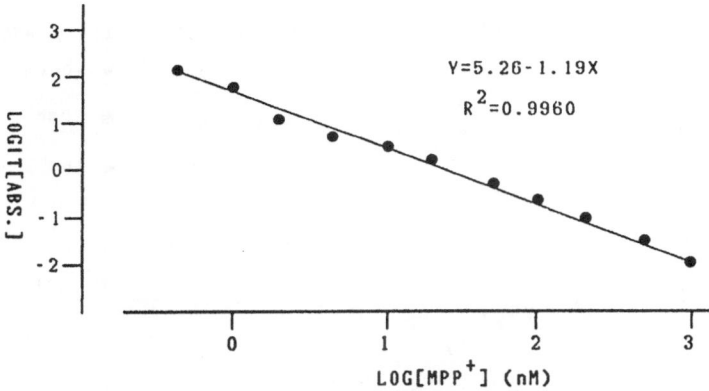

Fig. 1. Standard curve of MPP+ in 2 M NH₄OAc assayed using an amplified ELISA assay.

using the enzyme amplified ELISA procedure as described by Carr (9). Basically, recognition and competition for MPP+ -antigen bound to plastic wells by antibodies to MPP+ was as performed earlier. However, the sensitivity of the alkaline phosphatase linked sheep-anti-rabbit IgG interaction with this complex was enhanced by the fact that alkaline phosphatase produced an activator (NAD) in a cycling redox reaction involving alcohol dehydrogenase and diaphorase in which p-iodonitrotetrazolium violet was converted to a purple formazan. The resulting sensitivity enhancement permits a standard curve to be constructed over the concentration range .5 nM to 1000 nM, or approximately 6 pg to 12,000 pg/well as shown in Fig. 1.

Secondly, the development of a cation column fractionation procedure which could separate MPP+ -like compounds from crude brain extracts established a basis for comparing various patient samples. The chromatographic profile of immunoreactivity is a useful means of comparing samples from different patients and brain regions. A weak cation exchange resin was selected because strong cation resins bound MPP+ so avidly that concentrated acids were required for elution. It is important to minimize acid and ion concentration in the fractionation procedure because the ELISA method is sensitive to both, requiring standard curves to be made using equivalent ionic strengths as the neutralized samples. The use of the cation column separation is demonstrated in Fig. 2, in which the immunoreactivity in a monkey brain homogenate was assayed. The monkey had received MPTP one day prior to sacrifice. No immunoreactive substance eluted in the water washes of the column, only in the NH₄OAc fractions.

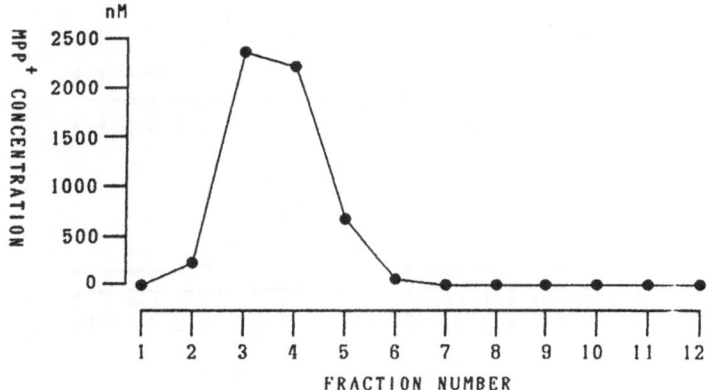

Fig. 2. MPP+ -immunoactivity in monkey brain extract from animal sacrificed one-day after MPTP injection as measured by ELISA assay of fractions eluted from cation resin with 2 M NH₄OAc.

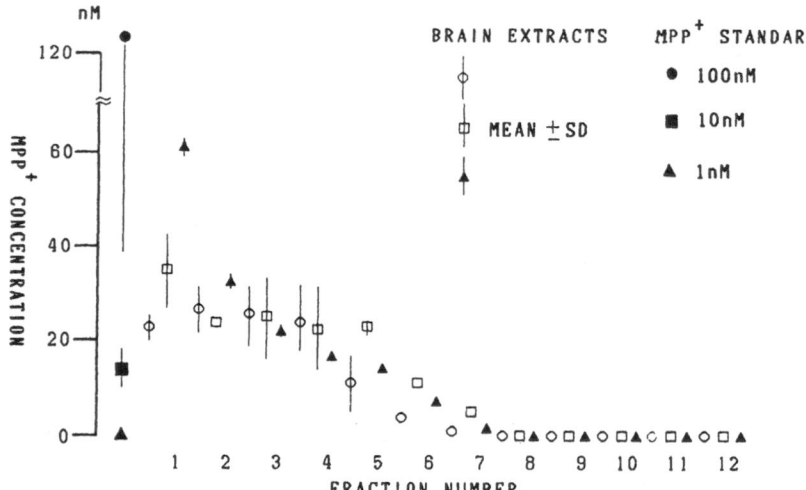

Fig. 3. MPP+-immunoactivity in human cortical tissue extract eluted from cation resin. Three aliquots of homogenate were extracted and analyzed in parallel, and standard concentrations of MPP+ (1, 10, 100 nM) were measured on the same ELISA plates.

The reproducibility of the screening method applied to a single human cortex homogenate processed in triplicate is shown in Fig. 3. Consistent concentrations of immunoreactive substances eluted with the first NH₄OAc fractions in all three samples. The early eluting immunoreactive substance has been found to derive from the resin itself, and resin preparation procedures have been modified to reduce this interference (Fig. 4). However, it is clear from the analysis of human brain doped with small amounts of MPP+ (Fig. 5), that background contributions are several orders of magnitude less than the signal generated by the presence of MPP+.

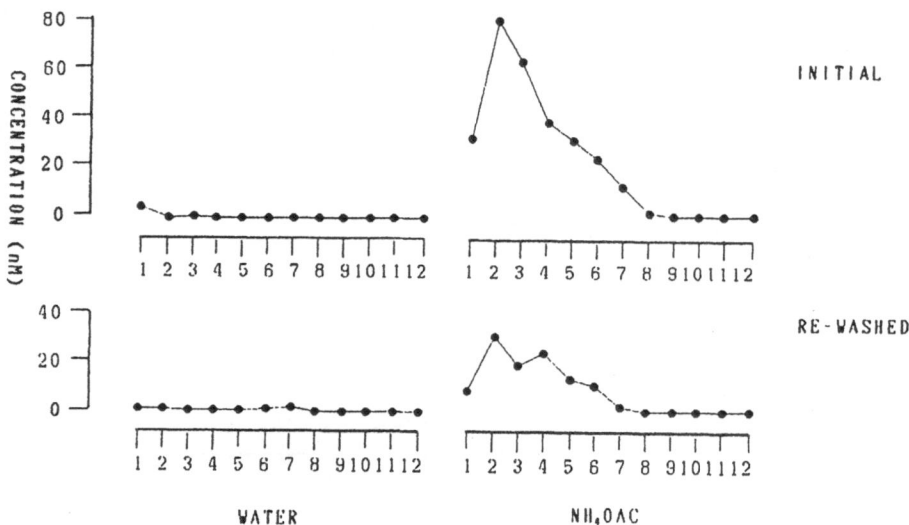

Fig. 4. Procedural blanks from Amberlite cation exchange resin demonstrated that nM concentrations of artifactual material elute with 2 M NH₄OAc, and can be substantially reduced by rewashing the resin with NH₄OAc and HCl.

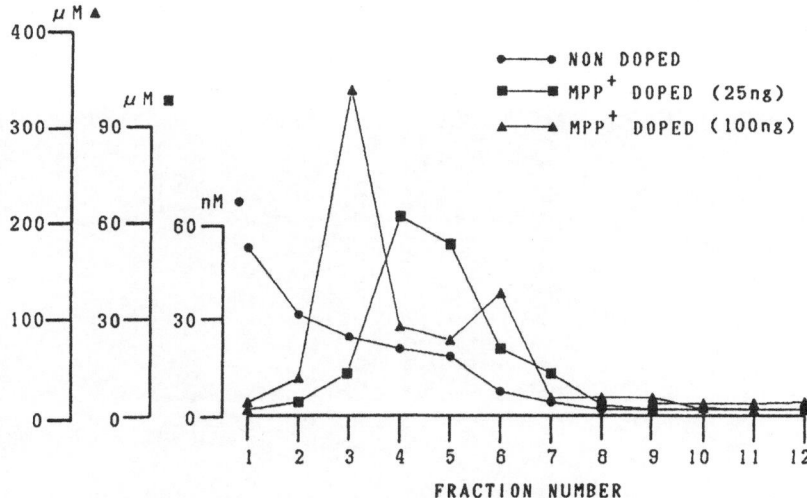

Fig. 5. MPP + -immunoreactivity in human cortical extracts eluted from cation resin with 2 M NH$_4$OAc. A non-doped extract exhibited only background immunoreactivity, whereas 200 mg homogenates doped with 25 or 100 ng MPP + exhibit robust peaks of activity.

There are many alternatives to the presentation and analysis of immunoassay data. Dilution curves for each sample are usually compared with the dilution curve of a standard to produce parallel curves for each fraction. However, because we are assaying for an immunoreactive substance which need not exhibit the same affinity for the MPP + -antibody as MPP + we have chosen to enhance the assay-to-assay reproducibility by measuring each cation column fraction with six replicates at a single concentration. Assuming that a 200 mg brain tissue extract contains 0.2-1.0 nmole of neurotoxin, then elution of this material as a chromatographic peak might spread it over several ml of eluant, with a final concentration in the .2-1.0 nmole/ml, or .2-1.0 micromolar. Because the standard curve extends to the picomolar range, there should be adequate sensitivity even to detect substances which only weakly cross-react with MPP +.

Cell Culture Assay of Brain Extracts

The cerebellar granule cell culture system produces a response that is easily detected in the presence of MPP + and its analogs. Healthy cells grown for 48 or 72 hours after the addition of MPTP, MPP +, BPTP (butylPTP), BPP +, p-amino-MPTP, and p-amino-MPP + can be visualized with fluorescein stain. As shown in Fig. 6, control cells, as well as those treated with the tetrahydropyridines MPTP, BPTP, and p-amino-MPTP are viable. However, equivalent concentrations of the pyridinium ions MPP +, BPP +, and p-amino-MPP + cause cell death. In dose response experiments, pyridinium compounds are toxic over 2-3 days at various concentrations. MPP + is the least potent and BPP + is the most potent. For example, p-amino-MPP + exerts its neurotoxic effects at a shorter time and at lower concentration (25 µM, 36 hrs). Tetrahydropyridines are essentially non-toxic until their concentrations exceed 100 µM.

Progress with the cerebellar granule cell culture system has only recently enabled the testing of some human brain extracts. At this time, procedures for producing concentrated extracts for screening valuable post-mortem samples is not satisfactory, because it is too easy to kill cells in culture for the wrong reasons (infection, foreign protein, salts, etc.). Nevertheless, when extraction procedures have been refined, the addition of chromatographic steps following the indication of a neurotoxic extract should provide an additional safeguard against artifactual observations.

We have purposely selected a cell culture for neurotoxin screening which does not require metabolic conversion of MPTP or MPTP-like substance to an oxidized quaternary amine for two reasons. First, the presence of a neurotoxin in human brain extracts should be detected with a minimum of specific enzyme or uptake system requirements which might preclude de-

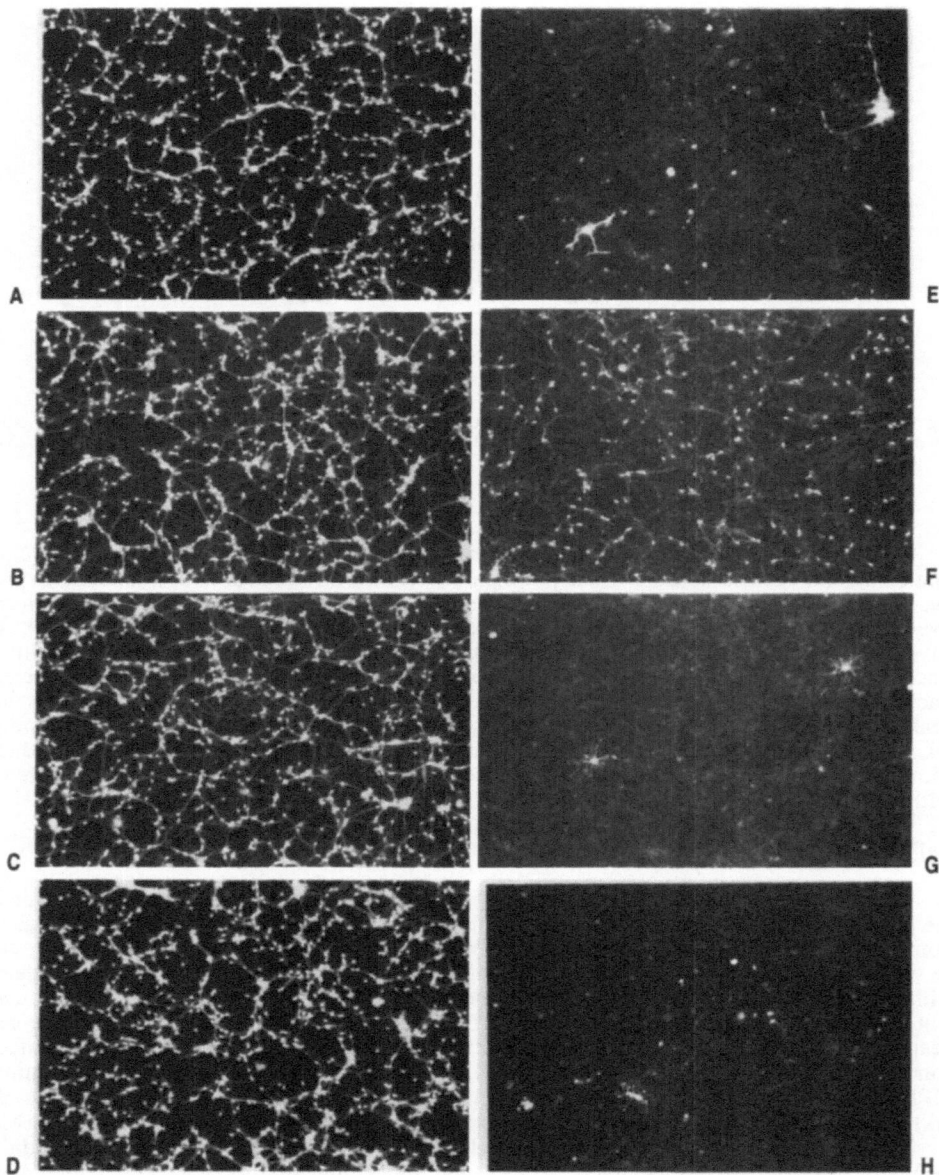

Fig. 6. Cerebellar granule cells were cultured in vitro, and at day 6 each compound was added to give a final concentration of 100μM and incubated for 48 hours, and stained with fluorescein. Photos are fluorescence micro-
graphs at 125X magnification. A, Control cells; B, MPTP; C, BPTP; D, p-amino-MPTP; E, EPP+; F,
MPP+; G, BPP+; H, p-amino-MPP+.

tection of a neurotoxin which could not enter the target cell. Secondly, we assume that the compound accumulated in human Parkinson's disease is in its neurotoxic form and does not require additional metabolic activation. Obviously, in order to screen for compounds like MPTP, a different cell culture system, for example, a co-culture of susceptible neurons and astrocytes, would be required for screening environmental samples or suspected agents with MPTP-like effects.

Conclusions

Two assay systems have been described for screening human brain extracts for neurotoxins related to MPP+ and MPTP. Procedures for the extraction and chromatographic separation of such low molecular weight compounds are being refined and tested. The detection of immuno- or toxin-reactivity will be correlated with dissected brain region in order to assure that the observations are relevant to idiopathic Parkinson's disease.

Acknowledgements

The receipt of tissue samples from the Brain Tissue Resource Center at Harvard (PHS grant number MH/NS 31862) and the National Neurological Research Bank at the VA Hospital Medical Center in Los Angeles is gratefully acknowledged. This project was supported in part (CJM) by a research grant from the National Parkinson Foundation. Helpful advice from Drs. R.I. Carr and L. Tamarkin assisted this research.

References

1. Duvoisin, R.C.: The cause of Parkinson's disease. In: Movement Disorders, Chap. 2, Marsden, C.D., Fahn, S. (Eds.), Butterworth Scientific, London, pp. 8-24, 1987.

2. Schwab, R.C., England, A.C., Jr.: Parkinson Syndromes due to various specific causes. In: Handbook Clin. Neurology, Chap. 9, Vinken, P.J., Bruyn, G.W. (Eds.), North Holland, pp. 227-247, 1968.

3. Davis, G.C., Williams, A.C., Markey, S.P. , Ebert, M.H., Caine, E.D., Reichert, C.M., Kopin, I.J.: Chronic parkinsonism secondary to intravenous injection of meperidine analogues. Psychiatry Research 1:249-254, 1979.

4. Langston, J.W., Ballard, P., Tetrud, J.W., Irwin, I.: Chronic parkinsonism in humans due to a product of meperidine-analog synthesis. Science 219:979-980, 1983.

5. Kopin, I.J., Markey, S.P.: MPTP toxicity: Implications for research in Parkinson's disease. Ann. Review Neurosci. 11:81-96, 1988.

6. Markey, S.P., Markey, C.J., Savitt, J.M., Bacon, J.P., Johannessen, J.N.: Search for environmental neurotoxins related to MPTP. Preparation of antibodies to MPTP and MPP+. In: MPTP - A Neurotoxin Producing A Parkinsonian Syndrome. Chap. 44, Markey, S.P., Castagnoli, N., Jr., Trevor, A.J., Kopin, I.J. (Eds.), Academic Press, Orlando, FL, pp. 487-493, 1986.

7. Johannessen, J.N., Savitt, J.M., Markey, C.J., Bacon, J.P., Weisz, A., Hanselman, D.S., Markey, S.P.: The development of amine substituted analogues of MPTP as unique tools for the study of MPTP toxicity and Parkinson's disease. Life Sci. 40:697-707, 1987.

8. Yang, S.C., Johannessen, J.N., Markey, S.P.: Metabolism of 14C-[phenyl]-MPTP in mouse and monkey implicates MPP+, and not bound metabolites, as the operative neurotoxin. Submitted for publication.

9. Carr, R.I., Mansour, M., Sadi, D., James, H., Jones, J.V.: A substrate amplification system for enzyme-linked immunoassays. Demonstration of its general applicability to ELISA systems for detecting antibodies and immune complexes. J. Immun. Methods 98: 201-208, 1987.

10. Gallo, V., Ciotti, M.T., Coletti, A., Aloisi, F., Levi, G.: Selective release of glutamate from cerebellar granule cells differentiating in culture. Proc. Natl. Acad. Sci. USA 79:7919-7923, 1982.

11. Novelli, A., Reilly, J.A., Lysko, P.G., Henneberry, R.C.: Glutamate becomes neurotoxic via the N-methyl-D-aspartate receptor when intracellular energy levels are reduced. Brain Res. (in press) 1988.

12. Niewola, Z., Walsh, S.T., Davies, G.E.: Enzyme linked immunosorbent asssay (ELISA) for paraquat. Int. J. Immunopharmacol. 5:211-218, 1983.

12

MPTP NEUROTOXICITY AND THE "BIOCHEMICAL" BLOOD-BRAIN BARRIER

Sami I. Harik, Naji J. Riachi, and Joseph C. LaManna

Department of Neurology. University Hospitals of Cleveland and
Case Western Reserve University School of Medicine

About 25 years ago, major breakthroughs were made in our understanding of the neurochemistry of the basal ganglia and the biochemical pathology of Parkinson's disease, which eventually led to the successful introduction of levodopa as replacement therapy for this malady. This dramatic success story, especially after the subsequent introduction of peripheral DOPA decarboxylase inhibitors and dopamine receptor agonists as adjuvants, was hailed as one of the best examples of how basic biomedical research can lead to rational therapies for human diseases. Meanwhile, the cause of Parkinson's disease, i.e. the progressive degeneration of nigrostriatal dopaminergic neurons, remained unknown. Furthermore, replacement therapy, which was once considered revolutionary, was found to be of little help in the treatment of subjects with advanced Parkinsonism and entirely ineffective in arresting the inexorable progression of Parkinson's disease. Another important fact that emerged in the last decade concerns the lack of evidence that Parkinson's disease is a genetic disorder, as evidenced by the absence of increased concordance among identical twins. This led to the belief that it may be caused by environmental factors (if it is not nature, it must be nurture).

Thus, it was not surprising that the discovery of 1-methyl-4-phenyl-2,3,6-tetrahydropyridine (MPTP), a specific and selective neurotoxin that causes degeneration of the nigrostriatal system, stimulated extensive research on the mechanisms of its toxicity and on the possible relation between MPTP (or other related toxins) and Parkinson's disease. It is now generally accepted that systemic MPTP induces in man and nonhuman primates a clinical, pathological and neurochemical state that closely resembles Parkinson's disease, or is at least, the best "model" of Parkinson's disease yet discovered (1-3). Thus, research on MPTP-induced neurotoxicity has important theoretical and practical implications directed at means for the prevention or arrest of nigrostriatal degeneration.

Although the mechanisms that underlie MPTP neurotoxicity are not entirely understood, a number of biological processes which are thought to explain the specific neurotoxicity of this agent have been defined: First, oxidation of MPTP by monoamine oxidase B (MAO-B) is necessary for its neurotoxicity, since pretreatment with specific MAO-B inhibitors (e.g deprenyl), but not with MAO-A inhibitors (e.g clorgyline), prevents MPTP neurotoxicity (4-8). This is the main rationale for ongoing clinical trials using deprenyl to arrest the progression of Parkinson's disease in subjects with early manifestations. Second, accumulation of the eventual product of MPTP oxidation, 1-methyl-4-phenylpyridinium (MPP +), by the high-affinity uptake process for dopamine, concentrates MPP + in neurons that possess this pump (9). This probably explains the specificity of MPTP toxicity to dopaminergic neurons, and has led to the finding that dopamine uptake inhibitors (e.g. mazindol and benztropine) decrease MPTP toxicity (10). Third, evidence now indicates that MPP + causes metabolic failure by interfering with mitochondrial oxidative metabolism (11-13).

One of the most intriguing observations regarding MPTP neurotoxicity is the marked variation among species in their susceptibility to systemic MPTP. For example. the green african monkey is susceptible to MPTP at < 0.5 mg/kg, daily. for 2-3 days, whereas the rat shows little evidence of toxicity when given 50 mg/kg/day for several weeks. This focused our attention on the biological basis for the marked species variations in systemic MPTP toxicity, which we thought may have immense heuristic and practical implications concerning not only the mechanisms of MPTP toxicity but also the treatment of Parkinson's disease and possibly other neurodegenerative disorders, if the reasons underlying resistance to MPTP toxicity were understood and could be conferred on susceptible animals.

There are several possible explanations for the differences among species. The first is neuromelanin; man and higher primates have melanin deposition in the substantia nigra (hence the name), while the substantia nigra of lower mammals is not pigmented. D'Amato et al. recently presented evidence that MPTP and MPP + bind to neuromelanin (14) and that chloroquin, which can displace these compounds from neuromelanin, may reduce MPTP toxicity in monkeys (15). The findings that older animals and pigmented strains of mice are sometimes more sensitive to systemic MPTP toxicity are often used to support the melanin hypothesis. The implication here is that neuromelanin accumulates with age, and that mouse strains with pigmented skin have more melanin in their brains too. We find no direct evidence to support these contentions. Furthermore, the melanin hypothesis does not explain why MPTP and its metabolites accumulate primarily in the striatum of monkeys (16), where there is no melanin, nor is it consistent with evidence suggesting that MPTP causes degeneration of dopaminergic substantia nigra neurons via a dying-back phenomenon following destruction of striatal dopaminergic terminals (17). Another possible explanation for differences among species is variation in regional brain MAO activity. For example, if MAO-B activity in the nigrostriatum of rats is meager, then MPP + will not be formed in sufficient concentrations to cause degeneration. Chemical assays of MAO activity indicates that MAO-B is ubiquitously distributed in the brain of the rat (18) and man (19) without major differences between the two species in overall activity. On the other hand, immunocytochemical methods show wide differences in cellular and regional MAO-B activity (20,21). Zimmer and Geneser have recently used this reasoning to explain known differences in susceptibility to MPTP between two strains of mice (22).

In attempting to understand the insensitivity of rats to MPTP toxicity, we found that direct microinfusion of MPTP into the rat substantia nigra produced selective destruction of dopaminergic neurons in the zona compacta, with the resultant depletion of dopamine and its metabolites in the ipsilateral striatum (23,24). This suggested to us that the rat's resistance to systemic MPTP may be due to failure of MPTP to reach the brain in sufficient concentrations, possibly because of a unique property of the rat blood-brain barrier and other organs that can metabolize MPTP and prevent it from reaching the brain. Because MPTP is a known substrate for enzymatic oxidation by MAO-B, and since brain capillaries are known to possess MAO activity (25-27), we reasoned that MAO activity in brain microvessels from different species may correlate inversely with their susceptibility to systemic MPTP neurotoxicity. To test our hypothesis, we examined isolated brain microvessels from several mammals, including on one extreme, humans which are sensitive to small doses of MPTP, and on the other, rats which are known to be resistant to MPTP toxicity. We also studied mitochondria-enriched preparations from the cerebral cortex and liver because of the importance of brain MAO in converting MPTP to the eventual toxin, MPP +, and in view of the importance of the liver in the detoxification of MPTP and numerous other toxins.

We assessed MAO by assaying specific [³H]pargyline binding, which is irreversible and stochiometric, and by measuring the enzymatic oxidation of MPTP and other known MAO substrates. The two molecular forms of MAO were estimated by performing [³H]pargyline binding in the presence and absence of either clorgyline or deprenyl, in addition to examination by SDS-PAGE after [³H]pargyline binding. Our results showed remarkable species differences in brain microvessel MAO, with rat microvessels having the highest MAO activity among all tissues that we tested, whereas human brain microvessel MAO activity was very low (18,28). Other species gave intermediate results, which in general, correlated well with their susceptibility to systemic MPTP toxicity. We have no explanation for the absence of reaction product in capillaries of rat brain when studied by immunocytochemical methods and specific antibodies against MAO (20).

To examine further the reasons for these differences among species and between strains of the same species, we concentrated our efforts on two strains of mice, CF 1 albino mice and C57 BL mice, because of the knowledge that the latter strain is more sensitive to MPTP (22,29,30). we measured MAO activity in the cerebral cortex, striatum and brain microvessels from the two strains and correlated the results with (i) in vitro metabolism of [³H]MPTP by isolated brain microvessels, (ii) regional brain levels of MPTP and MPP+ after the systemic administration of [³H]MPTP, and (iii) in vivo toxicity of systemic MPTP in the two strains of mice.

Adult male mice, weighing 25-30 g. of both strains were used. The toxicity of systemic MPTP was determined by injecting 20 mg/kg subcutaneously, twice daily (6 h apart), for 2 consecutive days. Control mice of both strains received vehicle solutions. The mice were killed one week later and their striata were assayed for dopamine and its metabolites, dihydroxyphenylacetic acid (DOPAC) and homovanillic acid (HVA), by HPLC with electrochemical detection. The results shown in Fig. 1 reveal that C57 BL mice are much more affected by MPTP than CF 1 albino mice.

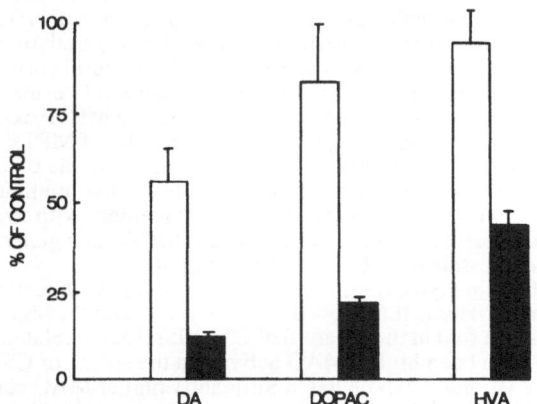

Figure 1. The effects of MPTP on striatal dopamine and its metabolites, DOPAC and HVA in adult male CF 1 (open columns) and C57 BL (filled columns) mice. The results are means ± SEM of 5 mice in each group. Dopamine, DOPAC and HVA in striata of control CF 1 mice were 13.37, 1.41 and 1.40 μg/g wet weight of tissue. The corresponding levels in C57 BL mice were 15.81, 1.89 and 1.76 μg/g. In CF 1 mice, MPTP induced a significant decrease in dopamine levels (p < 0.01), but the reduction in DOPAC and HVA levels were not significant. In C57 BL mice, MPTP caused a significant decrease in the levels of dopamine. DOPAC and HVA (p < 0.01). This figure is reproduced from reference 38.

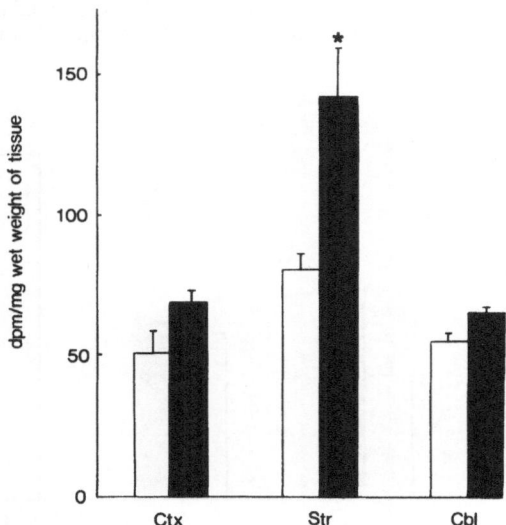

Figure 2. Regional brain [³H] content in CF 1 (open columns) and C57 BL (closed columns) mice 1 hr after the administration of MPTP (5 mg/kg, ip) containing a tracer of [³H]MPTP (100 μg Ci/kg). The results are means ± SEM of 3 mice in each group. Differences between the two strains of mice were significant only in the striatum at p < 0.005 (student t test, two-tailed).

In vivo studies of the tissue distribution and metabolism of [³H]MPTP were performed by injecting MPTP (5 mg/kg) containing a tracer of [³H]MPTP (100 μCi/kg). intraperitoneally. Mice were killed 1 hour later and samples taken from the frontoparietal cortex, striatum and cerebellum bilaterally. Samples from one half of the brain were dissolved and counted for their ³H content, and samples from the other half were extracted, and the extracts analyzed by HPLC to determine the nature and concentration of the tritiated compounds (5,31). The results depicted in Fig. 2 show significantly higher counts in the striata of C57 BL mice. Although ³H counts were slightly higher in the cerebral cortex and cerebellum of C57 BL mice, the differences here did not attain statistical significance. Analysis of the radioactive compounds revealed higher levels of MPTP and MPP+ in the striata of C57 BL mice (results not shown). These findings are consistent with prior results of higher levels of radiolabeled MPTP and its metabolites in the striata of animals that are more susceptible to MPTP toxicity (5,16,31), and strongly suggest that the striatum is the major site of MPTP accumulation and damage.

Results of regional brain MAO activity show that the cerebral cortex of C57 BL mice can oxidize MPTP and bind pargyline much better than the cerebral cortex of CF 1 albino mice, consistent with the findings of Zimmer and Geneser (22). The MPTP oxidation results are presented in Fig. 3. However, the striatum, which is the major site of MPTP toxicity and accumulation of its metabolites, had similar MPTP oxidizing activity in the two strains. On the other hand, brain microvessels from CF 1 mice have about twice the amount of MAO activity as those from C57 BL mice (Fig. 3). These results are in agreement with our prior demonstration that isolated brain microvessels from CF 1 mice oxidize MPTP, and generate MPP+ more vigorously than brain microvessels from C57 BL mice (Fig. 4).

The findings of the in vivo experiments do not correlate with results of MAO activity in the cerebral cortex and striatum. If the low dopamine levels and the high counts of [³H]MPTP and its metabolites that we find in the striatum of C57 BL mice are related to tissue MAO activity, then there should have been higher MAO activity in the striata of C57 BL mice. Also, the cerebral cortex of C57 BL mice, which has significantly higher MAO activity, did not contain higher counts of [³H]MPTP and its metabolites in vivo. Only brain microvessel activity correlated well (inversely) with MPTP neurotoxicity and with in vivo deposition of MPTP and its metabolites in the various regions of the brain. In our opinion, the best explanation for the differences among the two strains of mice in their susceptibility to MPTP neurotoxicity is that brain microvessels of CF 1 white mice have higher MAO activity and are, thus, capable of oxidizing MPTP to MPP+ in brain capillary endothelium. The formed MPP+ being highly charged and water-soluble, cannot easily traverse biological membranes into the brain. In our opinion, the activity of neuronal and glial MAO do not play an important role in determining the differences between the two strains of mice.

Thus, we hypothesize that MPTP is enzymatically metabolized in brain microvessels of rats and other animals that are resistant to systemic MPTP, most likely by MAO-B, to yield

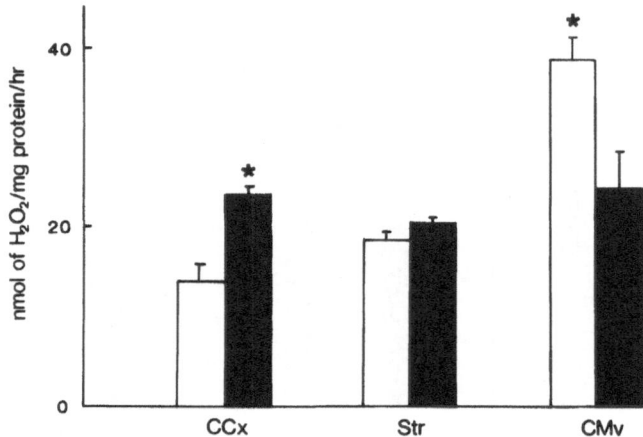

Figure 3. MPTP oxidation by tissues from CF 1 (open columns) and C57 BL (closed columns) mice. The concentration of MPTP in the incubation mixture was near saturating at 1 mM. The results are means ± SEM of 5 separate experiments. Results in the two strains of mice were significantly different in both the cerebral cortex and in cerebral microvessels at p < 0.005 (student t test, two-tailed).

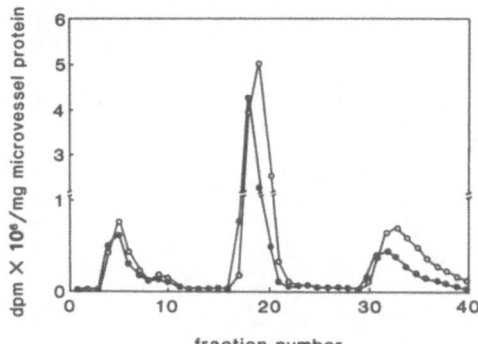

Figure 4. High performance liquid chromatography elution profile of radioactive compounds in extracts of 63 μg of brain microvessel protein from CF 1 mice (open symbols) and 85 μg of brain microvessel protein from C57 BL mice (closed symbols) that were each incubated for 1 hr with 1 μCi of [^{3}H]MPTP. We have determined in parallel experiments that authentic [^{3}H]MPTP+ and [^{3}H]MPP+ elute in fractions 17-21 and 30-40, respectively. This figure is taken from reference 31.

MPP+ and possibly other polar metabolites. It is known that MPP+ when injected systemically does not cause neurotoxicity, presumably because it cannot cross the blood-brain barrier (32,33), in contradistinction to its severe toxicity when instilled directly into the brain (23,24) or into the cerebral ventricles (34). The fate of the MPP+ formed in brain capillary endothelium is not known. Several scenarios could be envisioned. In one, MPTP is retained within endothelial cells possibly resulting in their death. However, since brain capillary endothelial cells can regenerate (35), the toxic effects of MPTP on these cells is less likely to be as devastating as on neurons. Another scenario is that MPP+ is transported unidirectionally across luminal membranes into the circulation.

To investigate the blood-brain barrier transport of MPTP, we performed preliminary experiments in anesthetized adult male Wistar rats by bolus injections of [^{14}C]butanol and [^{3}H]MPTP into the right atrium of the heart. We used a withdrawal syringe attached to a femoral artery as an artificial organ with a known arterial blood flow and no venous outflow, in a manner similar to that of the indicator-fractionation method (36). Rats were decapitated at intervals ranging from 10-120 sec after the bolus injection into the right atrium. [^{14}C]butanol was used as a lipid soluble tracer which is known to be almost completely extracted by the brain on the first pass, but which is known to leave the brain quickly thereafter. On the other hand, [^{3}H]MPTP was expected to be also completely extracted on the first pass but unlike butanol, was likely to be trapped in brain endothelium. Immediately after decapitation, the brain was quickly dissected and regional samples obtained for dual isotope counting.

The results of these experiments are briefly summarized in Table 1 and in Fig. 5. In Table 1, it is to be noted that although the ^{3}H:^{14}C ratio in the injectate was about 2.5, the isotope ratio in the arterial sample decreased to about 1.5 even when the time interval between the bolus injection and decapitation (with simultaneous cessation of arterial sample withdrawal) was 10 sec. This indicates that MPTP was preferentially removed from the pulmonary circula-

TABLE 1

[^{3}H]:[^{14}C] RATIOS IN ARTERIAL SYRINGE AND FRONTAL CEREBRAL CORTEX, AS FUNCTION OF THE INTERVAL FROM INTRA-ATRIAL BOLUS INJECTION TO DECAPITATION

Time Interval	Arterial Sample	Frontal Cortex
10 sec	1.6	1.4
20 sec	1.4	1.9
30 sec	1.5	2.3
60 sec	1.9	4.9
120 sec	1.7	10.3

[^{3}H]MPTP and [^{14}C]butanol, in a ratio ranging between 2.5 and 3, was injected in a bolus of 200 μl of buffer solution into the right atrium of the heart, and the rats were decapitated at intervals. The ratio of [^{3}H]MPTP to [^{14}C]butanol in the arterial sample, collected at a known and constant rate in that interval, and in the frontal cerebral cortex at decapitation, are presented. The ratio of MPTP to butanol in the brain increased with time. In arterial blood, the ratio of MPTP to butanol is lower than in the injectate, indicating differential increased uptake of MPTP by the pulmonary circulation.

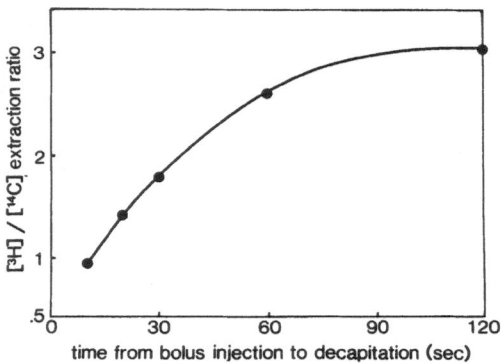

Figure 5. The extraction fraction ratio of [³H]MPTP to [¹⁴C]butanol in the parietal cortex is plotted as a function of the interval from bolus injection of 200 μl of buffer solution containing both tracers in the right atrium to decapitation. The ratio at 10 sec is ∼ 1, indicating similar blood-to-brain extraction of MPTP and butanol. With longer intervals, the ratio of MPTP to butanol in the brain increased, indicating differential brain-to-blood loss of butanol over MPTP.

tion presumably by capillary endothelium. The ratio of ³H:¹⁴C in the frontal cerebral cortex increased as a function of time indicating that MPTP is taken up to a similar extent as butanol on the first pass through the brain, but that in contradistinction to butanol, it does not leave the brain compartment (which includes brain endothelium) easily. The ratio of ³H:¹⁴C in the parietal cerebral cortex as a function of time between bolus injection and decapitation is depicted in Figure 5. There were little differences among the frontal cerebral cortex, parietal cerebral cortex, hippocampus, striatum, and cerebellum. The lack of regional differences speaks for the uniform effect of the cerebral circulation, which is the most likely place where [³H] MPTP is trapped in these short time periods of our study.

The important message from these preliminary experiments is that MPTP acts more or less like a chemical microsphere which seems to be trapped within the brain. Although it is probably just as lipid soluble as butanol, given their similar extraction by the brain when rats were decapitated 10 sec after the bolus injection. It is obvious however, that MPTP, unlike butanol, does not leave the brain easily. For that reason, calculation of cerebral blood flow using MPTP as the indicator is much less sensitive to changes in the time interval between bolus injection and decapitation. unlike the case when butanol is used as the indicator.

Taken together, our results are best explained by the existence of "biochemical" blood-brain barrier consisting of enzymes in brain endothelium which can metabolize lipid-soluble substances that are not impeded by the "physical" attributes of the blood-brain barrier. The concept of a "biochemical" blood-brain barrier, which was first proposed by Swedish investigators (37), was initially based on MAO and DOPA-decarboxylase, both of which play an important role in regulating the entry of amines and their precursors from blood-to-brain. If this "biochemical" blood-brain barrier proves to be the reason why certain animals are resistant to MPTP neurotoxicity, and if MPTP toxicity has anything to do with Parkinson's disease, then enzyme induction in brain endothelial cells may be the optimal means of preventing or arresting Parkinson's disease.

References

1. Davis, G.C., Williams, A.C., Markey. S.P., Ebert, M.H., Caine, E.D., Reichert, C.M., and Kopin, I.J.: Chronic parkinsonism secondary to intravenous injection of meperidine analogues. Psychiat. Res. 1, 249-254, 1979.

2. P. Langston, J.W., Ballard, P., Tetrud, J.W., and Irwin, I.: Chronic parkinsonism in humans due to a product of meperidine-analog synthesis. Science 219, 979-980, 1983.

3. Burns, R.S., Chiueh, C.C., Markey, S.P., Ebert, M.H., Jacobowitz, D.M., and Kopin, I.J.: A primate model of Parkinsonism: Selective destruction of dopaminergic neurons in the pars compacta of the substantia nigra by N-methyl-4-phenyl-1,2,3,6-tetrahydropyridine. Proc. Natl. Acad. Sci. USA 80, 4546-4550, 1983.

4. Chiba, K., Trevor, A., and Castagnoli, N. Jr.: Metabolism of the neurotoxic tertiary amine, MPTP, by brain monoamine oxidase. Biochem. Biophys. Res. Commun. 120, 574-578, 1984.

5. Markey, S.P., Johannessen, J.N., Chiueh, C.C., Burns, R.S. and Herkenham, M.A.: Intraneuronal generation of a pyridinium metabolite may cause drug-induced parkinsonism. Nature 311:464-466, 1984.

6. Heikkila, R.E., Manzino, L., Cabbat, F.S. and Duvoisin, R.C.: Protection against the dopaminergic neurotoxicity of 1 methyl-4-phenyl-1,2,5,6-tetrahydropyridine by monoamine oxidase inhibitors. Nature 311:467-469. 1984.

7. Langston, J.W., Irwin, I., Langston, E.B., and Forno, L.S.: Pargyline prevents MPTP-induced parkinsonism in primates. Science 225, 1480-1482, 1984.

8. Chiba, K., Peterson, L.A., Castagnoli, K.P., Trevor. A.J. and Castagnoli, N., Jr.: Studies on the molecular mechanism of bioactivation of the selective nigrostriatal toxin 1-methyl-4-phenyl-1,2,3,6-tetrahydropyridine. Drug Metab. Dispos. 13:342-347, 1985.

9. Javitch, J.A., D'Amato, R.J., Strittmatter, S.M., and Snyder, S.H.: Parkinsonism-inducing neurotoxin N-methyl-4-phenyl-1,2,3,6-tetrahydropyridine: uptake of the metabolite N-methyl-4-phenylpyridine by dopamine neurons explains selective toxicity. Proc. Natl. Acad. Sci. USA 82:2173-2177, 1985.

10. Ricaurte, G.A., Langston. J.W., DeLanney, L.W., Irwin, I.. and Brooks. J.D.: Dopamine uptake blockers protect against the dopamine-depleting effect of 1-methyl-4-phenyl-1,2,3,6-tetrahyropyridine (MPTP) in the mouse striatum. Neurosci. Lett. 59:259-264, 1985.

11. Nicklas, W.J., Vyas, I., and Heikkila, R.E.: Inhibition of NADH-linked oxidation in brain mitochondria by 1-methyl-4-phenylpyridine, a metabolite of the neurotoxin 1-methyl-4-phenyl-1,2,5,6-tetrahyropyridine. Life Sci. 36:2503-2508, 1985.

12. Vyas, I., Heikkila, R.E., and Nicklas, W.J.: Studies of the neurotoxicity of 1-methyl-4-phenyl-1,2,3,6-tetrahyropyridine. Inhibition of NAD-linked substrate oxidation by its metabolites 1-methyl-4-phenylpyridinium. J. Neurochem. 46:1501-1507, 1986.

13. Hollinden, G.E., Sanchez-Ramos, J.R., Sick, T.J. and Rosenthal, M.: MPP+ increases extracellular potassium in rat striatal slices: Preliminary evidence that consequences of MPP+ neurotoxicity are spread beyond dopaminergic terminals. Neurosci. Abstr. 13:1502, 1987.

14. D'Amato, R.J., Lipman Z.P., and Snyder, S.H.: Selectivity of the parkinsonian neurotoxin MPTP: Toxic metabolite MPP+ binds to neuromelanin. Science 231:987-989, 1986.

15. D'Amato, R.J., Alexander, G.M., Schwartzman, R.J., Kitt, C.A., Price, D.L., and Snyder, S.H.: Evidence for neuromelanin involvement in MPTP-induced neurotoxicity. Nature 327:324-326, 1987.

16. Johannessen, J.N., Chiueh, C.C., Burns, R.S. and Markey. S.P.: Differences in the metabolism of MPTP in the rodent and primate parallel differences in sensitivity to its neurotoxic effect. Life Sci. 36:219-224, 1985.

17. Elsworth, J.D., Deutch, A.Y., Redmond, D.E. Jr., Sladek, J.R., and Roth, R.H.: Differential responsiveness to 1-methyl-4-phenyl-1,2,3,6-tetrahydropyridine toxicity in sub-regions of the primate substantia nigra and striatum. Life Sci. 40, 193-202, 1987.

18. Kalaria, R.N. and Harik, S.I.: Blood-brain barrier monoamine oxidase: Enzyme characterization in cerebral microvessels and other tissues from six mammalian species, including human. J. Neurochem. 49:856-864, 1987.

19. Kalaria, R.N., Mitchell, M.J., and Harik, S.I.: Monoamine oxidases of the human brain and liver. Submitted to Brain.

20. Levitt, P., Pintar, J.E., and Breakefield, X.O.: Immunocytochemical demonstration of monoamine oxidase B in brain astrocytes and serotonergic neurons. Proc. Natl. Acad. Sci. U.S.A. 79:6385-6389, 1982.

21. Westlund, K.N., Denney, R.M., Kochersperger, L.M., Rose, R.M., and Abell, C.W.: Distinct monoamine oxidase A and B populations in primate brain. Science 230:181-183, 1985.

22. Zimmer, J. and Geneser, F.A.: Difference in monoamine oxidase B activity between C57 black and albino NMRI mouse strains may explain differential effects of the neurotoxin MPTP. Neurosci. Lett. 78:253-258, 1987.

23. Sayre, L.M., Arora, P.K., Iacofano, L.A. and Harik, S.I.: Comparative toxicity of MPTP, MPP+, and 3,3-dimethyl-MPDP+ to dopaminergic neurons of the rat substantia nigra. Eur. J. Pharmacol. 124:171-174, 1986.

24. Harik, S.I., Schmidley, J.W., Iacofano, L.A., Blue, P., Arora, P.K., and Sayre, L.M.: On the mechanisms underlying 1-methyl-4-phenyl-1,2,3,6-tetrahydropyridine neurotoxicity: The effects of perinigral effusion of 1-methyl-4-phenyl-1,2,3,6-tetrahydropyridine, its metabolite and their analogs in the rat. J. Pharmacol. Exp. Ther. 241:669-676. 1987.

25. Lai, F.M. and Spector, S.: Studies on the monoamine oxidase and catechol-O-methyltransferase of the rat cerebral microvessels. Arch. Int. Pharmacodyn. 233:227-234, 1978.

26. Hardebo, J.E., Emson, P.C.. Falck, B., Owman, C., and Rosengren, E.: Enzymes related to monoamine transmitter metabolism in brain microvessels. J. Neurochem. 35:1388-1393, 1980.

27. Lasbennes, F., Sercombe, R., and Seylaz, J.: Monoamine oxidase activity in brain microvessels determined using natural and artificial substrates: Relevance to the blood-brain barrier. J. Cereb. Blood Flow Metab. 3:521-528, 1983.

28. Kalaria, R.J., Mitchell, M.J. and Harik, S.I.: MPTP neurotoxicity: correlation with blood-brain barrier monoamine oxidase activity. Proc. Natl. Acad. Sci. U.S.A. 84:3321-3525, 1987.

29. Heikkila, R.E., Hess, A., and Duvoisin, R.C.: Dopaminergic neurotoxicity of 1-methyl-4-phenyl-1,2,5,6-tetrahydropyridine in mice. Science 224:1451-1453, 1984.

30. Sundstrom, E., Stromberg, I., Tsutsumi. T., Olson, L., and Jonsson G.: Studies on the effect of 1-ethyl-4-phenyl-1,2,3,6-tetrahydropyridine (MPTP) on central catecholamine neurons in C57BL/6 mice. Comparison with three other strains of mice. Brain Res. 405:26-38, 1987.

31. Riachi, N.J., Harik, S.I., Kalaria, R.N., and Sayre, L.M.: On the mechanisms underlying 1-methyl-4-phenyl-1,2,3,6-tetrahydropyridine neurotoxicity. II. Susceptibility among mammalian species correlates with the toxin's metabolic patterns in brain microvessels and liver. J. Pharmacol. Exp. Ther. 244:443-448, 1988.

32. Fuller, R.W., and Hemrick-Luecke, S.K.: Depletion of norepinephrine in mouse heart by 1-methyl-4-phenyl-1,2,3,6-tetrahydropyridine (MPTP) mimicked by 1-methyl-4-phenylpyridinium (MPP+) and not blocked by deprenyl. Life Sci. 39:1645-1650, 1986.

33. Johannessen, J.N., Adams, J.D., Schuller, H.M., Bacon, J.P. and Markey. S.P.: 1-methyl-4-phenylpyridine (MPP+) induces oxidative stress in the rodent. Life Sci. 38:743-749, 1986.

34. Sundstrom, E., Goldstein, M., and Jonsson, G.: Uptake inhibition protects nigrostriatal dopaminergic neurons from neurotoxicity of 1-methyl-4-phenylpyridinium (MPP+) in mice. Eur. J. Pharmacol. 131:289-292, 1986.

35. Rakic, P.: Limits of neurogenesis in primates. Science 227:1054-1056, 1985.

36. LaManna, J.C. and Harik, S.I.: Regional comparisons of brain glucose influx. Brain Res. 326:299-305, 1985.

37. Bertler, A., Falck, B., Owman, C., and Rosengren, E.: Localization of monoaminergic blood-brain barrier mechanisms. Pharmacol. Rev. 18:369-385, 1966.

38. Riachi, N.J. and Harik, S.I.: Differences in monoamine oxidase activity at the blood-brain barrier, not in brain, explain best the mouse strain differences in susceptibility to systemic 1-methyl-4-phenyl-1,2,3,6-tetrahydropyridine neurotoxicity. Submitted to Life Sci.

13

BIOCHEMISTRY OF THE NEUROTOXIC ACTION OF MPTP AND WHAT IT MAY TEACH US ABOUT THE ETIOLOGY OF IDIOPATHIC PARKINSONISM

Thomas P. Singer, Rona R. Ramsay, and Kathleen A. McKeown

Departments of Biochemistry/Biophysics and Pharmacy,
University of California, San Francisco. California 94143 and
Molecular Biology Division, Veterans Administration Medical Center,
San Francisco, California 94121

The past four years have been an exciting period in the history of parkinsonian research. Thanks to the efforts of an extraordinary number of investigators who have entered the field and a close interaction among neurologists, pharmacologists, biochemists, cell biologists, and chemists, we have progressed in a short time span from the first report that MPTP causes neurological symptoms close to those seen in Parkinsonism patients to an understanding of the main events leading to nigrostriatal cell death initiated by MPTP. One purpose of this paper is to summarize the biochemical events involved in this process, emphasizing recent, largely unpublished data and pointing out unresolved questions. Our second aim is to give an overview of what is being done to provide evidence for the hypothesis, which is gaining increasing acceptance, that idiopathic Parkinsonism is caused by slow-acting environmental neurotoxins. We will point out the battery of relatively simple in vitro tests for screening such potential neurotoxins and how they may eventually facilitate elimination of the disease. The original studies to be summarized represent, in part, collaboration with our colleagues, A. Trevor and N. Castagnoli at our University and in part, collaboration with Dr. R. Heikkila's laboratory.

Figure 1 represents the working hypothesis that we and many others have adopted to explain the selective toxicity of MPTP to nigrostriatal neurons. MPTP + passes the blood-brain barrier and is oxidized by monoamine oxidase (MAO)-B in the astrocytes to MPDP +, the dihydropyridinium form (1-3). MPDP + is then oxidized in part by MAO A or B, in part spontaneously, to MPP +, the pyridinium form (3), rather than undergoing chemical disproportio-

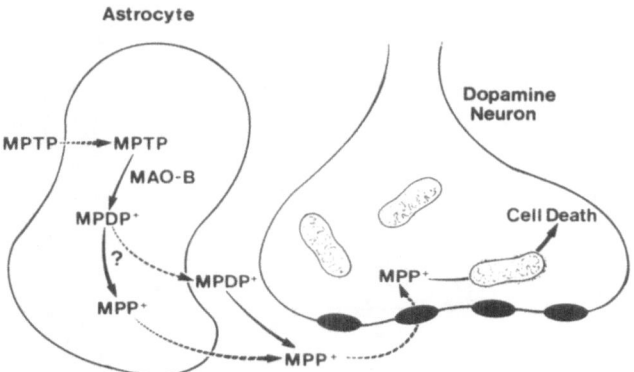

Figure 1. Schematic representation of the events from the study of MPTP into the brain to nigrostriatal cell destruction.

nation, as was believed at one time, since disproportionation does not occur to any significant extent at physiological pH. As shown on this scheme, it is not clear whether the second 2-e oxidation step to yield MPP + occurs in the astrocytes, in other cells, or even in the extracellular fluid. The neurotoxic bioactivation product, MPP + is then collected and concentrated by the synaptic dopamine reuptake system (4). This event, more than any other, may be responsible for the remarkable selectivity of MPP + for nigrostriatal cells. Once inside the dopamine terminus, MPP + causes destruction of the neuron. The hypothesis first voiced by Nicklas et al. (5) that the target of MPP + is the initial segment of the mitochondrial NADH oxidase chain is gaining ever increasing acceptance because so many experimental observations are best explained by it. Evidence for alternative hypotheses has not come to light.

It is not known, however, whether the mitochondria attacked by MPP + are those at the nerve terminus or in the stroma, which they might reach via the microtubules. It has been learned, however, (6,7) that positively charged MPP + molecules are rapidly and selectively concentrated across the inner membrane barrier by an energized pumping system. Inside the mitochondria, MPP + then combines non-covalently with NADH dehydrogenase (8,9) at the junction where ubiquinone (Q) reacts. This is also the combining site of barbiturates, rotenone, and of piericidin A (10,11) (Figure 2). Electron transport to Q is thus blocked, NADH oxidation and consequent oxidative phosphorylation cease and, when ATP becomes depleted, cell function deteriorates and cell death follows.

The Formation of MPP +

In our studies of the bioactivation of MPTP and its congeners, the basic assumption has been that for a tetrahydropyridine to be toxic to dopaminergic neurons, it must be first oxidized by MAO-A or B at appreciable rates. Along with several other laboratories, we have tested a large number of MPTP analogs for oxidation by the two types of MAO. The main purpose of such studies is to define the structural requirements for in vivo neurotoxicity although, as a spin-off, we are also learning totally unexpected features of the substrate specifics of MAO-A and B.

The data in Tables I and II represent the latest results, using compounds synthesized in Dr. Heikkila's laboratory, and are expressed as turnover number/K_m. This representation takes into account not only the maximum potential rate of oxidation of a given compound, but also its apparent affinity for the enzyme, so that it is a good basis for comparing activities at the low concentrations likely to prevail in cells. It is interesting that alkyl substitution in the 2' position of the aromatic ring progressively increases the reactivity of HPTP analogs with MAO-A as the alkyl-substituent is lengthened. For example, the 2'-isopropyl derivative is oxidized by MAO-A nearly as well as the best conventional substrate, kynuramine. Halogen substitution at the 3' position, but not at the 4', also increases the reactivity with MAO-A. In contrast, lengthening

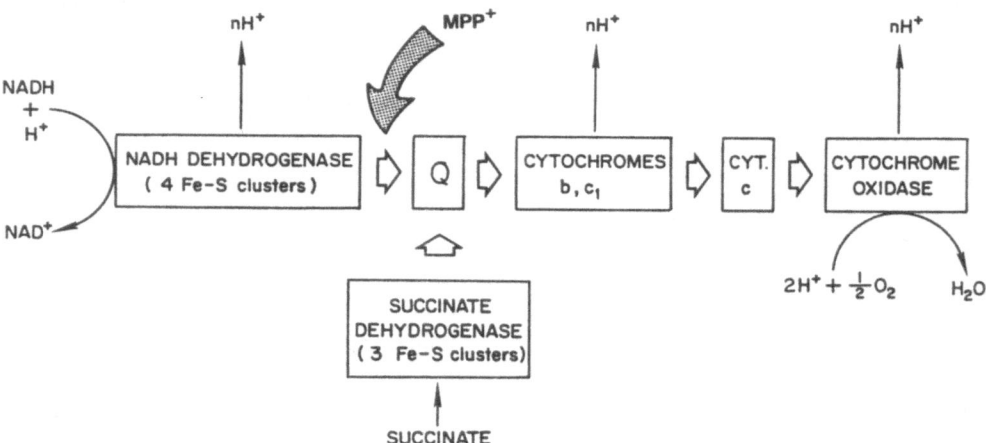

Figure 2. Site of inhibition of NADH oxidation by MPP + in the mitochondrial respiratory chain. Note that this is also site of inhibition by barbiturates, rotenone, and piericidin A.

Table I

Oxidation of MPTP and its Analogs by MAO A

Substrate	Polarographic Assay			Spectrophotometric Assay		
	Turnover Number	Km (mM)	Turnover Number/Km	Turnover Number	Km (mM)	Turnover Number/Km
Kynuramine	154	0.12	1280	120	0.15	857
MPTP	20	0.14	143	14	0.14	100
2'-Me-MPTP	83	0.14	593			
2'-Et-MPTP	53	0.077	688			
2'-O-Me-MPTP	23	0.045	509			
2'-Cl-MPTP	22	0.055	392			
2'-isoPr-MPTP	95	0.084	1127			
2'6'-diMe-MPTP	103	0.21	488			
3'-Me-MPTP	13	0.17	76			
3'-F-MPTP	25	0.064	391			
3'-Cl-MPTP	34	0.060	563			
3'-Br-MPTP	36	0.12	302			
4'-Cl-MPTP	46	0.667	69			
4'-Me-MPTP	11.1	0.19	58			
Et(4Ph)TP	7	0.25	28			
MeCTP	54	0.2	270			
Me(4Bz)TP	17	0.17	100			
PPTP	24	0.33	73			

All assays were at 30° and initial rates at V_{max} were measured with pure MAO B from beef liver and MAO A from human placenta. 2'-isoPr-MPTP = 2'-isopropyl MPTP; MeCTP = 1-methyl-4-cyclohexyl-tetrahydropyridine; Me(4Bz)TP =1-methyl-4-benzyl-tetrahydropyridine. Unpublished data of Youngster, S., Heikkila, R.E., McKeown, K., and Singer, T.P.

the alkyl substituent at 2' beyond a -CH$_3$ group progressively decreases the reactivity with MAO-B. Substitution at 3' enhances reactivity with MAO-B;thus the turnover number/K$_m$ ratio for the 3'-Br derivative is twice that for benzylamine, the fastest known substrate. The substitution of a benzyl group for the phenyl at the 4' position of MPTP yields an even more strikingly efficient substrate.

This data, using pure enzymes, correlates well with the in vivo findings of Dr. Heikkila's laboratory (13,14). As he and his colleagues have reported, the neurotoxic action of MPTP to black mice is completely blocked by deprenyl, a MAO-B inhibitor, but preventing the neurotoxicity of 2'-CH$_3$-MPTP requires clorgyline and deprenyl, i.e., both MAO-A and MAO-B inhibitors; in the case of 2'-ethyl MPTP, clorgyline alone blocks most of the neurotoxicity. These are just a few examples of the excellent correlation which has been achieved between in vivo data on neurotoxicity to C57 black mice and biochemical studies with pure enzymes.

While enzyme studies have thus become a primary tool for screening potential neurotoxins, there are some questions they cannot answer. One is the site of MPDP+ oxidation to MPP+. Does it occur in the glia, in the extracellular fluid, or in other types of cells? This

Table II

Oxidation of MPTP and its Analogs by MAO B

Substrate	Polarographic Assay			Spectrophotometric Assay		
	Turnover Number	Km (mM)	Turnover Number/Km	Turnover Number	Km (mM)	Turnover Number/Km
Benzylamine	415	0.39	1064	600	0.38	1580
MPTP	204	0.39	523	197	0.30	657
2'-Me-MPTP	357	0.28	1275			
2'-Et-MPTP	227	0.77	295			
2'-O-Me-MPTP	198	0.85	233			
2'-Cl-MPTP	257	0.19	1353			
2'-isoPr-MPTP	51	1.0	51			
2',6'-diMe-MPTP	161	0.77	209			
3'-Me-MPTP	130	0.20	650			
3'-F-MPTP	198	0.22	900			
3'-Cl-MPTP	215	0.19	1132			
3'-Br-MPTP	169	0.083	2036			
3'-O-Me-MPTP	151	0.16	944			
4'-Me-MPTP	169	0.49	345			
4'-Cl-MPTP	119	0.20	595			
4'-F-MPTP	93	0.22	423			
Et(4Ph)TP	34	0.17	200			
MeCTP	344	0.50	688			
Me(4-Bz)TP	222	0.083	2680			
Me(4-t-Bu)TP	183	2.0	92			
PPTP	24.5	0.8	31			

All assays were at 30° at V_{max} with respect to substrate with the pure MAO B from beef liver and MAO A from human placenta. Abbreviations as in Table I. Et(4Ph)TP = 1-ethyl-4-phenyl-tetrahydropyridine; Me(4-t-Bu)TP = 1-methyl-4-t-buytl-tetrahydropyridine. Unpublished data of Youngster, S., Heikkila, R.E., McKeown, K., and Singer, T.P.

question is being studied with cultures of the appropriate cells in collaboration with Dr. Hefti's laboratory.

There seemed to be one instance of a contradiction between in vivo effects and studies with pure enzymes. We have reported (2,3) that MAO-A also oxidizes MPTP all the way to MPP+, albeit considerably more slowly than MAO-B, suggesting that even in the presence of MAO-B blocker MPTP should elicit signs of neuronal damage, although more slowly than in an untreated animal. Yet, in vivo deprenyl alone suffices to block the toxicity of MPTP. The answer to this paradox is seen in Table III. The products of the oxidation of MPTP and its analogs, the pyridinium form (Table III) and the dihydropyridinium form (3,15), are excellent competitive inhibitors of MAO-A, but poor inhibitors of the B form. In many cases, the inhibition is too low to be measured (Table III, N.D.). Thus, soon after processing of MPTP starts, inhibi-

Table III

K_i Values for Inhibition of MAO A and B by MPP+ Analogs

Compound	K_i	
	MAO-A μM	MAO-B μM
MPP+	3.0	230
EPP+	6.8	N.D.
PPP+	38.0	N.D.
2'-Me-MPP+	4.8	> 100
2'-Et-MPP+	4.0	> 100
2'-O-Me-MPP+	1.6	250
2'-Cl-MPP+	1.6	N.D.
3'-Me-MPP+	1.5	206
3'-O-Me-MPP+	2.5	N.D.
3'-Cl-MPP+	4.0	360
3'-Br-MPP+	2.8	285
3'-F-MPP+	10.3	N.D.
4'-Me-MPP+	1.6	N.D.
4'-F-MPP+	10.6	N.D.
MCP+	7.0	N.D.
Me-4-BzP+	70.0	N.D.

Assays were spectrophotometric at 30°, derived from the initial rate of benzylamine or kynuramine oxidation. Noncompetitive inhibition was small and could not be compared for kinetic reasons with other values. N.D. = not determined, because of very high K_i value and high absorbance of the inhibitor at 250 nm where benzylamine oxidation is measured.

tion turns off the A enzyme, while the B form continues acting. Irreversible mechanism-based or "suicide" inhibition, although observed with both forms of MAO (15), similarly fails to block the action of MAO-B.

Path of MPP+ to the Mitochondrial Target

Uncertainty still surrounds the passage of the highly reactive MPP+ molecule to the dopaminergic synapse. There is little doubt, however, that it gets there and there seems to be a consensus that the next key event is its uptake by the dopamine reuptake system (4). This important event may be studied in vitro with synaptosome preparations, using labeled MPP+. Although such preparations are heavily contaminated with mitochondria (which also concentrate MPP+ selectively), as well as other inclusions which bind this reactive electrophile, the specific synaptic binding may be studied by the use of mazindol or of other reuptake inhibitors, which do not effect these other processes.

Although we are strongly inclined that the eventual cause of nigrostriatal cell death is damage to the mitochondrial energy generating system, "oxidative stress," an alternate hypothesis, is still frequently considered. Based on the structural analogy between MPP+ and the herbicide paraquat, it proposes that MPP+, like paraquat, undergoes a reduction-oxidation cycle in the cell, generating toxic oxidation products. The underlying assumption, namely that

MPP+ can be reduced under intracellular conditions, has been shown to be incorrect, however (16). Moreover, no persuasive evidence for this idea has ever been reported.

In contrast, the events thought to occur in mitochondria have been demonstrated in vitro and add up to a plausible and coherent picture. Moreover, direct evidence for it has recently emerged from two disparate lines of investigation. Christie-Pope et al. (17) have been studying the events occurring in cultures of dopaminegic neurons from embryonic dog brains on exposure to MPP+ and found that the first visible damage is disintegration of the mitochondria. Sanchez-Ramos and Hefti (18) have found that in dissociated neuronal cultures from rodents MPP+, but not MPTP, causes cell death and that it can be largely prevented by the barbiturate pentothal. This is as it should be, since both compounds combine at the same site in the NADH oxidizing system reversibly and thus should compete with each other (Figure 2).

Although initially skeptical, we have become outspoken advocates of the mitochondrial hypothesis on the basis of subsequent evidence found in this and other laboratories. We readily confirmed the initial observation of Nicklas et al. (5) that preincubation of rat liver or brain mitochondria with 0.2 to 0.5 mM MPP+ prevents the ADP-stimulated oxidation (i.e., state 3) of NAD+ linked substrates but not of succinate (Figure 3), pinpointing the site of blockade in

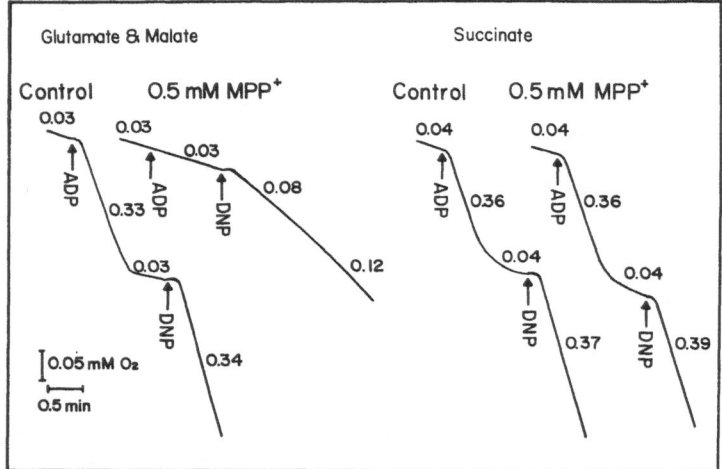

Figure 3. Inibition of the oxidation of NAD+ -linked substrates in intact liver mitochondria by brief preincabation with 0.5 mM MPP+. Note that succinate oxidation is not affected, hence blockade is in the Complex I segment of the respiratory chain.

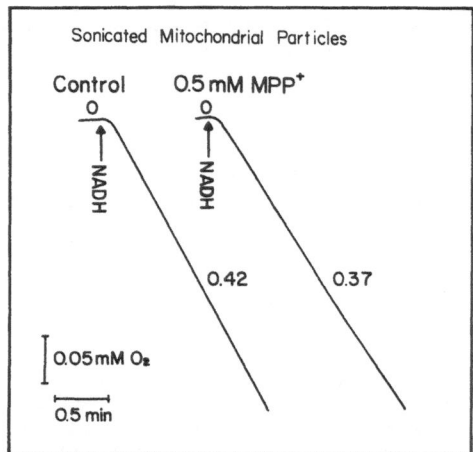

Figure 4. Failure of 9.5 mM MPP+ to abolish NADH oxidation in inverted mitochondria.

the NADH dehydrogenase region of the respiratory chain. However, the concentrations of MPP+ required for this seemed too high compared with the level found in post mortem samples of substantia nigra from MPTP-treated animals. Moreover, neither the highly purified, soluble enzyme nor Complex I, its membraneous counterpart, were inhibited significantly by 0.5 mM MPP+.

In order to rule out the possibility that the lack of major effects of 0.5 mM MPP+ on Complex I or soluble NADH dehydrogenase was due to some subtle modification, we subjected intact mitochondria to brief (10 sec) sonication so as to invert the inner membrane and then repeated the measurement of NADH oxidase activity with and without MPP+ treatment. Again, at 0.5 mM concentration, inhibition was trivial (Figure 4). Having eliminated alternative possibilities, we were left with the conclusion that there must be some mechanism for concentrating MPP+ from the cytoplasm into the mitochondria so as to achieve the high levels (10-20 mM) required for extensive blockade of NADH dehydrogenase and that this effect disappeared on inverting the inner membrane. The idea was tested by using [^{3}H] MPP+ and measuring its rate of uptake into mitochondria. Fig. 5 (solid circles) shows that a very rapid and extensive concentration of MPP+ occurred in the mitochondria. In similar experiments, starting with as low as 50 μM external concentration, an internal concentration of > 20 mM was achieved in about 30 minutes, a concentration more than enough to block the NADH oxidation completely.

The uptake of MPP+ was energy dependent (Figure 5), driven by the potential gradient of the membrane, since nigericin, a specific blocker of the proton gradient stimulated the uptake (triangles), while valinomycin + K+ which dissipates the electrical gradient (Figure 5, squares), prevented the uptake, as did the respiratory inhibitors rotenone and antimycin (Figure 5, open circles). Uncouplers also prevented the uptake of MPP+ as would be expected. Figure 6 shows that the addition of the uncoupler 2,6-dinitrophenol during the uptake process (10 min in Figure 6) causes immediate efflux of MPP+ with the concentration gradient. The energized uptake resembled the carrier-mediated uptake of metabolites across the inner membrane in many respects. Thus, the temperature coefficient was high (Figure 7), with an energy of activation of 72 Kjoules, in the range reported for metabolite carriers, as was the K_m value for MPP+. Although these and other characteristics are compatible with the concept of an energy-dependent carrier, they do not prove its existence. In fact, these features are also reminiscent of Pressman's (20) earlier observations on the energized uptake of hydrophobic molecules with a shielded (+) charge (e.g., alkylguanidines) into mitochondria, which is not thought to involve a carrier system. There are notable differences between the uptake of the two types of

Figure 5. Dependence of the uptake of MPP+ on the brain membrane potential. The uptake of [^{3}H]-MPP+ (0.5 mM) by mitochondria was monitored in aliquots removed at the intervals shown. Solid circles, control. Triangles, with 8μM nigericin, which dissipates the H+ gradient. Solid squares, with 4.4mg/ml of valinomycin (in the presence of K+), which dissipates the transmembrane potential. Open circles, with 10μM antimycin A + 10 μg/ml rotenone.

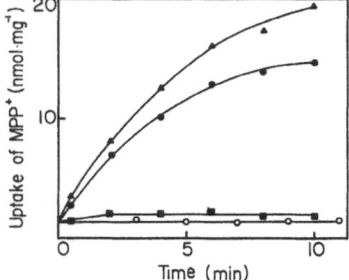

Figure 6. Effect of temperature on the energized uptake of [^{3}H]-MPP by mitochondria.

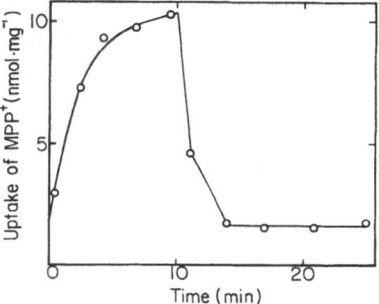

molecules, however, the most important of which is virtual lack of structural specificity in the uptake of the alkylguanidine, as against a high degree of specificity in the uptake of MPP +. We have studied the rates of uptake of MPP + analogs by two techniques. With radioactive analogs their uptake was measured directly, while with unlabeled analogs competitive inhibition of the accumulation of [³H] MPP + was measured. We have reported (21) that some positively charged pyridinium compounds, such as 1-methyl-2-phenylpyridinium, do not seem to be taken up by mitochondria, nor do uncharged analogs, such as 4-phenylpyridine, interfere with the uptake of [³H] MPP +. Figure 8 illustrates the effect of a series of unlabeled MPP + analogs on the uptake of MPP +. It is clear that their effectiveness as inhibitors varies widely, indicating probably different rates of uptake of these compounds. While the question of whether the energydependent uptake of MPP + into mitochondria is carrier mediated remains open, the process can be readily distinguished from the synaptic uptake of MPP + in that the mitochondrial pump is not affected by reuptake inhibitors. In mixtures of mitochondria and synaptosomes, as usually present in synaptosomal preparations, MPP + uptake with and without mazindol or dopamine provides a measure of the contribution of each process to total MPP + uptake (6).

At the high concentration of MPP + reached inside the mitochondria (20 mM or more) NADH oxidation in Complex I or in inner membrane preparations is nearly completely blocked. The predictable result of this in vivo is rapid ATP depletion and ultimate cell death. MPP + is noncovalently bound to the dehydrogenase (as are barbiturates and rotenone), so that on dilution activity is rapidly recovered (8). The techniques previously used to localize the binding site of barbiturates and rotenone (10,11) indicate that MPP + is bound in the same vicinity, i.e., at the junction with ubiquinone (Figure 2). We plan to localize more closely the binding site on the complex NADH dehydrogenase molecule, using photoaffinity labeled MPP +.

If we accept that NADH dehydrogenase is the likely target of MPP +, measurement of the I_{50} (or K_i) values in Complex I or inner membrane preparations provides a convenient screening technique for related compounds of potential neurotoxicity. For rapid onset of neurotoxic action and consequent dopamine depletion MPP + analogs must also be concentrated by the synaptic reuptake system and by mitochondria. However, pyridinium diffusion into nigrostriatal cells and their mitochondria might suffice for slow acting potential neuroto-

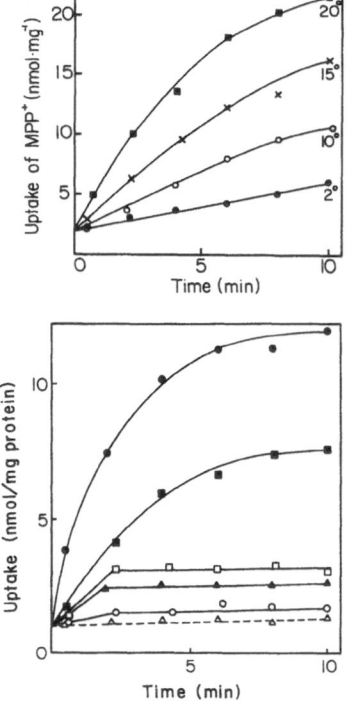

Figure 7. Effect of 10 μM 2,4-DNP (added as 10 min) on the uptake of MPP + by mitochondria.

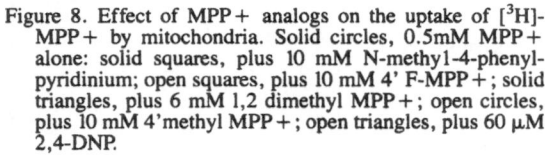

Figure 8. Effect of MPP + analogs on the uptake of [³H]-MPP + by mitochondria. Solid circles, 0.5mM MPP + alone: solid squares, plus 10 mM N-methyl-4-phenyl-pyridinium; open squares, plus 10 mM 4' F-MPP +; solid triangles, plus 6 mM 1,2 dimethyl MPP +; open circles, plus 10 mM 4'methyl MPP +; open triangles, plus 60 μM 2,4-DNP.

Table IV

Inhibition of Mitochondrial Respiration and NADH Oxidation by MPP+ Analogs

	NADH Oxidase Activity		Mitochondrial Respiration on NAD+ -Linked Substrates	
	IC_{50} (mM)	Relative Activity	K* $(M^{-1}min^{-1})$	Relative Activity
MPP+	4.0 ± 1.4	100	0.629	100
2'-Me-MPP+	4.1	98	0.547	87
2'-Cl-MPP+	0.7	57	0.448	71
2'-O-Me-MPP+	1.2	330	1.05	167
3'-Me-MPP+	1.7	235	1.56	248
3'-Br-MPP+	2.3	174	0.631	100
3'-Cl-MPP+	3.0	133	0.574	91
3'-O-Me-MPP+	2.6	154	0.564	90
3'-F-MPP+	5.5	73	0.355	56
4'-Me-MPP+	0.4	1000	2.50	397
4'-F-MPP+	2.0	200	0.200	32
PPP+	2.4	167	< MPP+	[~60]
MCP+	4.8	83	0.626	100
M4BzP+	3.0	133	< MPP+	[~50]
M3BzP+	4.8	83	< MPP+	[~50]
N-DiMe-PTP+	0.7	571	<<< MPP+	[<5]
MPPyrimidinium+	0.7	571	<<< MPP+	[<5]
MPyrP+	7.4	54	<< MPP+	[~20]
DP+	0.03	13300	>> MPP+	[~10000]

*K = Empirical expression relating inhibitor concentration to the time required to reach 50% inhibition of respiration, calculated from the slope of Dixon plots of [I] versus $1/t_{0.5}$. PPP+ = N-propyl-4-pyridinium; MCP+ = N-methyl-cyclohexylpyridinium; N-DiMe-PTP+ =N-dimethyl-4-phenyl-tetrahydropyridinium; MPPyrimidinium+ = N-methyl-4-phenyl-pyrimidinium; MPyrP+ = N-methyl-4pyrimidyl-pyridinium; DP+ = N-dodecylpyridinium.

xins, but they must still be inhibitors of NADH dehydrogenase. If this proposal has merit, the assay of NADH dehydrogenase activity in suitable preparations can be a primary tool in screening slow acting neurotoxins targeted to dopamine cells.

We reported recently (21) that among the pyridinium currently tested, 4-phenyl-pyridinium have the lowest I_{50} value (about 1mM), and is thus nearly an order of magnitude more inhibitory than MPP+. It is only moderately inhibitory to mitochondrial respiration (22), probably because it is not concentrated by the mitochondrial uptake system, but remains nevertheless a candidate for chronic cumulative neurotoxicity since it can run freely through the blood-brain barrier and across the inner membrane of the mitochondria. Moreover, it is known to be an environmental pollutant and the constituent of some spices (23). The corresponding tetrahydropyridines, such as MPTP are even more inhibitory, but again these compounds are not pumped into mitochondria and are thus not respiratory inhibitors. We have also examined a series of MPP + analogs synthesized in Dr. Heikkila's laboratory, comparing their action as inhibitors of NADH dehydrogenase in membranous NADH dehydrogenase preparations and as inhibitors of respiration (Table IV). It is interesting to note that all compounds which inhibits respiration in mitochondria also inhibit NADH dehydrogenase in isolated membranes, as it should be, if the site of action is as shown in Figure 2. On the other hand, several potent inhibitors of NADH dehydrogenase have very little effect on mitochondrial respiration (e.g. N-dimethyl-4-phenylpyridinium, and N-dodecyl-4-phenyl pyrimidinium). These possible candidates for slow acting neurotoxins. Attention is also called to the enormous effects of DP + (N-dodecylpyridinium) in both types of assays. It will be very interesting to test its Parent tetrahypropyridine for neurotoxicity and processing by MAO. In some, we have available a battery of readily responsible and fairly simple biochemical tests (oxidation by MAO-A and B, uptake of the pyridiuium by synaptosomes and by mitochondrial, blockade of mitochondrial respiration and inhibition of NADH dehydrogenase in purified membrane preparations, which might prove to be useful in screening tetrahydropyridines, and structurally related compounds for their potential in eliciting either rapid onset or slow progression of nigrostriatal degeneration. This may be the first and perhaps the easiest step toward defining the structural requirements for compounds to be reputative idiological aspects of parkinsonism.

Acknowledgements

This investigation was supported by Program Project HL 162521 and Grant NS-23066 for the NIH, Grant No. DMB8718741 from the National Science Foundation, and by the Veterans Administration.

References

1. Chiba, K., Trevor, A. and Castagnoli, N., Jr. (1984). Metabolism of the neurotoxic testiary amine, MPTP, by brain monoamine oxidase. Biochem. Biophys. Res. Commun. 120: 574-578.

2. Salach, J.I., Singer, T.P., Castagnoli, N., Jr. and Trevor, A.J. (1984). Oxidation of the neurotoxic amine 1-methyl-4-phenyl-1,2,3,6-tetrahydropyridine (MPTP) by monoamine oxidases A and B and suicide inactivation of the enzymes by MPTP. Biochem. Biophys. Res. Commun. 125: 831-825.

3. Singer, T.P., Salach, J.I., Castagnoli, N., Jr. and Trevor, A. (1986). Interactions of the neurotoxic amine 1-methyl-4-phenyl-1,2,3,6-tetrahydropyridine with monoamine oxidases. Biochem. J. 235: 785-789.

4. Javitch, J.A., D'Amato, R.J., Strittmatter, S.M. and Snyder, S.H. (1985). Parkinsonism-inducing neurotoxin, N-methyl-4-phenyl-1,2,3,6- tetrahydropyridine: uptake of the metabolite N-methyl-4-phenylpyridine by dopamine neurons explains selective toxicity. Proc. Natl. Acad. Sci. 82: 2173-2177.

5. Nicklas, W.J., Vyas, I. and Heikkila, R.E. (1985). Inhibition of NADH-linked oxidation in brain mitochondria by 1-methyl-4-phenylpyridinium, a metabolite of the neurotoxin 1-methyl-4-phenyl-1,2,3,6-tetrahydropyridine. Life Sci. 36: 2503-2508.

6. Ramsay, R.R., Dadgar, J., Trevor, A. and Singer, T.P. (1987). Energy-driven uptake of N-methyl-4-phenylpyridine by brain mitochondria mediates the neurotoxicity of MPTP. Life Sci. 39: 581-588.

7. Ramsay, R.R. and Singer, T.P. (1986). Energy-dependent uptake of N-methyl-4-phenylpyridinium, the neurotoxic, metabolite of 1-methyl-4-phenyl-1,2,3,6-tetrahydropyridine, by mitochondria. J. Biol. Chem. 261: 7585-7587.

8. Ramsay, R.R. Kowal, A.T., Johnson, M.K., Salach, J.I., and Singer, T.P. (1987). The inhibition site of MPP +, the neurotoxic bioactivation product of 1-methyl-4-phenyl-1,2,3,6-tetrahydropyridine is near the Q-binding site of NADH dehydrogenase. Arch. Biochim. Biophys. 259: 645-649.

9. Singer, T.P., Castagnoli N., Ramsay, R.R., and Trevor, A.J. (1987). Biochemical events in the development of parkinsonism induced by 1-methyl-4-phenyl-1,2,3,6-tetrahydropyridine. J. Neurochem. 49: 1-8.

10. Horgan, D.J., Singer, T.P. and Casida, J.E. (1968). Studies on the respiratory chain-linked reduced nicotinamide adenine dinucleotide dehydrogenase. XIII. Binding sites of rotenone, piericidin A, and amytal in the respiratory chain. J. Biol. Chem. 243: 834-841.

11. Palmer, A., Horgan, D.J., Tisdale, H., Singer, T.P. and Beinert, H. (1968). Studies on the respiratory chain-linked reduced nicotinamide adenine dinucleotide dehydrogenase. XIV. Location of the sites of inhibition of rotenone, barbiturates, and piericidin by means of electron paramagnetic resonance spectroscopy. J. Biol. Chem. 243: 844-847.

12. Youngster, S.K., Heikkila, R.E., McKeown, K.A., and Singer, T.P. (to be published).

13. Youngster, S.K., Sonsalla, P.K. and Heikkila, R.E. (1987). Evaluation of the biological activity of several analogs of the dopaminergic neurotoxin 1-methyl-4-phenyl-1,2,3,6-tetrahydropyridine. J. Neurochem. 48: 992-934.

14. Youngster, S.K., (1987). Ph.D. Thesis, New Jersey Medical College.

15. Singer, T.P., Salach, J.I. and Crabtree, D. (1985). Reversible inhibition and mechanism-based irreversible inactivation of monoamine oxidases by 1-methyl-4-phenyl-1,2,3,6-tetrahydropyridine (MPTP). Biochem. Biophys. Res. Commun. 127: 707-712.

16. Sayre, L.M., Arora, P.K., Feka, S.C., and Urbach, F.L. (1986). Mechanism of induction of Parkinson's Disease by 1-methyl-4-phenyl-1,2,3,6-tetrahydropyridine (MPTP). Chemical and electrochemical characterization of a geminal-dimethyl-blocked analogue of a postulated toxic metabolite. J. Am. Chem. Soc. 108: 2464-2466.

17. Christie-Pope, B., Burns, R.S. and Whetsell, W.O., Jr. (1987). Ultrastructural alterations induced by 1-methyl-4-phenylpyridine (MPP+) in organotypic cultures of canine substantia nigra and rat mesencephalon. Neurosci. Soc. Abstr., p.788.

18. Sanchez-Ramos, J.R., Barrett, J.N., Goldstein, M, Weiner, W.J. and Hefti, F. (1986). MPP+, but not MPTP is toxic to dopamine neurons in cultures of dissociated rat mesencephalic neurons. Neurosci. Lett. 72 215-220. Neurosci. Lett. 72 215-220.

19. Sanchez-Ramos, J.R. and Hefti, F., personal communication.

20. Pressman, B.C. (1963). The effects of guanidine and alkylguanidines on the energy transfer reactions of mitochondria. J. Biol. Chem. 238: 401-409.

21. Ramsay, R.R., McKeown, K.A., Johnson, E.A., Booth, R.G., and Singer, T.P. (1987). Inhibition of NADH oxidation by pyridine derivatives. Biochem. Biophys. Res. Commun. 146: 53-60.

22. Ramsay, R.R., Salach, J.I., Dadgar, J., and Singer, T.P. (1986). Inhibition of mitochondrial NADH dehydrogenase by pyridine derivatives and its possible relation to experimental and idiopathic parkinsonism. Biochem. Biophys. Res. Commun. 135 269-275.

23. Snyder, S.H. and D'Amato, R.J. (1985). Predicting parkinson's disease. Nature 317: 198-199.

14

A BIOLOGICAL EVALUATION OF SOME 2'-SUBSTITUTED ANALOGS OF MPTP

Stephen K. Younster and Richard E. Heikkila

Department of Neurology
University of Medicine and Dentistry of New Jersey
Robert Wood Johnson Medical School, Piscataway, NJ 08854

INTRODUCTION

1-Methyl-4-phenyl-1,2,3,6-tetrahydropyridine (MPTP) has been shown to produce a Parkinsonian syndrome in man (Davis et al., 1979; Langston et al., 1983) and to destroy the nigrostriatal dopaminergic pathway in monkeys (Burns et al., 1983; Langston et al., 1984a) and mice (Heikkila et al., 1984a; Hallman et al., 1985). The discovery that administration of such a simple chemical substance can reproduce the behavioral and pathological symptoms of Parkinson's disease has led some investigators to suggest that idiopathic Parkinson's disease may be caused by a compound present either in the environment or as an endogenous biological constituent (Snyder and D'Amato, 1986). Searches have been conducted for compounds structurally similar to MPTP which might be responsible for producing idiopathic Parkinson's disease and thousands of compounds, either synthesized in the laboratory or isolated from natural sources, have been identified which share certain structural features with MPTP (Markey and Schmuff, 1986). Our laboratory has been involved in the synthesis and evaluation of the neurotoxic potential of numerous analogs of MPTP with the intent of defining the structural requirements for production of MPTP-like neurotoxicity so that we will be able to predict the likelihood that any particular compound will be an MPTP-like neurotoxin.

Information regarding the mechanism of MPTP-induced neurotoxicity is consistent with the premise that MPTP is a protoxin bioactivated to the neurotoxic species 1-methyl-4-phenyl-pyridinium ion (MPP+) by monoamine oxidase oxidase B (MAO-B) in the brain. MAO-B catalyzes the oxidation of MPTP to 1-methyl-4-phenyl-3,4-dihydropyridinium ion (MPDP+), which subsequently oxidizes to MPP+ (Chiba et al., 1984; Chiba et al., 1985; Markey et al., 1984). This process is necessary for the neurotoxic activity of MPTP, since pretreatment of experimental animals with inhibitors of MAO-B prevents MPTP-induced neurotoxicity (Heikkila et al., 1984b; Langston et al., 1984b) and also prevents, or greatly diminishes, the amount of MPP+ formed in the brain (Markey et al., 1984). MPP+, but neither MPDP+ nor MPTP, is a good substrate for the dopamine (DA) carrier and the neurotoxicity of MPTP in experimental animals is prevented by preadministering an inhibitor of DA uptake (Javitch et al., 1985; Mayer et al., 1986; Gessner et al., 1985). These results are consistent with the conclusion that MPP+ is generated outside of the catecholaminergic neurons, which contain predominately the A-form of MAO and very little MAO-B, and is then concentrated by the DA transporter present in the plasma membrane of nigrostriatal dopaminergic neurons. MPP+ has been shown to be more toxic to dopaminergic neurons than MPTP when the compounds are stereotaxically injected into the substantia nigra, median forebrain bundle or the caudate of rats (Heikkila et al., 1985; Bradbury et al., 1985) or incubated with cultured rat mesencephalic dopaminergic neurons (Mytilineou et al., 1985). Also, MPP+, but neither MPDP+ nor MPTP, is an effective inhibitor of complex I of the mitochondrial respiratory chain (Nicklas et al., 1985; Vyas et al., 1986; Ramsey et al., 1986), which could greatly deplete cells of ATP and may be the actual

Table 1

Effects of pretreatment with MAO inhibitors on the in vivo neurotoxicity induced by 2'Me-MPTP

Pretreatment	Treatment	n	Neostriatal DA Levels (ug/g tissue)
—	—	5	16.8 ± 1.2
Clorgyline	—	5	18.9 ± 1.0
Deprenyl	—	5	16.9 ± 1.0
Both	—	5	18.9 ± 1.6
—	2'Me-MPTP	4	7.6 ± 1.4 (−55%)
Clorgyline	2'Me-MPTP	4	7.9 ± 1.2 (−58%)
Deprenyl	2'Me-MPTP	4	6.7 ± 0.7 (−61%)
Both	2'Me-MPTP	5	17.2 ± 1.2

Clorgyline, deprenyl, or the combination of clorgyline and deprenyl ("Both" in Table) were administered (each at 2.5 mg/kg, i.p.) to male Swiss Webster (CF-W) mice on day 1. On day 2, mice were given 2 injections (113 unmol/kg/injection, i.p.) with a 6 hr interval between injections. On day 16, mice were killed and DA levels were determined as described (Sonsalla et al., 1987). Data are mean values ± S.D. Numbers in parentheses indicate percentage changes from appropriate controls for statistically significant effects.

mechanism whereby MPP+ kills cells. In addition to synthesizing MPTP analogs and testing them for neurotoxic potential, we have been attempting to determine the reasons for the neurotoxicity, or lack of neurotoxicity, of the analogs in light of the mechanistic findings described above. As such, we test the MPTP analogs for their abilities to be oxidized by MAO to the corresponding dihydopyridinium and pyridinium ions and we screen the pyridinium compounds for their capacities both to be transported by the DA carrier and to inhibit Complex I of the respiratory chain. The studies have produced a considerable amount of information regarding the structural requirements for these various activities and our results with MPTP analogs containing substituents in the 2'position are the subject of this report.

Results and Discussion

We have previously shown that 1-methyl-4-(2'methylphenyl)-1,2,3,6-tetrahydropyridine (2'Me-MPTP) is a more potent nigrostriatal dopaminergic neurotoxin than MPTP in mice (Youngster et al., 1986; Sonsalla et al., 1987). In the experiment described in Table 1, 2'Me-MPTP administration reduced neostriatal DA by 55% relative to control DA content. This decrement in neostriatal DA is an indication of the neurotoxic activity of 2'Me-MPTP (see Sonsalla et al., 1987 for further evidence of neurotoxicity). Prevention of 2'Me-MPTP-induced neurotoxicity is not accomplished by inhibiting MAO-B or MAO-A alone, but, rather, requires inhibition of both MAO-A and MAO-B together (see Table 1, Kindt et al., in press Heikkila et al., in press). These results indicate that, unlike MPTP which is bioactivated to a neurotoxin by MAO-B but not MAO-A (Heikkila et al., 1984b), 2'Me-MTPT can be bioactivated to a neurotoxin by either MAO-A or MAO-B. Metabolic studies have shown that 2'Me-MPTP is oxidized to 1-methyl-4-(2'methylphenyl)pyridinium ion (2'Me-MPP+) both in vitro and in vivo (Kindt et al:, in press; Heikkila et al., in press; unpublished results). In accord with the neurotoxicity studies, inhibition of both MAO-A and MAO-B is required to significantly diminish the levels of 2'Me-MPP+ in the brains of mice following the peripheral administration of 2'Me-MPTP. This suggests that 2'Me-MPP+ by is the ultimate toxic species. Consistent with the in vivo data, the in vitro kinetic data for the oxidation of 2'Me-MPTP by MAO (Table 4) indicate that 2'Me-MPTP is a good substrate for both MAO-A and MAO-B. MPTP is oxidized rapidly and selectively by MAO-B (Table 4), consistent with the in vivo role of MAO-B, but not MAO-A, in bioactivating MPTP to a neurotoxin (Heikkila et al., 1984b). In contrast to the considerable differences between 2'Me-MPTP and MPTP, the pyridinium metabolites 2'Me-MPP+ and MPP+ are quite similar. Both compounds have comparable abilities to be transported by the DA carrier, as indicated by their abilities to cause the mazindol-inhibitable release of preaccumulated [^{3}H] DA from neostriatal synaptosomes (see Table 5; Sonsalla et al., 1987). Also, other studies have shown that 2'Me-MPP+ and MPP+ are comparable inhibitors of Complex I in the mitochondria (Kindt et al., 1987; Nicklas et al., 1987).

Table 2

Effects of pretreatment with MAO inhibitors on the in vivo neurotoxicity induced by 2'Et-MPTP

Pretreatment	Treatment	n	Neostriatal DA Levels (ug/g tissue)
—	—	5	13.7 ± 0.7
Clorgyline	—	5	14.0 ± 1.6
Deprenyl	—	5	13.5 ± 0.3
Both	—	5	13.9 ± 0.3
—	2'Et-MPTP	8	$6.9 \pm 2.1 \, (-50\%)$
Clorgyline	2'Et-MPTP	9	$11.5 \pm 1.3 \, (-18\%)$
Deprenyl	2'Et-MPTP	8	$5.7 \pm 2.7 \, (-58\%)$
Both	2'Et-MPTP	8	13.6 ± 1.1

Clorgyline, deprenyl, or the combination of clorgyline and deprenyl ("Both" in Table) were administered (each at 2.5 mg/kg, i.p.) to male Swiss-Webster mice on day 1. On days 2 and 3, 2'Et-MPTP was administered one injection per day (170 unmol/kg, i.p.). Mice were killed on days 18 and 19 and neostriatal levels of DA were determined as described (Sonsalla et al., 1987). Data are mean values $\pm$ S.D. Numbers in parentheses indicate significant differences from appropriate controls.

Table 3

Effects of MPTP and 2'nPr-MPTP on neostriatal levels of DA and DOPAC

Treatment	n	Neostriatal DA Levels (ug/g tissue)
—	4	13.9 ± 1.1
MPTP	4	$7.1 \pm 1.5 \, (-49\%)$
2'nPr-MPTP	4	13.4 ± 1.3

Male Swiss-Webster (CF-W) mice were injected (i.p.) 4 times at 2 hr intervals with MPTP at a dose of 113 umol/kg/injection. The mice died when we attempted to administer 2'nPr-MPTP using this dosing regimen, but they survived 2 injections (113 umol/kg/injection) 6 hr apart, which was the paradigm used in this experiment. MPTP is not neurotoxic to mice when administered using this paradigm (Youngster et al., 1986). Mice were killed 4 days later and neostriatal levels of DA were determined as described (Sonsalla et al. 1987). Data are mean values $\pm$ S.D. Numbers in parentheses indicate percentage changes from appropriate controls for statistically significant effects.

The interesting results obtained with 2'Me-MPTP prompted us to synthesize and test 1-methyl-4-2'ethylphenyl)-1,2,3,6-tetrahydropyridine (Et-MPTP) and 1-methyl-4-(2'n-prop-ylphenyl)-1,2,3,6-tetrahydropyridine (2'nPr-MPTP). As shown in Table 2, 2'Et-MPTP is primarily bioactivated to a neurotoxin by MAO-A since selective inhibition of MAO-A by clorgyline attenuates the DA-depleting effects of 2'Et-MPTP from 50% to 18%, and inhibition of MAO-B has no effect on 2'Et-MPTP-induced neurotoxicity. The MAO-catalized oxidation of 2'Et-MPTP-produces a pyridinium compound, 2'Et-MPP+, both in vitro and in vivo (Heikkila et al., in press; unpublished results). Also, consistent with the neurotoxicity studies, metabolic studies have shown that clorgyline pretreatment, but not deprenyl pretreatment, significantly reduces the amount of the pyridinum formed in the brains of mice receiving 2'Et-MPTP peripherally. The in vitro data for the MAO-catalyzed oxidation of 2'Et-MPTP and 2'Me-MPTP (Table 4) are consistent with the greater role for MAO-A in the bioactivation of 2'Et-MPTP relative to 2'Me-MPTP. 2'Et-MPTP has a somewhat higher Vmax than 2'Me-MPTP for MAO-A while the Km's of the 2 compounds for MAO-A are similar. Also, 2'Et-MPTP has a lower Vmax and higher Km than 2'Me-MPTP for MAO-B, consistent with the smaller role for MAO-B in the bioactivation of 2'Et-MPTP relative to 2'Me-MPTP. The [^{3}H]DA release studies (Table 5) indicate that 2'Et-MPP+ is transported by the DA carrier similarly to 2'Me-MPP+. Also, 2'Et-MPP+ has been shown to be a similar inhibitor of Complex I to 2'Me-MPP+ (Nicklas et al., 1987).

Table 4

Kinetic constants for the oxidation of MPTP and analogs by MAO-A and MAO-B

Compound	MAO type	Km (uM)	Vmax (nmoles H_2O_2/hr/g tissue)
MPTP	A	MD[a]	ND[a]
	B	54 ± 16	484 ± 91
2'Me-MPTP	A	17 ± 13	388 ± 152
	B	38 ± 4	2268 ± 470
2'Et-MPTP	A	24 ± 3	579 ± 7
	B	69 ± 11	602 ± 77
2'nPr-MPTP	A	93[b]	800[b]
	B	ND[a]	ND[a]

Mitochondrial preparations from the whole brains of Swiss Webster mice were used as the source of MAO. The MAO-A activity was determined in the presence of 0.25 uM deprenyl and the MAO-B activity in the presence of 0.25 uM clorgyline, concentrations which are selective and complete for inhibition of MAO-B and MAO-A, respectively. The method used for the determination of MAO activity was based on H_2O_2 formation and is described in Youngster et al., 1987. Lineweaver-Burk plots of the data were used to determine the Km and Vmax values shown, which represent the means $\pm$ S.D. for three separate experiments.
[a] ND-the activity was too low to determine with the method utilized
[b] values are the average of two determinations

Table 5

Abilities of pyridiniums to cause release of previously accumulated [^{3}H]DA from mouse neostriatal synaptosomes

Compound	EC_{50} (uM)
MPP+	0.83 ± 0.03
2'Me-MPP+	0.53 ± 0.05
2'Et-MPP+	0.97 ± 0.08
2'nPr-MPP+	>10

Data represents the mean $EC_{50} \pm$ S.D. values for 3-5 separate experiments for each compound (see Sonsalla et al., 1987 for experimental details). EC_{50} values are the concentrations of the pyridinium compounds that cause the release of 50% of preaccumulated [^{3}H]DA from a mouse (male Swiss-Webster CF-W) neostriatal synaptosomal preparation. The EC_{50} values were obtained by linear regression of a plot of log (concentration) vs. percentage release (expressed as a percent of control release) using 3-5 concentrations of pyridiniums which produced between 20% and 80% release. The release caused by each compound was blocked by 10 uM mazindol.

We synthesized 2'nPr-MPTP anticipating that this compound would be a neurotoxin bioactivated exclusively by MAO-A. However, as shown in Table 3, 2'nPr-MPTP administration to mice had no effect on the levels of DA, indicating that this compound is not neurotoxic under these experimental conditions. In vitro studies (Table 4) indicate that 2'nPr-MPTP is oxidized rapidly and selectively by MAO-A and, furthermore, preliminary metabolic studies have shown that substantial amounts of the pyridinium metabolite 2'nPr-MPP+ are formed in the brains of mice injected with 2'nPr-MPTP peripherally (unpublished results). Also in some preliminary studies, we have found that 2'nPr-MPP+ is a potent inhibitor of Complex I. However, compared to the other pyridinium metabolites described in this report, 2'nPr-MPP+ is a very weak substrate for the DA carrier (Table 5), and it seems likely that this explains the lack of neurotoxicity of 2'nPr-MPTP.

The results of the study have revealed some interesting structure-activity relationships. Clearly, MPTP analogs having 2'substituents should be considered potentially neurotoxic. We should mention that although the addition of a methyl or ethyl group may seem to be a trivial change in the molecule, MPTP analogs have been synthesized and screened with the methyl substitute in every other possible position of the molecule (Fries et al., 1986; Youngster et al.,

1987), and only the analogs with the methyl in either the 2'position are neurotoxic to mice (3'Me-MPTP is a less potent neurotoxin than MPTP in mice, unpublished results). The results also demonstrate how increasing the size of the 2'substituent increases the Vmax for the oxidation of the resulting compounds by MAO-A. Accordingly, two of these analogs, 2'Me-MPTP and 2'Et-MPTP, can be bioactivated to neurotoxins by MAO-A, suggesting the possibility that putative environmental or endogenous compounds responsible for idiopathic Parkinson's disease may be bioactivated by the A-form of MAO, as well as the B-form (eg. Lewin, 1986). The effects that the 2'substitutions have on the oxidation of the compounds by MAO-B seem more complex, the methyl group greatly enhancing activity and the propyl group eliminating activity. Equally interesting are the results concerning the effects the 2'substituents have on the activities of the MPP+ analogs as substrates for the DA carrier. The minimal effect of the methyl and ethyl substituents and drastic reduction in activity induced by the propyl substituent points to an important structural constraint in this area of the molecule for transport by the DA carrier. Finally, the results are consistent with our hypothesis that in order for a tetraydropyridine MPTP analog to be an MPTP-like neurotoxin, it must be metabolized by MAO to a pyridinium compound which is both a good substrate for the DA carrier and an inhibitor of mitochondrial respiration. The neurotoxic analogs 2'Me-MPTP and 2' and 2'Et-MPTP fulfill the criteria, whereas the non-neurotoxic analog 2'nPr-MPTP does not.

References

Burns, R.S., C.C. Chiueh, S.P. Markey, M.H Ebert, D.M. Jacobowitz and I.J. Kopin, 1983, A primate model of parkinsonism: selective destruction of dopaminergic neurons in the pars compacta of the substantia nigra by N-methyl-4-phenyl-1,2,3,6-tetrahydropyridine, Proc. Natl. Acad. Sci., U.S.A. 80:4546-4550.

Bradbury, A.J., Costall, B., Domeney, A.M., Jenner, P., Kelly, M.E., Marsden, C.D, and Naylor, R.J., 1986, MPP+ is neurotoxic to the nigrostriatal dopamine pathway, Nature 319:56-7.

Chiba, K., Petersen, K.P., Castagnoli, K.P., Trevor, A.J., and Castagnoli, N. Jr., 1985, Studies on the molecular mechanism of bioactivation of the selective nigrostriatal toxin 1-methyl-4-phenyl-1,2,3,6-tetrahydropyridine, Drug Metab. Disp. 13:342-347.

Chiba, K. A., Trevor, A.J., and Castagnoi, N. Jr., 1984, Metabolism of the neurotoxic tertiary amine, MPTP, by brain monoamine oxidase, Biochem. Biophys. Res. Commun. 120:674-578.

David, G.C., Williams, A.C., Markey,, S.P., Ebert, M.H., Caine, E.D, Reichert, C.M., and Kopin, I.J, 1979, Chronic parkinsonism secondary to intravenous injection of meperidine analogues, Psychiatry Res. 1:249-254.

Fries, D.S, Vries, J.D., Hazelhoff, B., and Horn, A.S., 1986, Synthesis and toxicity toward nigrostriatal dopamine neurons of 1-methyl-4-phenyl-1,2,3,6-tetrahydropyridine (MPTP) analogues, J. Med. Chem. 29:424-427.

Gessner, W., Brossi, A., Shen, R., and Abell, C.W., 1985, Further insight into the mode of action of the neurotoxin 1-methyl-4-phenyl/1,2,3,6-tetrahydropyridine (MPTP), FEBS Lett. 183:345-348.

Hallman, H., Lange, J., Olson, L., Stromberg, I. and Jonsson, G. 1985, Neurochemical and histochemical characterization of neurotoxic effects of 1-methyl-4-phenyl-1,2,3,6-tetrahydropyridine on brain catechalamine neurons in the mouse, J. Neurochem. 44:117-127.

Heikkila, R.E., Hess, A., and Duvoisin, R.C., 1984a, Dopaminergic neurotoxicity of 1-methyl-4-phenyl-1,2,3,6-tetrahydropyridine (MPTP) in mice, Science, 224: 1451-1453.

Heikkila, R.E., Kindt, M.V. Sonsalla, P.K. Giovanni, A., Youngster S.K., McKeown, K.A. and Singer, T.P. in press, Proc. Natl. Acad. Sci., U.S.A.

Heikkila, R.E., Manzino, L., Cabbat, F.S., and Duvoisin, R.C., 1984b, Protection against the dopaminergic neurotoxicity of 1-methyl-4-pheny-1,2,3,6-tetrahydropyridine by monoamine oxidase inhibitors, Nature 311:467-469.

Heikkila, R.E., Nicklas, W.J., and Duvoisin, R.C., 1985, Dopaminergic toxicity after the stereotaxic administration of the 1-methyl-4-phenylpyridinium ion (MPP+) to rats, Neurosci. Lett. 59:135-140.

Javitch, J.A., D'Amato, R.J, Strittmatter, S.M., and Snyder, S.H. 1985, Parkinsonism-inducing neurotoxin, N-methyl-4-phenyl-1,2,3,6-tetrahydropyridine: uptake of the metabolite N-methyl-4-phenylpyridine by dopaine neurons explains selective toxicity, Proc. Natl. Acad. Sci., U.S.A. 82:2173-2177.

Kindt, M.V., Heikkila, R.E., and Nicklas, W.J., 1987, Mitochondrial and metabolic toxicity of 1-metyl-4-(2'methylphenyl)-1,2,3,6-tetrahydropyridine, J. Pharmacol. Exp. Ther. 242:858-863.

Kindt, M.V. Youngster, S.K. Sonsalla, P.K., Duvoisin, R.C., and Heikkila, R.E., in press Role for monoamine oxidase-A (MAO-A) in the bioactivation and nigrostriatal dopaminergic neurotoxicity of the MPTP analog, 2'Me-MPTP, Eur. J. Pharmacol.

Langston, J.W., Ballard, P., Tetrud, J.W., and Irwin, I. 1983, Chronic parkinsonism in humans due to a product of meperidine-analog synthesis, Science 219:979-980.

Langston, J.W., Forno, L.S., Rebert, C.S., and Irwin, I., 1984a, Selective nigral toxicity after systemic administration of 1-methyl-4-phenyl-1,2,3,6-tetrahydropyridine (MPTP) in the squirrel monkey, Brain Res. 292:390-394.

Langston, J.W., Irwin, I., Langston, E.B., and Forno, L.S., 1984b, Pargyline prevents MPTP-induced parkinsonism in primates, Science 225:1480-1482.

Lewin R., 1985, Clinical trials for Parkinson's disease? Science 230:527.

Markey, S.P., Johannessen, J.N., Chiueh, C.C., Burns, R.S., and Herkenham, M.A., 1984, Intraneuronal generation of a pyridinium metabolite may cause drug-induced parkinsonism, Nature 311:464-467.

Markey, S.P. and Schmuff, N.R., 1986, The pharmacology of the parkinsonian syndrome producing neurotoxin MPTP (1-methyl-4-phenyl-1,2,3,6-tetrahydropyridine) and structurally related compounds, Med. Res. Rev. 6:386-429.

Mayer, R.A., Kindt, M.V., and Heikkila, R.E., 1986, Prevention of the nigrostriatal toxicity of 1-methyl-4-phenyl-1,2,3,6-tetrahydropyridine by inhibitors of 3,4-dihydroxyphenylethylamine transport, J. Neurochem, 47:1073-1079.

Mytilineou, C., Cohen, G., and Heikkila, R.E., 1985, 1-Methyl-4-phenylpyridine (MPP+) is toxic to mesencephalic dopamine neurons in culture, Neurosci. Lett. 57:19-24.

Nicklas, W.J. Vyas, I., and Heikkila, R.E., 1985, Inhibition of NADH-linked oxidation in brain mitochondria by 1-methyl-4-phenylpridine, a metabolite of the neurotoxin, 1-methyl-4-phenyl-1,2,3,6-tetrahydropyridine, Life Sci. 36, 2503-2508.

Nicklas, W.J., Younster, S.K., Kindt, M.V., and Heikkila, R.E., 1986, MPTP, MPP+, and mitochondrial function, Life Sci 40:721-729.

Ramsay, R.R., Salach, J.I., Dadgar, J. and Singer, T.P. 1986, Inhibition of mitrochondrial NADH dehydrogenase by pyridine derivatives and its possible relation to experimental and idiopathic parkinsonism, Biochem. Biophys. Res. Commun. 135:743-748.

Snyder, S.H. and D'Amato, 1986, MPTP, a neurotoxin relevant to the pathophysiology of Parkinson's disease, Neurology 36:250-258.

Sonsalla, P.K., Youngster, S.K., Kindt, M.V., and Heikkila, R.E., 1987, Characteristics of 1-methyl-4-(2'-methylphenyl)-1,2,3,6-tetrahydropyridine-induced neurotoxicity in the mouse, J. Pharmacol. Exp. Ther. 242:850-857.

Vyas, I., Heikkilla, R.E., and Nicklas, W.J., 1986, Studies on the neurotoxicity of 1-methyl-4-phenyl-1,2,3,6-tetrahydropyridine. Inhibition of NAD-linked substrate oxidation by its metabolite, 1-methyl-4-phenyl-pyridinium, J. Neurochem. 46:1501-1507.

Youngster, S.K., Duvoisin, R.C., Hess, A., Sonsalla, P.K., Kindt, M.V., and Heikikila, R.E., 1986, 1-Methyl-4-(2'-methyphenyl)-1,2,3,6-tetrahydropyridine (2'Me-MPTP) is a more potent dopaminergic neurotoxin than MPTP in mice, Eur. J. Pharmacol. 122:283-287.

Youngster, S.K., Sonsalla, P.K., and Heikkila, R.E., 1987, Evaluation of the biological activity of several analogs of the dopaminergic neurotoxin MPTP, J. Neurochem. 48:929-934.

15

HYDROGEN PEROXIDE PRODUCTION IN DOPAMINE NEURONS: IMPLICATIONS FOR UNDERSTANDING PARKINSON'S DISEASE

Gerald Cohen and Mary Beth Spina

Department of Neurology, Neurobiology Center
and Graduate School of Biological Sciences
Mount Sinai School of Medicine of the City University of New York
New York, N.Y. 10029, U.S.A.

INTRODUCTION: DOPAMINE TURNOVER AND H_2O_2 PRODUCTION

The turnover of monoamine neurotransmitters in brain should, in theory, be associated with an increased flux of hydrogen peroxide. The H_2O_2 arises as a result of the oxidative deamination of the neurotransmitters by monoamine oxidase (MAO). The enzymatic reaction with dopamine as substrate is depicted in Equation 1 below:

$$\text{dopamine} + O_2 + H_2O \xrightarrow{\text{MAO}} \text{3,4-dihydroxyphenylacetaldehyde} + H_2O_2 + NH_3 \tag{1}$$

The aldehyde product of the reaction is further metabolized in brain to the familiar acidic products, 3,4-dihydroxyphenylacetic acid (DOPAC) and 3-methoxy-4-hydroxyphenylacetic acid (homovanillic acid, HVA). DOPAC arises predominantly from the presynaptic turnover of dopamine. HVA, on the other hand, is formed by postsynaptic deamination of the dopamine metabolite, 3-0-methyldopamine, as depicted in Equation 2:

$$\text{3-0-methyldopamine} + O_2 + H_2O \xrightarrow{\text{MAO}} \text{3-methoxy-4-hydroxyphenylacetaldehyde} \\ + H_2O_2 + NH_3 \tag{2}$$

and by postsynaptic 0-methylation of DOPAC. The aldehyde product in Equation 2 is oxidized by brain aldehyde dehydrogenase to form HVA.

Consideration of Equations 1 and 2 leads to a conclusion that the formation of DOPAC and HVA in brain is accompanied, mole-for-mole, by the formation of H_2O_2. Similarly, the catabolism of serotonin and norepinephrine in brain (to 5-hydroxyindoleacetic acid and 3-methoxy-4-hydroxyphenylglycol, respectively) should be accompanied, mole-for-mole, by the generation of H_2O_2.

H_2O_2 is an oxidant and a potential cellular toxin. H_2O_2-mediated cell damage is seen in phagocytosis, drug-induced hemolytic anemia, and with the selective cellular toxins, paraquat and alloxan. In neural systems, H_2O_2 production has been linked to the destructive effects of 6-hydroxydopamine, 6-aminodopamine, and 5,7-dihydroxytryptamine on catecholamine neurons. Therefore, it seems apparent that the turnover of monoamines can be associated with an oxidative stress. Increased turnover of monoamines should be accompanied by an increased flux of H_2O_2 and, hence, increased probability for oxidative damage to monoamine systems.

DOPAMINE TURNOVER AND H_2O_2 PRODUCTION IN PARKINSON'S DISEASE

In Parkinson's disease, an observed rise in the ratio HVA/dopamine in brain autopsy specimens has been interpreted as indicating an increased turnover of dopamine (e.g., Hornykiewicz & Kish, 1986). In rats with lesions of substantia nigra neurons, increased levels of

DOPAC and HVA were reported in the remaining nerve terminals in the striatum when the extent of the lesion was greater than 60-80% (e.g., Hefti et al., 1980; Altar et al., 1987). The lesion experiments with rats parallel the situation in Parkinson's disease because the classic symptoms of tremor and rigidity emerge only after 80% or more of the nigrostriatal neurons have been lost. Therefore, an increased turnover of dopamine in Parkinson's disease should be associated with: (1) increased production of H_2O_2, and (2) increased oxidative stress within the surviving dopaminergic neurons. Increased release, reuptake, and presynaptic turnover of dopamine may contribute to the progressive destruction of nigrostriatal neurons seen in Parkinson's disease (Cohen, 1983).

We were interested in exploring ways to detect the production of H_2O_2 in dopamine nerve terminals. In an earlier study, production of H_2O_2 was detected in vivo within the rat brain parenchyma (as distinguished from the brain microvasculature) by utilizing 3-amino-1,2,4-triazole, an H_2O_2-dependent inhibitor of catalase (Sinet, Heikkila & Cohen, 1980). However, the H_2O_2 was peroxisomal in origin. The observed H_2O_2-production was unaffected by treatment with reserpine to provoke the turnover of monoamines, or treatment with pargyline to inhibit MAO. These results were in keeping with the known peroxisomal localization of both catalase and a number of H_2O_2-generating enzymes.

MAO, however, is located in the outer membrane of mitochondria. It is known that H_2O_2 generated by MAO in other tissues is removed by glutathione peroxidase (GSH peroxidase), and not by catalase (e.g., Oshino & Chance, 1977). In broken cell preparations of brain, the oxidative deamination of dopamine or serotonin by MAO is coupled directly to the oxidation of GSH to glutathione disulfide (GSSG) (Maker et al., 1981). Therefore, we turned our attention to the formation of GSSG as an index of H_2O_2 production during the turnover of dopamine (Spina & Cohen, 1988).

The reaction catalyzed by GSH peroxidase is shown in Equation 3. Once formed, GSSG can be reduced by GSSG reductase, which utilizes NADPH as cofactor (Equation 4):

$$H_2O_2 + 2GSH \xrightarrow{\text{GSH peroxidase}} GSSG + H_2O \tag{3}$$

$$GSSG + NADPH + H+ \xrightarrow{\text{GSSG reductase}} 2GSH + NADP+ \tag{4}$$

GSSG reductase normally prevents a marked accumulation of GSSG in tissues. Hence, during the production of H_2O_2 by MAO, we can anticipate an increase in the steady-state level of GSSG, but not the total conversion of GSH to GSSG. A significant rise in the level of GSSG would reflect the presence of an oxidative stress.

REDUCED AND OXIDIZED GLUTATHIONE IN BRAIN

GSH is ubiquitously distributed in relatively high concentration in body tissues. In brain, levels are in the range 1-3 mM, compared to a high in the range of 10 mM in liver. The

TABLE 1

LEVELS OF GSSG AND GSH IN RODENT AND PRIMATE BRAIN

SOURCE OF TISSUE	GSSG (mM) (mean ± SEM)	GSH (mM)	% GSSG
Rat frontal cortex (n = 4)	0.0038 ± 0.0019	1.45 ± 0.02	0.5
Monkey striatum (n = 3)	0.0081 ± 0.0032	3.10 ± 0.58	0.5
Human parietal cortex at autopsy (n = 5)	0.0068 ± 0.0020	1.18 ± 0.09	1.2

GSH was measured by high performance liquid chromatography. GSSG was measured spectrophotometrically (Tietze, 1969) with dithionitrobenzoate (DTNB) and GSSG reductase. The % GSSG is defined as 100% x 2GSSG/(GSH + 2GSSG). Data are from Slivka, Spina & Cohen (1987).

levels of GSSG, on the other hand, are quite low, constituting 1% or less of the total glutathione. Levels of GSH and GSSG in rodent and primate brain are shown in Table 1. The ratio GSSG/GSH reflects, in part, the redox status of the tissue.

Earlier reports of very high GSSG levels (up to 1 mM) in autopsy and biopsy specimens of human brain appear incompatible with the viability of neural tissue; these values probably reflect an analytic difficulty, as discussed elsewhere (Slivka, Spina & Cohen, 1987). A reported increase in GSSG in the substantia nigra from 89.3% of the total glutathione in control subjects to 100% in Parkinson's disease (Perry, Godin & Hansen, 1982) is similarly suspect.

EXPERIMENTAL DESIGN

The goal of these experiments (Spina & Cohen, 1988) was to detect the production of H_2O_2 associated with the oxidative deamination of dopamine within dopamine neurons. A synaptosomal preparation (P2 pellet) of rat striatum was incubated with L-dopa. Under these circumstances, dopamine is formed by the action of the aromatic amino acid decarboxylase (Equation 5):

$$\text{L-dopa} \xrightarrow{\text{dopa-decarboxylase}} \text{dopamine} + CO_2 \tag{5}$$

Reserpine phosphate (10 μM) was added to prevent the storage of dopamine in synaptic vesicles and, thereby, facilitate the turnover of the newly-synthesized dopamine. Lesion studies by Hefti, Melamed & Wurtman (1981) had shown that 80-85% of the dopa decarboxylase in the striatum is localized to the terminals of dopamine neurons. Hence, the synthesis and turnover of dopamine within the synaptosomal preparation was highly localized to the dopamine nerve terminals. In some experiments, reserpinized synaptosomes were incubated with dopamine, which was used in place of L-dopa.

The experimental expectation was that the catabolism of dopamine within the terminals would give rise to H_2O_2 and, subsequently, to the oxidation of GSH. In the presence of added glucose, which sustains a flow of reducing equivalents to recycle GSSG back to GSH (Equation 4), only a fraction of the GSSG produced would remain within the tissue preparation. However, a measurable rise in GSSG would reflect the presence of a significant oxidative stress.

A problem in attempting to target dopamine terminals for study is that they constitute only 1%, or less, of the mass of the striatum. Hence, only 1% or less of the striatum participates in an experiment conducted in vivo. The isolation of a P2 pellet results in a partial enrichment for in vitro studies; but, it would not be reasonable to look for significant reduction in total GSH, even if all of the GSH within the population of dopamine nerve terminals were to be oxidized to GSSG. Therefore, we concentrated on the measurement of GSSG, which could signal a change within a small tissue compartment.
Incubation with L-dopa and reserpine targets the dopamine nerve terminals population for study.

EXPERIMENTAL METHODS

A P2 pellet was isolated from rat striatum by a standard procedure: The tissue was homogenized in cold 0.32M sucrose and subjected to two centrifugations; a low-speed spin to remove cell debris and erythrocytes, followed by centrifugation at 12,000 x g for 40 minutes. The isolated pellet contained synaptosomes and free mitochondria. The contribution of free mitochondria to H_2O_2 production was assessed in experiments in which catalase was added to the medium. The pellet was resuspended in the medium and replicate aliquots were used. In some experiments, slices of striatum (1.0 x 0.4 x 0.4 mm) were used in place of the synaptosomal pellet.

The tissue preparations were incubated in a modified Krebs-Ringer phosphate buffer at pH 7.4 and 37°C. The buffer contained 5.5 mM glucose, 0.1 mM ascorbate, and 0.1 mM diethylenetriaminepentaacetate (DTPA) as a metalchelator. Samples were preincubated for 15 or 30 minutes with 10 μM reserpine before L-dopa (or dopamine) was added. Inhibitors of MAO or dopadecarboxylase were also preincubated before the addition of L-dopa. At the termination of the experiment, the pellet was isolated again by high-speed centrifugation and analyzed for GSSG.

TABLE 2

FORMATION OF GSSG IN SLICES OF RAT STRIATUM AFTER
INCUBATION WITH t-BUTYL HYDROPEROXIDE OR H_2O_2

PEROXIDE ADDED	GSSG IN TISSUE	INCREASE IN GSSG
(μM)	(μM $\pm$ SEM)	(μM)
hydrogen peroxide		
0 (Control)	1.0 $\pm$ 0.2	
100 μM	21.8 $\pm$ 1.3	20.8
t-butyl hydroperoxide		
0 (Control)	6.0 $\pm$ 0.2	
10 μM	10.9 $\pm$ 0.7	4.9
50 μM	23.6 $\pm$ 0.7	17.6
100 μM	38.5 $\pm$ 1.2	32.5

Tissue slices (25 mg) were incubated in 2 mL of medium for 30 minutes at 37°C in the presence of 5.5 mM glucose. For experiments with H_2O_2, 50 μM azide wa added to inhibit catalase in tissue or within entrapped erythrocytes. Result are for n = 6 (t-butyl hydroperoxide) or n = 4 (hydrogen peroxide) per group.

TABLE 3

INCREASE IN GSSG AFTER INCUBATION OF SYNAPTOSOMES WITH RESERPINE + L-DOPA

EXPT.	TISSUE GSSG CONTROL (pmol/mg	RES $\pm$ L-DOPA striatum $\pm$ SEM)	INCREASE IN GSSG (pmol/mg)	(%)
1	19.8 $\pm$ 0.0	25.3 $\pm$ 0.9 *	5.5	27.8
2	21.0 $\pm$ 0.4	28.0 $\pm$ 0.6 *	7.0	33.3
3	21.9 $\pm$ 0.8	31.8 $\pm$ 1.2 *	9.9	45.2

Synaptosomes were preincubated at 37°C in the presence of 10 μM reserpine phosphate for 15 min. Subsequently, 1 mM L-dopa was added and the incubatio: was continued for an additional 45 min. Values are for n = 3 per group. * p < 0.01

GSSG was measured by a modification of the enzymatic recycling procedure of Tietze (1969). The method is based on the reduction of dithionitrobenzoate (DTNB) to thionitrobenzoate (TNB) by GSSG reductase in the presence of GSSG. GSH was first removed by reaction with N-ethylmaleimide (Adams et al., 1983). The rate of TNB formation was monitored spectrophotometrically at 412 nm. The rate was compared to standards of known GSSG concentrations.

MODEL EXPERIMENTS

In preliminary experiments, slices of striatum were exposed to H_2O_2 or to an organic peroxide, t-butyl hydroperoxide, and the formation of GSSG was measured. t-Butyl hydroperoxide is a good substrate for GSH peroxidase (Equation 6):

$$\text{t-butyl hydroperoxide} + 2GSH \xrightarrow{\text{GSH peroxidase}} \text{t-butyl alcohol} + GSSG + H_2O \tag{6}$$

Earlier workers (e.g., Sies, 1985) had used t-butyl hydroperoxide and H_2O_2 to study the formation and efflux of GSSG from intact perfused liver.

In experiments with brain slices (Table 2), a significant rise in GSSG was observed when 100 μM H_2O_2 was added. Similarly, incubation with t-butyl hydroperoxide evoked a rise in GSSG, which was incremented with increasing concentration of added peroxide over the range 10-100 μM. These results illustrate a rise in the steady-state level of GSSG within the tissue during exposure to peroxides.

EXPERIMENTS WITH L-DOPA OR DOPAMINE

The results shown in Table 3 are representative experiments that illustrate the rise in GSSG when suspensions of synaptosomes are incubated with reserpine and 1 mM L-dopa. Results are expressed in terms of the original wet weight of striatum. In each experiment, a significant rise in tissue GSSG was seen, constituting an increase of 28-45% over the basal level. No GSSG was observed in the medium in these experiments.

Table 4 compares pooled results for multiple experiments with 1 mM L-dopa or for 10 μM dopamine during incubation with 10 μM reserpine phosphate. Dopamine could be substituted for the L-dopa, but, in general, L-dopa gave a more vigorous response.

In another set of experiments, the concentration of L-dopa was varied from 40 μM to 1 mM. A significant rise in GSSG was observed with 40 μM L-dopa (+ 3.4 pmoles/mg), which increased progressively at 200 μM L-dopa (+ 4.9 pmoles/mg) and at 1 mM L-dopa (+ 9.5 pmoles/mg). Therefore, the rise in GSSG was dependent upon the concentration of L-dopa added.

In still other experiments, the effect of reserpine alone or L-dopa alone was studied. Reserpine alone produced only a small rise in GSSG, which was not always significant in individual experiments. The poor response to reserpine, by itself, can be attributed to a relative absence of dopamine and formation of only trace amounts of DOPAC, as observed by high performance liquid chromatography with electrochemical detection. On the other hand, 1 mM L-dopa, by itself, accounted for most of the increment in GSSG seen when this concentration of L-dopa was combined with reserpine.

DRUG MANIPULATION OF GSSG FORMATION

Inhibitors of MAO were tested in combination with reserpine and L-dopa. If GSSG accumulation is dependent upon the oxidative deamination of dopamine, MAO inhibitors should prevent GSSG accumulation. When synaptosomes were preincubated with either pargyline or clorgyline, the rise in GSSG was markedly suppressed (Table 5). In other experiments with 10 μM dopamine used in place of the 1 mM L-dopa, similar strong inhibition of GSSG accumulation was seen in the presence of clorgyline. These results show that MAO activity was responsible for the accumulation of GSSG in the tissue.

Similarly, inhibitors of dopa decarboxylase inhibitors were tested. When the formation of dopamine is suppressed by inhibition of the decarboxylase, the rise in GSSG should also be inhibited. Table 5 shows the results of representative experiments with NSD-1055 and carbidopa. Both of these decarboxylase inhibitors suppressed the rise in GSSG, but not as effec-

TABLE 4

COMPARISON OF THE EFFECTS DOPAMINE (10 μM) AND L-DOPA (1 mM) ON TISSUE GSSG LEVELS IN THE PRESENCE OF RESERPINE

EXPERIMENT	GSSG INCREASE OVER BASAL LEVELS (pmol/mg ± SEM)	% INCREASE
1 mM L-Dopa (n = 11 expts)	7.0 ± 0.7 *	38.0 ± 4.5
10 μM Dopamine (n = 11 expts)	4.1 ± 0.5 *	22.9 ± 2.6

Experiments were similar in design to Table 3.
* p < 0.001 in paired t-test with respective controls.

TABLE 5

GSSG LEVELS IN THE PRESENCE OF INHIBITORS OF MONOAMINE OXIDASE
AND DOPA DECARBOXYLASE

| DRUG STUDIED | GSSG FORMED | | % INHIBITION |
| | (-)DRUG | (+)DRUG | |
	(pmoles/mg $\pm$ SEM)		
MAO Inhibitors			
Clorgyline (10 μM)	7.1	0.8 *	89 %
Pargyline (10 μM)	6.7	0.8 *	88 %
Decarboxylase Inhibitors			
NSD-1055 (250 μM)	9.3	3.1 *	68 %
Carbidopa (250 μM)	9.2	3.6 *	61 %

Experiments were conducted as in Table 3. The MAO inhibitors were preincubated for 15 min; the dopa decarboxylase inhibitors were preincubated for 30 min.
* p < 0.01

TABLE 6

GSSG LEVELS IN THE PRESENCE AND ABSENCE OF CATALASE IN THE MEDIUM

| EXPT | INCREASE IN GSSG LEVELS | | % INHIBITION |
| | (-) CATALASE | (+) CATAL±SE | |
	(pmoles/mg $\pm$ SEM)		
1	10.4	5.4	48 %
2	8.9	4.0	55 %
3	8.0	4.5	44 %

Experiments were conducted as in Table 3. Catalase was added to a final concentration of 2 ug/mL.

tively as the MAO inhibitors. It is concluded that the decarboxylation of L-dopa, followed by oxidative deamination of dopamine, plays a dominant role in the rise in steady-state level of GSSG within the tissue.

EFFECT OF CATALASE

The levels of tissue catalase are particularly low in brain. Moreover, it has been established that H_2O_2 generated by mitochondria is detoxified by GSH peroxidase. Therefore, the oxidation of GSH to GSSG is the dominant pathway for the detoxification of H_2O_2 during the turnover of monoamine neurotransmitters by isolated synaptosomes.

However, the addition of catalase to the medium provides a useful experimental probe because free mitochondria in the medium can also participate in generating H_2O_2. H_2O_2 generated in the medium can penetrate into the tissue to induce a rise in GSSG. This phenomenon was demonstrated earlier by the addition of reagent H_2O_2 to tissue slices (Table 2). Catalase added to the medium can scavenge the H_2O_2 produced by free mitochondria and, thereby, prevent external H_2O_2 from affecting GSSG levels.

In the experiments shown in Table 6, the formation of GSSG was compared in the presence and absence of added catalase. Catalase suppressed the rise in GSSG by 45-56% in 3 experiments. Therefore, approximately half of the rise in GSSG can be ascribed to the enzymatic activity associated with free mitochondria, and half to events taking place within the synaptosomes.

GSSG LEVELS IN VIVO

Some experiments were conducted with mice in vivo (Table 7). A combination of reserpine and L-dopa evoked a significant rise in the steady-state level of GSSG in the striatum. Therefore, the phenomenon that was studied in some detail with synaptosomes in vitro has a parallel with events that take place in the brain in vivo.

SIGNIFICANCE OF RESULTS

These experiments show that the turnover of dopamine can give rise to an oxidant stress, which is signaled by an elevation in the steady-state level of oxidized glutathione. The rise in GSSG was observed both in vitro with isolated synaptosomes (Table 3) and in vivo within the intact striatum of mice (Table 7).

It must be emphasized that the dopaminergic innervation of the striatum in a normal rodent constitutes 1% or less of the total mass of tissue. Therefore, the observed rise in GSSG, if properly attributed to dopaminergic terminals, represents a relatively larger rise within a small compartment.

Although the oxidation of GSH by GSH peroxidase is correctly and commonly looked upon as a detoxification step for a primary toxin, H_2O_2, the accumulation of GSSG may evoke secondary responses: First, if the loss of GSH is large enough, further detoxification of H_2O_2 will be impaired. As a consequence, the dopamine nerve terminals will be exposed to progressive oxidative damage from a natural metabolic product of neurotransmitter catabolism. Alternatively, even if the rise in GSSG is not massive, smaller elevations can still evoke physiologic sequelae. A literature exists (e.g., Gilbert, 1982) to indicate that, in other organs, GSSG levels in the range 50-200 μM can suppress the activity of certain sulfhydryl-dependent enzymes or serve as a "third messenger" to alter cellular function. Therefore, the potential consequences of elevated GSSG levels require closer scrutiny.

Do the changes we have seen represent very large increments within a small compartment comprised of dopamine terminals? Or is GSSG accumulation spread more diffusely across a variety of cell types? If changes are diffuse, they may be small enough to be readily accommodated, and no particular consequences may follow. If they are sufficient to evoke "third messenger" and other responses, important alterations in physiologic parameters may be evoked in affected cell types; such alterations may turn out to be either beneficial or harmful, and cannot be predicted in advance.

But, if the rise in GSSG seen with reserpine and L-dopa is indeed large and localized to the dopamine nerve terminals, and if this experimentally-induced change mirrors events that take place within the basal ganglia in Parkinson's disease, the potential ramifications may be detrimental to surviving dopamine neurons. We suggest that detailed consideration needs to be given to the role of increased presynaptic metabolism of dopamine (increased turnover) in evoking and maintaining an oxidant stress in Parkinson's disease. The presence of such a stressor mechanism within a partially degenerated nigrostriatal tract may provide some explanation for the relentless and progressive deterioration of remaining nigrostriatal neurons after the symptomatology of Parkinson's disease has emerged.

TABLE 7

INCREASED LEVELS OF GSSG IN VIVO

EXPT	GSSG IN THE STRIATμM		RISE IN GSSG
	CONTROLS	DRUG TREATED	
	(pmol/mg striatum $\pm$ SEM)		(pmol/mg)
1	3.8 ± 0.1	10.5 ± 0.1	$6.7*$
2	15.3 ± 0.3	22.0 ± 0.5	$6.7*$

Mice received i.p injections of reserpine (10 mg/kg), followed 1 hour later by L-DOPA (300 mg/kg). The striatum was assayed at 1 hr after L-dopa (n = 3-4 mice per group).
* p < 0.001

REFERENCES

Adams, J.D., Lauterberg, B.H., and Mitchell, J.R., Plasma glutathione and glutathione disulfide in the rat: regulation and response to oxidative stress. J. Pharmacol. Exp. Ther. 227: 749-754, 1983.

Altar, C.A., Marien, M.R., and Marshall, J.F., Time course of adaptations in dopamine biosynthesis, metabolism, and release following nigrostriatal lesions: implications for behavioral recovery from brain injury. J. Neurochem. 48: 390-399, 1987.

Cohen, G., The pathobiology of Parkinson's disease: biochemical aspects of dopamine neuron senescence. J. Neural Trans. Suppl. 19: 89-103, 1983.

Gilbert, H.F., Biological disulfides: the third messenger? J. Biol. Chem. 257: 12086-12091, 1982.

Hefti, F., Melamed, E., and Wurtman, R.J., Partial lesions of the dopaminergic nigrostriatal system in the rat brain: Biochemical characteristics. Brain Res. 195: 123-137, 1980.

Hefti, F., Melamed, E., and Wurtman, R.J., The site of dopamine formation in rat striatum after L-dopa administration. J. Pharmacol. Exp. Ther. 217: 18-197, 1981.

Hornykiewicz, 0. and Kish, S.J., Biochemical pathophysiology of Parkinson's disease. In Advances in Neurology, 45: 19-34, 1986.

Maker, H.S., Weiss, C., Silides, D.J., and Cohen, G., Coupling of dopamine oxidation (monoamine oxidase activity) to glutathione oxidation via the generation of hydrogen peroxide in rat brain homogenates. J. Neurochem. 36: 589-593, 1981.

Perry, T.L., Godin, D.V., and Hansen, S., Parkinson's disease: a disorder due to nigral glutathione deficiency? Neurosi. Lett. 33: 305-310 (1982).

Oshino, N. and Chance, B., Properties of glutathione release observed during reduction of organic hydroperoxide, demethylation of aminopyrine and oxidation of some substances in perfused rat liver, and their implications for the physiological function of catalase. Biochem. J. 162: 509-525, 1977.

Sies, H., Hydroperoxides and thiol oxidants in the study of oxidative stress in intact cells and organs. In H.Sies (Ed.), Oxidative Stress, Academic, London, 73-90, 1985.

Sinet, P.M., Heikkila, R.E., and Cohen, G., Hydrogen peroxide production by rat brain in vivo. J. Neurochem. 34: 1421-1428, 1980.
Slivka, A., Spina, M.B., and Cohen, G., Reduced and oxidized glutathione in human and monkey brain. Neurosci. Lett. 74: 112-118, 1987.

Spina, M.B. and Cohen, G., Exposure of striatal synaptosomes to L-Dopa increases levels of oxidized glutathione, Trans. Amer. Soc. Neurochem., in press, 1988.

Tietze, F., Enzymic method for quantitative determination of nanogram amounts of total and oxidized glutathione: applications to mammalian blood and other tissues. Analyt. Biochem. 27: 502-522, 1969.

16

STUDIES ON THE TOXICITY OF MPTP TO DOPAMINE NEURONS IN TISSUE AND CELL CULTURES

Catherine Mytilineou, Linda K. Friedman and Peter Danias

Department of Neurology, Mount Sinai School of Medicine
New York, N.Y. 10029

1-Methyl-4-phenyl-1,2,3,6-tetrahydropyridine (MPTP) is a neurotoxin with high specificity for the dopamine (DA) neurons of the nigrostriatal pathway of humans (Davis et al., 1979), monkeys (Burns et al., 1983) and mice (Heikkila et al., 1984a). In humans, the lesion which results from MPTP toxicity produces symptoms similar to idiopathic parkinsonism (Langston et al., 1983). The etiology of idiopathic parkinsonism remains still unknown and it is hoped that study of MPTP induced parkinsonism can elucidate the causes of this disease. Although it is possible that idiopathic parkinsonism bears no relationship with MPTP intoxication, knowledge of the cellular and molecular events that occur during MPTP induced neuronal degeneration would be a significant contribution to our understanding of neuronal susceptibility to environmental and/or endogenous neurotoxins.

Our laboratory has used organotypic and dissociated cell cultures from embryonic rat brain to study the toxicity of MPTP to monoamine neurons (Mytilineou and Cohen, 1984; Friedman and Mytilineou, 1987). For the purpose of our studies, we define neurotoxicity as damage to DA neurons, which is expressed either by a reduction in the high affinity uptake for DA or by the disappearance or morphological disintegration of neurons stained with antibodies against tyrosine hydroxylase (T-OH), the rate limiting enzyme in catecholamine biosynthesis. A reduction in the high affinity uptake, measured after washout of the toxins, represents damage or degeneration of neuronal fibers, the area where the high affinity uptake pump is predominantly localized. T-OH immunocytochemistry is a reliable marker for visualization of catecholamine neurons. By using both uptake measurement and T-OH immunocytochemistry to determine neurotoxicity, we can obtain a fairly accurate picture of the localization and the extent of damage to the neurons. The methods used for the establishment of cultures, uptake measurements and immunocytochemistry have been described in previous publications (Mytilineou and Cohen, 1984; Friedman and Mytilineou, 1987).

1. CHARACTERISTICS OF MPTP TOXICITY TO DA NEURONS IN CULTURE

In order to effectively use the tissue culture model in studies of the mechanism of MPTP neurotoxicity, it is important to determine whether MPTP affects the DA neurons in culture in a way similar to that in vivo. Experiments in our laboratory have demonstrated that exposure to MPTP induces damage to cultured DA neurons, which can result in neuronal degeneration (Figure 1; also see Mytilineou and Cohen, 1984; Friedman and Mytilineou, 1987). The concentration of MPTP necessary to induce toxicity in culture is in the μM range (1-10μM), a range similar to that used in experiments with monkeys. For example, in squirrel monkeys given repeated injections of 0.5 to 3.0 mg/kg MPTP to produce degeneration of nigrostriatal DA neurons (Langston et al., 1984a), the average whole-body concentration of MPTP, after each injection, would be between 3 and 17 μM, similar to the concentrations in our experiments.

The toxicity of in vivo MPTP administration depends on its bioactivation by monoamine

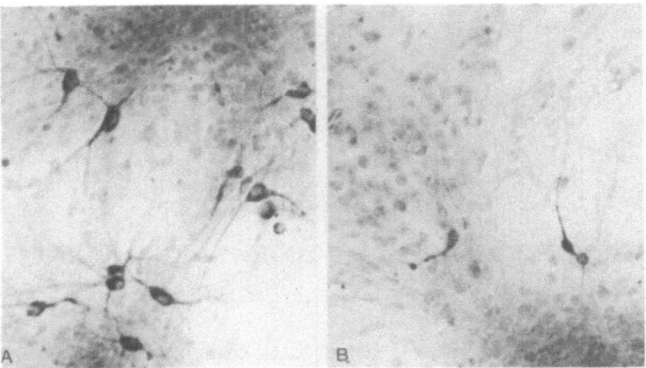

Figure 1. Photomicrographs from dissociated mesencephalic cultures stained with antibodies to T-OH. A. Control culture,demonstrating several T-OH positive neurons. B. Culture treated with 5 μM MPTP for 7 days. Two T-OH positive neurons in the field appear severely damaged. X 400.

oxidase (MAO) B. MPTP oxidation by MAO B results in the formation of the dihydropyridine MPDP + which is subsequently converted to the highly polar pyridinium ion MPP + (Chiba et al., 1984; Castagnoli et al., 1985). Treatment with the MAO B inhibitors pargyline and deprenyl protects monkeys (Cohen et al., 1984; Langston et al., 1984b) and mice (Heikkila et al. 1984b) from the toxicity of MPTP. Similarly, pretreatment of the cultures with pargyline or deprenyl protects the DA neurons from MPTP toxicity (Mytilineou and Cohen, 1984; Cohen and Mytilineou, 1985). Thus it appears that like in vivo, the toxicity of MPTP in culture depends on its bioactivation by MAO B. Furthermore, MPP +, the product of MPTP oxidation by MAO B is a very potent, specific dopaminergic neurotoxin in culture (Mytilineou et al., 1985; Sanchez-Ramos et al., 1986)

A characteristic of MPTP toxicity in culture is that it requires extended period of exposure to produce its toxic effect to DA neurons (Table 1). Exposure to MPTP for 24 hours does not result in a significant reduction of [³H]DA uptake by the cultures, while 4 and 7 day exposure produces a decrease in the [³H]DA uptake by 40 and 75% respectively.

One reason for the requirement of prolonged exposure to MPTP to induce toxicity in cultures, could be a slow rate of MPTP metabolism, because of low levels of MAO B activity. To examine this possibility, we measured the levels of MAO B in the cultures during the period that we expose them to MPTP. As shown in Table 2, at 1 week in vitro (the time when we apply MPTP to the cultures), the activity of MAO B is very low, compared to the levels of MAO B in the midbrain of the adult rat (4.8% of adult levels). MAO B activity increases significantly by the 2nd week in vitro, but it still remains below adult levels (30.3%). Consequently, a low rate of MPTP metabolism should be expected in the culture preparation. This was confirmed by measuring the amount of MPTP metabolized by the cultures during exposure to 5μM [³H]MPTP for 4 days. [³H]MPTP was separated by ether extraction from its metabolic product, MPP +, which is highly soluble in water at pH 12 (Glover et al., 1986; Mytilineou and Friedman, submitted for publication). As seen in Figure 2, MPTP metabolism proceeds at a rather slow rate in the cultures, and the amount of MPTP metabolized increases with increasing

Figure 2. Metabolism of MPTP by mesencephalic cell cultures during a 4 day exposure to 7.5 nmoles MPTP. The amount of MPTP metabolized was measured at 24 hour intervals. The values represent the means ± SEM from 5 separate cultures for each day.

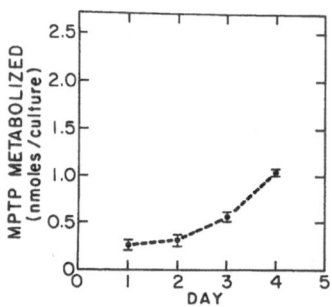

TABLE 1

EFFECT OF EXPOSURE TIME ON THE TOXICITY OF 10μM MPTP
TO DOPAMINE NEURONS IN EXPLANT CULTURES

Exposure Time (10 μM MPTP)	3H-DA Uptake % of Control	
24 hours	86.2 ± 13.8	N.S.
4 days	61.1 ± 6.7	p<0.05
7 days	25.7 ± 4.7	p<0.001

Explant cultures were treated with 10 μM MPTP for the period indicated. The cultures were washed with drug free medium for 24 hours before measurement of [^{3}H]DA uptake.

TABLE 2

MONOAMINE OXIDASE A AND B ACTIVITIES IN DISSOCIATED
MESENCEPHALIC CULTURES

	7 DAY OLD CULTURES	14 DAY OLD CULTURES
MAO-A		
nmoles/mg protein/hour	18.6 ± 0.6	36.9 ± 2.4
% of adult MAO-A	30.8	61.2
MAO-B		
nmoles/mg protein/hour	1.16 ± 0.18	7.4 ± 0.11
% of adult MAO-B	4.8	30.3

MAO activity was measured in homogenates from dissociated mesencephalic cultures using [^{3}H]5-hydroxytryptamine and [^{14}C]benzylamine as substrates for MAO-A and MAO-B activity respectively. Blank values were obtained by the addition of 1 μM clorgyline and 1 pM deprenyl. Adult rat MAO activity was measured in the ventral midbrain containing the substantia nigra.

TABLE 3

RELATIONSHIP BETWEEN MPTP METABOLISM AND TOXICITY TO DA NEURONS

TYPE OF CULTURE	MPTP Metabolized (nmoles/culture)	% Reduction of [^{3}H]DA Uptake
Low Cell Density (1X10^6 cells)	1.04 ± 0.04	56.6 ± 6.2
High Cell Density (2X10^6 cells)	2.02 ± 0.11	90.1 ± 0.8

Cultures were treated with 7.5 nmoles MPTP and the water soluble metabolites were separated from MPTP by ether extraction at pH 12 after 4 days incubation. Although the cultures with the high density of cells metabolized more MPTP/culture, when expressed as nmoles/mg protein the values were 2.44 ± 0.09 for the low cell density and 2.42 ± 0.13 for the high cell density cultures. MPTP toxicity was studied after exposure of cultures to 10 μM MPTP for 7 days. [^{3}H]DA uptake was measured 24 hour after removal of MPTP. Values are expressed as the percent reduction in [^{3}H]DA uptake. Uptake by the control cultures was 3.71 ± 0.1 pmoles [^{3}H]DA/culture/10 min.

the exposure time. This suggests that the prolonged exposure to MPTP required to produce toxic changes to DA neurons, could be related to a slow rate of MPTP metabolism in the cultures, due to the low levels of MAO B activity.

It has been proposed that MPP+, formed by the oxidation of MPTP, escapes in the extracellular space and then enters the DA neurons by the monoamine uptake pump (Javitch et al., 1985). In the cultures, the entire volume of the feeding medium corresponds to the extracellular space in vivo. Consequently, a much greater dilution of MPP+ would be expected to occur in the cultures when compared to the in vivo situation. Thus, prolonged exposure to MPTP would also be necessary in order to accumulate toxic levels of MPP+ into the feeding medium. We have shown that in dissociated cell cultures incubated with [³H]MPTP, more than 99% of the water soluble metabolite can be recovered from the feeding medium (Mytilineou and Friedman, submitted for publication). Thus, the low levels of MAO B activity in the cultures, combined with the dilution of MPP+ in the feeding medium, could explain the requirement of several-day exposure to MPTP to induce toxicity.

The relationship between the ratio of tissue to feeding medium volume and the toxicity of MPTP to DA neurons is shown in Table 3. In this experiment we doubled the number of cells plated in one group of cultures, but retained a constant volume of feeding medium. Thus, the group of cultures with the greater number of cells would have a higher number of MAO molecules per dish and should synthesize greater total amount of MPP+. Since the volume of feeding medium was the same in both sets of cultures, the cultures with the higher cell number should have higher concentrations of MPP+ in the feeding medium and consequently should suffer greater damage to DA neurons. As seen in Table 3, the water soluble MPTP metabolite in the feeding medium was greater in the cultures with the higher number of plated cells, although the rate of MPTP metabolism per mg of protein was the same for both groups of cultures. Furthermore, as was predicted, the toxicity to DA neurons was greater in the cultures with the higher number of plated cells, both at 1 and 10 µM MPTP concentration.

2. INTERACTIONS OF MPTP WITH MAO. IMPLICATIONS ON TOXICITY

If one examines the dose response curve of MPTP toxicity to DA neurons in culture, an interesting phenomenon becomes apparent (Figure 3). Although at the lower range of concentrations (from 0.1 to 5 µM) the toxicity increases with increasing concentrations of MPTP, further increase in the concentration (10 to 200 µM) results in attenuation of toxicity (Friedman and Mytilineou, 1987). Since it has been shown that MPTP, in addition to being a substrate for MAO B, is also an inhibitor of MAO B activity (Fuller and Hemrick-Luecke, 1985; Salach et

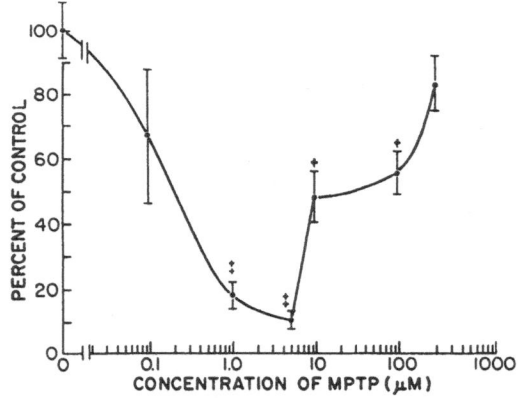

Figure 3. Uptake of [³H]DA by mesencephalic neurons in culture after exposure to different concentrations of MPTP for 7 days.

Treatment with MPTP began on the 8th day in vitro. The cultures were incubated with drug free medium for 24 hours before measurement of [³H]DA uptake. The values are the average ± SEM of [³H]DA uptake of 9-16 cultures. They are expressed as % of the uptake by the control (untreated) cultures. The value for the control cultures was: 0.54 + 0.07 pmoles [³H]DA/culture.† Significantly different from control p<0.01; ‡ p<0.001; Student's t-test. (From Friedman and Mytilineou, 1987).

TABLE 4

EFFECT OF MPTP ON MONOAMINE OXIDASE ACTIVITY IN CULTURES

Culture Treatment	MAO Activity nmoles/mg protein/hour		% Control	
	MAO A	MAO B	MAO A	MAO B
Untreated	41.9±7.1	4.3±0.7	100	100
Untr. + 5µM MPTP	34.2±5.6	3.7±1.0	82	86
Untr. + 100µM MPTP	8.4±1.1	2.3±0.5	20	53
5µM MPTP, 7 days	47.2±6.5	3.0±0.5	113	70
5µM, 7d. + 5µM MPTP	38.4±5.3	2.8±0.4	92	65
100µM MPTP, 7days	29.8±4.1	1.5±0.2	71	35
100µM, 7d. + 100µM MPTP	8.2±1.3	0.5±0.1	20	12

In the untreated cultures MPTP was added during the MAO assay. Cultures treated for 7 days with MPTP were rinsed twice with buffer and then assayed for MAO activity in the presence or absence of MPTP.

al., 1984; Tipton et al., 1986), we have examined whether the reduced toxicity at the high concentrations of MPTP could be the result of inhibition of MAO B activity. We treated cultures with either the most toxic concentration of MPTP (5 µM), or a concentration that results in significant diminution of toxicity (100µM) and determined the levels of MAO A and B activities at the end of a 7 day exposure. MAO activity was assayed by a modification of the method of Youdim (1975) using [^{3}H]serotonin and [^{14}C]benzylamine as substrates for MAO A and B respectively. The results of this experiment are presented in Table 4. When untreated (control) cultures were exposed to 5 µM MPTP during the 10 min. incubation period of the assay for MAO activity, there was a slight but not significant inhibitory effect on MAO activity. In cultures that were pretreated with 5 µM MPTP for 7 days and analyzed without MPTP present during the assay, there was no inhibitory effect on MAO A, but there was an approximately 30% inhibition of MAO B. Addition of 5 µM MPTP during the assay did not produce a significant further inhibitory effect. However, when 100 µM MPTP was added to untreated cultures, it resulted in a potent inhibition of both MAO A and B activities. MAO A activity was reduced by 80%, while MAO B activity was reduced by 47%. In cultures that were treated with 100 µM MPTP for 7 days and the toxin removed prior to the assay, both MAO A and B activities were depressed, but this time the effect was more pronounced for the B form of the enzyme. Under these conditions the activity of MAO A was reduced by 29%, while MAO B activity was reduced by 65%. When 100 µM MPTP was added to these cultures during the assay, both MAO A and B activities were suppressed even further, by 80 and 88% respectively. Thus, it appears that the continuous presence of MPTP in the cultures can have a profound inhibitory effect on MAO activity, particularly at the higher concentrations. The inhibitory effect of MPTP on MAO activity observed in cultures is consistent with published results showing that MPTP is mainly a competitive, partially reversible inhibitor of MAO A and a slowly acting irreversible inactivator of MAO B (Tipton et al. 1986). The potent inhibitory effect of 100 µM MPTP on both MAO A and B activity in cultures could explain the reduced toxicity observed at this concentration. Although MPTP is a better substrate for MAO B it can also be oxidized by the A form of the enzyme (Salach et al., 1984; Singer et al., 1985). However, under prolonged exposure to 100 µM MPTP, both MAO A and B activities are inhibited by at least 80%. Under these conditions, very little oxidation of MPTP to MPP+ would be expected. On the other hand, at the concentration of MPTP that is most toxic to the DA neurons in culture (5 µM), there was no

Figure 4. Effect of MPTP on the uptake of [³H]MPP+ by mesencephalic neurons in culture.

The values are the mean ± SEM from 14-16 cultures per group and are expressed as % of [³H]MPP+ uptake by the controls (no MPTP present in the incubation medium). [³H]MPP+ uptake by the control cultures was 65±4 fmoles/culture. † significantly different from control p< 0.01; ‡ p<0.001; Student's t-test. (From Friedman and Mytilineou, 1987).

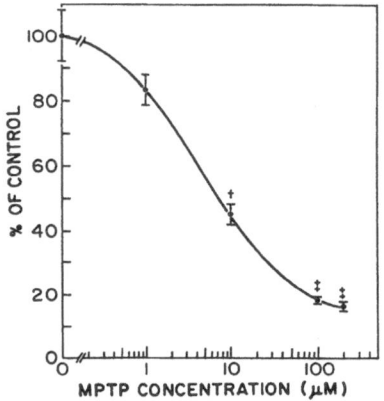

significant inhibitory effect on MAO A, and a 30-35% inhibition of MAO B was produced only after prolonged exposure. Thus it appears that MPTP, through its interaction with MAO can, at high concentrations, suppress its own toxicity to DA neurons.

3. INHIBITION OF THE HIGH AFFINITY UPTAKE OF MPP+ BY MPTP

Although MPTP is a very poor substrate for the high affinity uptake pump of catecholamine neurons (Javitch et al., 1985; Chiba et al., 1985), it has been shown that it can be an effective inhibitor of MPP+ uptake by the DA neurons (Javitch and Snyder, 1985). We therefore examined the possibility that, in addition to the inhibition of MAO activity, high concentrations of MPTP might prevent MPP+ uptake by the DA neurons and thus protect them from toxicity. In the culture system, where a large part of MPTP remains unmetabolized in the feeding medium, such an inhibitory effect could modify significantly the intraneuronal levels of MPP+. When we incubated mesencephalic cultures with [³H]MPP+ in the presence of increasing concentrations of MPTP (Figure 4), MPP+accumulation was inhibited by about 50% at 10 µM and by more than 80% at 100 and 200 µM MPTP (Friedman and Mytilineou, 1987).

Thus, at the concentrations that result in reduced toxicity to DA neurons, MPTP is a very potent inhibitor of MPP+ uptake. The above data indicate that the MPTP molecule can, at a particular range of concentrations, interfere with its own bioactivation to MPP+ by inhibiting MAO activity and, at the same time, can prevent the accumulation of the neurotoxic product by the DA neurons. These properties of MPTP are probably responsible for the U-shape of the dose-response curve of MPTP toxicity in culture. The same properties of the MPTP molecule could also explain why repeated injections of low doses of MPTP produce larger depletion of striatal DA levels in mice than the same total amount of MPTP administered in a single injection (Sonsalla and Heikkila, 1986).

4. STUDIES ON THE MECHANISM OF MPTP NEUROTOXICITY

The experiments described above were conducted in order to define the conditions necessary to induce toxicity to DA neurons in culture by exposure to MPTP. The goal of our research is to use the tissue culture model to study the mechanisms of neuronal degeneration produced by MPTP. One of the first questions we addressed was whether the neuronal DA content and the rate of DA metabolism by the mesencephalic dopaminergic neurons are related to the toxic effect of MPTP. The intraneuronal concentration of the monoamine neurotransmitters has been shown to modify the toxicity of methamphetamine (Schmidt et al., 1985) and 6-hydroxydopamine (Sachs et al., 1976). In our experiments we exposed cultures to L-DOPA under conditions that significantly increase both the levels and the rate of metabolism of DA (Mytilineou, 1987). MPTP induced toxicity, determined by the remaining [³H]DA uptake, was not influenced by the administration of L-DOPA to the cultures (Table 5), indicating that DA content and metabolism are probably not directly related to MPTP neurotoxicity. The same conclusion has been reached from in vivo experiments, where decrease in the levels of DA by administration of alpha-methyltyrosine, an inhibitor of tyrosine hydroxylase, or administration of L-DOPA had no effect on the toxicity of MPTP to the striatal DA neurons in mice (Fuller and Hemrick-Luecke, 1985; Schmidt et al., 1985).

The involvement of oxygen reactive species in the toxicity of MPTP has been suggested. Hydrogen peroxide, superoxide radical or hydroxyl radical could be formed during the oxidation of MPTP to MPDP+ by MAO oxidase (Irwin, 1986). In addition, MPP+ itself could cause oxidative stress, as shown by the increased levels of glutathione disulfide (GSSG) in the blood of rats, following peripheral administration of MPP+ (Johannessen et al., 1986). Attempts to protect from MPTP toxicity by antioxidants or free radical scavengers in vivo, have

TABLE 5

THE EFFECT OF L-DOPA TREATMENT ON THE TOXICITY OF MPTP
TO DA NEURONS IN EXPLANT CULTURES

TREATMENT	[^{3}H]DA Uptake (% Control)
Control	100.0 ± 8.4
MPTP (10 μM)	68.1 ± 5.3
L-DOPA (100μM)	97.3 ± 12.0
L-DOPA (100 μM) + MPTP (10 μM)	66.2 ± 7.9

Explant cultures were exposed to 100 μM L-Dopa in the presence of 0.2 mg/ml ascorbic acid. Control cultures were treated with ascorbic acid alone. MPTP (10 μM) was added to the cultures in the presence or absence of L-DOPA. Treatment was repeated for two 24 hour periods in order to keep the levels and metabolism of DA high (data from Mytilineou, 1987).

TABLE 6

EFFECT OF HYDROXYL-RADICAL SCAVENGERS ON MPTP AND MPP+ TOXICITY

TREATMENT	[^{3}H]DA UPTAKE (% of Control)
MPTP (5 μM)	34.7 ± 3.7
ETHANOL (50 mM) + MPTP	34.1 ± 1.5
BUTANOL (17 mM) + MPTP	29.5 ± 2.9
DMSO (13 mM) + MPTP	30.6 ± 2.4
MPP+ (1 μM)	34.9 ± 4.1
ETHANOL (50 mM) + MPP+	42.3 ± 4.1
BUTANOL (17 mM) + MPP+	41.8 ± 1.5
DMSO (13 mM) + MPP+	43.7 ± 2.1

Cultures were exposed to the hydroxyl-radical scavengers for 24 hours prior to the addition of MPTP or MPP+. The scavengers were present during the treatment with the toxins. [^{3}H]DA uptake was measured 24 hours after removal of drugs. Control uptake was 0.37 pmoles [^{3}H]DA/culture/10 min.

TABLE 7

EFFECT OF N-ACETYLCYSTEINE AND CYSTEAMINE ON MPTP AND MPP+
TOXICITY TO DA NEURONS IN CULTURE

TREATMENT	[³H]DA UPTAKE (% of Control)
MPTP (5μM)	25.4 ± 1.5
NAC (1mM) + MPTP (5μM)	26.8 ± 3.8
CAM (50μM) + MPTP (5μM)	22.0 ± 2.6
MPP+ (1μM)	25.5 ± 5.3
NAC (1mM) + MPP+ (1μM)	28.9 ± 2.0
CAM (50μM) + MPP+ (1μM)	23.5 ± 1.8

Cultures were exposed to 1mM N-actylcysteine (NAC) or 50μM cysteamine (CAM) 24 hours prior to the exposure to 5μM MPTP or 1μM MPP+. MPTP treatment lasted for 6 days and MPP+ for 3 days. [³H]DA uptake was measured 24 hours after removal of the toxins. Control cultures accumulated 0.79 ± 0.08 pmoles [³H]DA/culture/10 min.

produced conflicting results. Partial protection from MPTP toxicity in mice has been reported after large doses of ascorbic acid, alpha-tocopherol, beta-carotene and N-acetylcysteine (Perry et al., 1985). However a different study (Martinovits et al., 1986) showed no protection of DA neurons from MPTP by treatment with ascorbic acid, alphatocopherol, or dimethylsulfoxide (DMSO).

In the tissue culture system, direct application of drugs to monolayers of brain cells bypasses the in vivo peripheral metabolic detoxification or other pharmacokinetics that could modify the direct action of the compounds. We therefore examined the effect of exposure of cultures to three known potent hydroxyradical scavengers (ethanol, N-butanol and DMSO), on the toxicity of MPTP and MPP+ to the DA neurons. The results of this experiment are shown in Table 6. Cultures were exposed to ethanol (50 mM), N-butanol (17 mM) or DMSO (13 mM), 24 hours before addition of MPTP (5 μM) or MPP+ (1 μM). The scavengers were present during the entire period of treatment with the neurotoxins. MPTP was present for 6 days and MPP+ for 3 days. Analysis for neurotoxicity was performed, 24 hours after removal of all drugs, by determining the [³H]DA uptake in the cultures. Exposure to ethanol, butanol or DMSO alone did not affect the [³H]DA uptake by the cultures. Treatment with MPTP or MPP+, according the protocols used, produced a similar decrease in the uptake of [³H]DA, by approximately 65%. The presence of the hydroxylradical scavengers 24 hours prior and during the exposure to MPTP or MPP+ did not modify the reduction in [³H]DA uptake produced by the toxins. We also examined the effect of the antioxidants and hydroxyl-radical scavengers cysteamine and N-acetylcysteine on the toxicity of MPTP or MPP+ to the DA neurons in culture (Table 7). Incubation of the cultures with 1 mM N-acetylcysteine or 50 μM cysteamine (high concentrations of cysteamine produced generalized toxicity to the cultures) again failed to modify the degree of damage to DA neurons produced by MPTP or MPP+. Failure to protect the cultures from MPTP or MPP+ toxicity by antioxidants and hydroxyl-radical scavengers does not necessarily prove the lack of involvement of these toxic species in the neuronal degeneration produced by the toxins. However these date do not lend support to such a hypothesis.

ACKNOWLEDGMENTS

Supported by NIH grants NS-11631, NS-18979, NS-23017, the American Federation for Aging Research and the American Parkinson Disease Association. We thank Julia Shen for expert technical assistance.

REFERENCES

Burns, R. S., Chiueh, C. C., Markey, S. P., Ebert, M. H., Jacobowitz, D. M. and Kopin, I. J. (1983) A primate model of parkinsonism: selective destruction of dopaminergic neurons in the pars compacta of the substantia nigra by N-methyl-4-phenyl-1,2,3,6-tetrahydropyridine. Proc. Natl. Acad. Sci. U.S.A., 80:4546-4550.

Castagnoli, N., Chiba, K. and Trevor, A.J. (1985) Potential bioactivation pathways for the neurotoxin 1-methyl-4-phenyl-1,2,3,6-tetrahydropyridine (MPTP). Life Sci. 36:225-230.

Chiba, K., Trevor, A. and Castagnoli, N., Jr. (1984) Metabolism of the neurotoxic tertiary amine, MPTP, by brain monoamine oxidase. Biochem. Biophys. Res. Commun. 120:574-578.

Chiba, K., Trevor, A.J. and Castagnoli, N., Jr. (1985) Active uptake of MPP+, a metabolite of MPTP, by brain synaptosomes. Biochem. Biophys. Res. Commun. 128:1228-1232, 1985.

Cohen, G., Pasik, P., .Cohen, B., Leist, A., Mytilineou, C. and Yahr, M. D. (1984) Pargyline and deprenyl prevent the neurotoxicity of 1-methyl-4-phenyl-1,2,3,6-tetrahydropyridine (MPTP) in monkeys. Eur. J. Pharmacol., 106:209-210.

Cohen, G. and Mytilineou, C. (1985) Studies on the mechanism of action of 1-methyl-4-phenyl-1,2,3,6-tetrahydropyridine (MPTP). Life Sci. 36:237-242.

Davis, G.C., Williams, A.C., Markey, S.P., Ebert, M.H., Jacobowitz, D.M. and Kopin, I.J. (1979) Chronic parkinsonism secondary to intravenous injection of meperidine analogues. Psychiat. Res. 1:249-254.

Friedman, L. and Mytilineou, C. (1987) The toxicity of MPTP to dopamine neurons in culture is reduced at high concentrations. Neurosci. Lett. 79:65-72.

Fuller, R. and Hemrick-Luecke, S.K. (1985) Mechanisms of MPTP (1-methyl-4-phenyl-1,2,3,6-tetrahydropyridine) neurotoxicity to striatal dopamine neurons in mice. Prog. Neuro-Psychopharmacol. Biol. Psychiat. 9:687-690.

Glover, V., Gibb, C. and Sandler, M. (1986) Monoamine oxidase B (MAO-B) is the major catalyst for 1-methyl-4-phenyl-1,2,3,6-tetrahydropyridine (MPTP) oxidation in human brain and other tissues. Neurosci. Lett. 64:216-220.

Heikkila, R. E., Hess, A. and Duvoisin, R. (1984a) Dopaminergic neurotoxicity of 1-methyl-4-phenyl-1,2,5,6-tetrahydropyridine in mice. Science, 224:1451-1453.

Heikkila, R. E., Manzino. L., Cabbat, F. S. and Duvoisin, R. C. (1984b) Protection against the dopaminergic neurotoxicity of 1-methyl-4-phenyl-1,2,5,6-tetrahydropyridine by monoamine oxidase inhibitors. Nature, 311: 467-469.

Irwin, I. (1986) The neurotoxin 1-methyl-4-phenyl-1,2,3,6-tetrahydropyridine (MPTP): A key to Parkinson's disease? Pharmaceut. Res. 3:7-11, 1986.

Javitch, J.A. and Snyder, S.H. (1985) Uptake of MPP(+) by dopamine neurons explains selectivity of parkinsonism-inducing neurotoxin, MPTP. Eur. J. Pharmacol. 106:445-456.

Javitch, J. A., D'Amato, R. J., Strittmatter, S. M. and Snyder, S. H. (1985) Parkinsonism-inducing neurotoxin, N-methyl-4-phenyl-1,2,3,6-tetrahydropyridine: Uptake of the metabolite N-methyl-4-phenylpyridine by dopamine neurons explains selective toxicity. Proc. Natl. Acad. Sci. U.S.A., 82:2173-2177.

Johannessen, J.N., Adams, J.D., Schuller, H.M., Bacon, J.P. and Markey, S.P. (1986) 1-Methyl-4-phenylpyridine (MPP+) induces oxidative stress in the rodent. Life Sci. 38:743-749.

Langston, J. W., Ballard, P., Tetrud, J. W. and Irwin, I. (1983) Chronic parkinsonism in humans due to a product of meperidine-analogue synthesis. Science 219:979-980.

Langston, J.W., Forno, L.S., Rebert, C.S. and Irwin, I. (1984a) Selective nigral toxicity after systemic administration of 1-methyl-4-phenyl-1,2,3,6-tetrahydropyridine (MPTP) in the squirrel monkey. Brain Res. 292: 390-394.

Langston, J. W., Irwin, I., Langston, E. B. and Forno, L. S. (1984b) Pargyline prevents MPTP-induced parkinsonism in primates. Science, 225, 1480-1482.

Martinovits, G., Melamed, E., Cohen, O., Rosenthal, J. and Uzzan, A. (1986) Systemic administration of antioxidants does not protect mice against the dopaminergic neurotoxicity of methyl-4-phenyl-1,2,3,6-tetrahydropyridine. Neurosci. Lett. 69:192-197.

Mytilineou, C. (1987) Tissue culture studies of aging in the nervous system: toxicity of MPTP to dopamine neurons. In Vernadakis, A. (ed.), "Model Systems of Development and Aging of the Nervous System". pp. 491-500, Martinus Nijhoff.

Mytilineou, C. and Cohen, G. (1984) 1-Methyl-4-phenyl-1,2,3,6-tetrahydropyridine destroys dopamine neurons in explants of rat embryo mesencephalon. Science, 225:529-531.

Mytilineou, C., Cohen, G. and Heikkila, R. E. (1985) 1-Methyl-4-phenylpyridine (MPP +) is toxic to mesencephalic dopamine neurons in culture. Neurosci. Lett. 57:19-24.

Perry, T.L., Yong, V.W., Clavier, R.M., Jones, K., Wright, J.M., Foulks, J.G. and Wall, R.A. (1985) Partial protection from the dopaminergic neurotoxin 1-methyl-4-phenyl-1,2,3,6-tetrahydropyridine by four different antioxidants in the mouse. Neurosci. Lett. 60:109-114.

Sachs, Ch., Jonsson,.G., Heikkila, R. and Cohen, G. (1975) Control of the neurotoxicity of 6-hydroxydopamine by intraneuronal noradrenaline in rat iris. Acta Physiol. Scand. 93:345-351.

Salach, J.I., Singer, T.P., Castagnoli, N. and Trevor, A. (1984) Oxidation of the neurotoxic amine 1-methyl-4-phenyl-1,2,3,6-tetrahydropyridine (MPTP) by monoamine oxidases A and B and suicide inactivation of the enzymes by MPTP. Biochem. Biophys. Res. Commun. 125:831-835.

Sanchez-Ramos, J.R., Barrett, J.N., Goldstein, M., Weiner, W.J. and Hefti, F. (1986) 1-Methyl-4-phenyl-pyridinium (MPP +) but not 1-methyl-4-phenyl-1,2,3,6 tetrahydropyridine (MPTP) selectively destroys dopaminergic neurons in cultures of dissociated rat mesencephalic neurons. Neurosci. Lett., 72:215-220.

Schmidt, C.J., Bruckwick, E. and Lovenberg, W. (1985) Lack of evidence supporting a role for dopamine in 1-methyl-4-phenyl-1,2,3,6-tetrahydropyridine neurotoxicity. Eur. J. Pharmacol. 113:149-150.

Schmidt, C.J., Ritter, J.K., Sonsalla, P.K., Hanson, G.R. and Gibb, J.W. (1985) The role of dopamine in the neurotoxic effects of methamphetamine. J. Pharmacol. Exp. Ther. 233:539-544.

Singer, T.P., Salach, J.I. and Crabtree, D. (1985) Reversible inhibition and mechanism-based irreversible inactivation of monoamine oxidases by 1-methyl-4-phenyl-1,2,3,6-tetrahydropyridine (MPTP). Biochem. Biophys. Res. Comm. 127:707-712.

Sonsalla, P.K. and Heikkila, R.E. (1986) The influence of dose and dosing interval on MPTP induced dopaminergic neurotoxicity. Eur. J. Pharmacol. 129:339-345.

Tipton, K.F., McCrodden, J.M. and Youdim, M.B.H. (1986) Oxidation and enzyme-activated irreversible inhibition of rat liver monoamine oxidase-B by 1-methyl-4-phenyl-1,2,3,6-tetrahydropyridine. Biochem. J. 240:379-383.

Youdim, M.B.H. (1975) Assay and purification of brain monoamine oxidase. In "Research methods in Neurobiology" Vol. 3,p.p. 167-207 (Marks, N. Rodnight, R. Eds.) New York, Plenum.

17

TOXICITY OF STRUCTURAL ANALOGS OF 1-METHYL-4-PHENYL
PYRIDINIUM (MPP+) AND RELATED COMPOUNDS ON DOPAMINERGIC
NEURONS IN CULTURE

Franz Hefti, Juan R. Sanchez-Ramos,
Patrick P. Michel, Simon Efange, and Berton C. Pressman

Departments of Neurology, Pharmacology, and Radiology
University of Miami, Miami, FL 33101

MPTP selectively destroys the nigrostriatal dopaminergic neurons in primates and the mouse and produces a neuropathological lesion in the primate similar to that seen in idiopathic Parkinson's disease (Heikkila, 1984: Langston, 1985; Langston et al., 1983). These findings led to the speculation that environmental toxins may cause idopathic Parkinson's disease. According to this hypothesis, exposure to a neurotoxin during a short period of time may reduce the number of dopaminergic neurons. so that an age-related slow degeneration of these cells will reduce, at an earlier time, the number of surviving cells to a level insufficient to support normal function. Alternatively, it is possible that long-term exposure to small quantities of a neurotoxin might produce similar effects as short-term exposure to high levels and gradually result in the appearance of clinical symptoms. At present, no chemical has been identified which could be responsible for causing idiopathic Parkinson's disease. However, in the modern industrial society, humans are exposed to a vast number of xenobiotics, which are novel chemical structures and to which no defense mechanism has developed during evolution. The discovery of MPTP illustrates that a simple chemical which was used as intermediate in chemical syntheses can turn out to be a major neurotoxin. It is therefore important to study xenobiotics on a large scale for their potential neurotoxicity. We have developed culture systems to study the mechanism of action by which MPTP destroys dopaminergic neurons and to search for neurotoxins which, similar to MPTP, produce toxic effects on dopaminergic neurons. Culture systems offer the advantage that relatively large number of compounds can be tested and that their effects can be assessed in absence of interference by the blood brain barrier or metabolizing enzymes.

Dopaminergic neurons in culture

Cultures were prepared from dissociated cells from the ventral tegmentum of the mesencephalon of rat embryos (gestational day E15 to E16). The area dissected contained the substantia nigra and the ventral tegmental area, i.e,, areas containing dopaminergic cell groups A8, A9 and A10 which give rise to the nigro-striatal and mesolimbic pathways, respectively. Cells were dissociated, plated in culture dishes and grown in a modified DMEM medium. The methods used to prepare the cultures has been described in detail elsewhere (Sanchez-Ramos et al., 1987, 1988). For the study of drug effects, cultures were grown for 7 days and MPTP or related compounds were added to the medium for 24 hours. After a further delay of 24 hours, dopaminergic neurons were visualized using cytofluorescence of accumulated catecholamines or tyrosine hydroxylase (TH) immunocytochemistry.

For cytofluorescence, cultures were incubated with 1.0 µM L-norepinephrine which, after accumulation within dopaminergic neurons, was condensed to a fluorescent product with glyoxylic acid (DeLaTorre, 1980). To visualize TH, cultures were fixed and incubated overnight with a rabbit antiserum to TH (provided by Dr. M. Goldstein). This antibody was visualized by further incubation with a biotinylated anti-rabbit IgG fraction followed by a complex of avidin and biotinylated horseradish peroxidase (Vectastain). The peroxidase was visualized using diaminobenzidine and hydrogen peroxide. The number of fluorescent or TH-positive cells was established by counting their cell bodies in several randomly chosen visual fields per dish. To assess for general toxicity, the total number of cell bodies in the same area was counted by means of phase contrast microscopy. The number of dopaminergic cells in these cultures ranged from 0.5 to 1.0% of the total number of cells.

Toxicity of MPP+

In confirmation of earlier findings obtained on explant cultures (Mytilineou and Cohen, 1984: Mytilineou et al., 1985), MPTP, in concentrations from 10 to 100 μM, was found to diminish markedly the number of dopaminergic neurons visualized with catecholamine fluorescence in the cultures. This finding suggested that MPTP had destroyed the dopaminergic cells. However, the same concentrations of MPTP failed to reduce the number of TH positive neurons in the cultures. This indicates that MPTP interfered with uptake or storage of norepinephrine used to label the dopaminergic neurons fluorescently, but did not actually destroy the

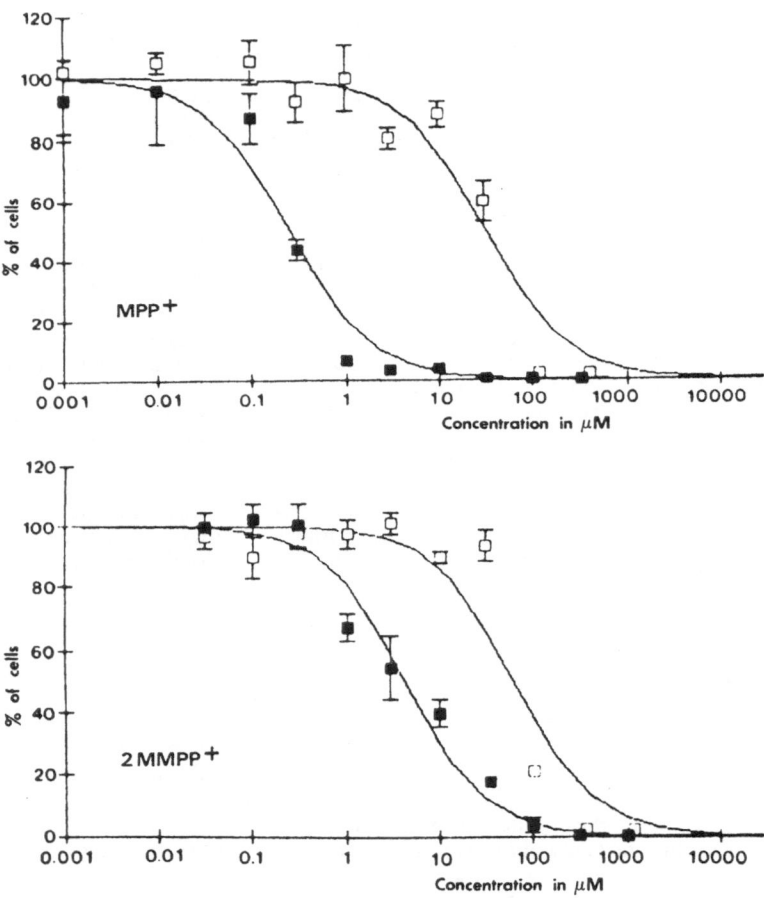

Figure 1: Specific toxicity of MPP+ and 2'-methyl-MPP+ (2'MMPP+) for dopaminergic neurons in cultures of dissociated mesencephalic cells. Cultures were grown for 7 days and then exposed for 24 hours to various concentrations of the toxins. After subsequent 24 hours free of toxin, they were fixed and processed for visualization of dopaminergic neurons. Filled squares represent the percent of surviving TH-positive neurons; open squares represent the percent of cells visualized with phase contrast microscopy, an index for unspecific toxicity.

dopaminergic neurons. Concentrations of MPTP higher than 300 μM were necessary to reduce the number of TH-positive neurons in culture. However, at these concentrations, MPTP not only decreased the number of TH-positive cells but also that of all other cells in culture, indicating that high concentrations of MPTP produce general toxic effects.

In contrast to the non-selective toxicity produced by MPTP, MPP+ selectively affected TH-positive neurons. Increasing concentrations of MPP+ in the cultures resulted, first, in disappearance of catecholamine cytofluorescence and, at higher concentrations, in disappearance of the TH immunoreactivity; still higher concentrations produced toxic effects on all cells present in the culture (Figure 1). The EC50 of the three effects were statistically different from each other. These findings confirm the selectivity of MPP+ for dopaminergic neurons. Administration of MPP+ at concentrations which selectively reduced the number of TH-positive cells resulted in characteristic morphological changes in those TH-positive cells which survived the treatment. Their cell bodies were often spherical, in contrast to the normal polygonal or spindle forms of TH positive neurons in control cultures. Moreover, cell processes tended to become fragmented and disappear. Occasionally, TH-immunoreactivity was visualized as a clump of particulate matter associated with fiber remnants. The observation of these morphological changes supports the view that application of MPP+ results in destruction of the dopaminergic neurons in culture.

Compounds known to block uptake of catecholamines were tested for their ability to prevent the selective toxic effects of MPP+ on dopaminergic neurons. A specific inhibitor of dopamine uptake, mazindol, was found to provide partial protection against the MPP+ induced loss of TH-positive neurons, whereas desipramine, an inhibitor of norepinephrine uptake was ineffective. The results confirmed that MPP+ enters dopaminergic neurons by means of the specific dopamine re-uptake mechanism. In earlier studies, selective inhibitors of dopamine uptake prevented MPTP-induced dopamine depletion in the mouse brain and in mouse striatal synaptosomes (Javich and Snyder, 1984: Pileblad and Carlsson, 1985).

The finding that MPP+, but not MPTP, is selectively toxic to TH positive neurons in rat nigral cell cultures is consistent with strong prior evidence that MPP+ is the active agent in the MPTP-induced destruction of dopaminergic neurons in vivo. MPTP is rapidly transformed in vivo to MPP+ (Heikkila et al., 1985; Markey et al., 1984). MAO-B inhibitors prevent both behavioral effects and the cell loss in the substantia nigra in the monkey, and the reduction of dopamine levels in rodents (Heikkila et al., 1984; Langston et al., 1984; Markey et al., 1984). Our data demonstrate that rat dopaminergic neurons in cultures of dissociated nigral cells are susceptible to low concentrations of MPP+. These findings therefore support the notion that MPTP's lack of effect in rats is due to inefficient conversion to MPP+ (Kalaria et al., 1987). Our findings do not support the speculation that the species differences reflect the amount of neuromelanin contained in dopaminergic neurons (D'Amato et al., 1987). MPP+ destroyed rat dopaminergic cells in cultures, even though these cells do not contain significant quantities of neuromelanin. It might also be argued that to the extent that neuromelanin complexes MPP+, it may increase overall MPP+ levels but actually decrease the availablitiy of MPP+ to injure other cell organelles like mitochondria.

Toxicity of MPTP- and MPP+-analogs

As with the parent compound, analogs of MPTP, lacking a positive charge, failed to produce selective decreases in the number of TH+ neurons. Compounds tested included MPTP an analog in which a methyl group was added to the phenyl ring at the 2' position (2'-methyl MPTP) and analogs in which the N-alkyl group was lengthened to N-ethyl (EPTP) and N-propyl (PPTP). These MPTP analogs were not toxic at the doses tested (0.1 to 100 μM). Substitution of a cyclohexyl group for the phenyl group of MPTP (1-methyl-4-cyclohexyl tetrahydropyridine or MCTP) also failed to reduce the number of TH+ neurons until a dose of 100 μM was attained which produced a generalized toxicity as evidenced by decreased numbers of all cell types. These findings confirm that uncharged MPTP analogs, which require conversion to pyridinium ions for toxicity, do not selectively affect cultured dopaminergic neurons.

Some of the analogs of MPP+ analyzed to date were toxic in a similar fashion to MPP+ (Table 1). MPP+ was the most toxic compound of the group of pyridinium analogs tested and produced a 50% decrease in the number of TH+ neurons (EC50) at a concentration of 0.3 μM. N-ethyl-phenyl pyridinium (EPP+) was the next most potent analog with an EC50 of 2.3 μM, while 2'- methyl-MPP+ had an EC50 of 4.2μM. N-propyl- phenyl pyridinium showed no selective toxicity. These findings indicate that the pyridinium metabolites of those compounds

Table 1

Toxicity of compounds tested on cultures of dissociated mesencephalic neurons. $EC_{50}TH$ is the concentration resulting in 50% loss of TH-positive neurons, $EC_{50}Phaco$ is the concentration resulting in a 50% loss of all cultured cells visualized by phase contrast microsopy. The ratio $EC_{50}Phaco/EC_{50}TH$ (R_{spec}) represents an index for selectivity for dopaminergic neurons. Values are means of at least two independent experiments with 20-40 cultures each. Concentrations are given in μM.

		$EC_{50}TH$	$EC_{50}Phaco$	R_{spec}
PYRIDINES				
MPP+	1-methyl-4-phenyl-pyridinium	0.30	32	107.00
EPP+	1-ethyl-4-phenyl-pyridinium	2.3	30	13.0
2'MMPP+	2'-methyl-MPP+	4.2	61	14.5
paraq.	paraquat	23	78	3.39
PPP+	1-propyl-4-phenyl-pyridinium	28	49	1.75
MBP+	1-methyl-4-benzyl-pyridinium	29	40	1.38
BP+	2,2'-bipyridinium	33	31	0.94
CMP+	2-chloro-1-methyl-pyridinium	55	58	1.05
MPTP	1-methyl-4-phenyl-1,2,3,6-tetrahydropyridine	660	1100	1.67
PP+	1-(4-pyridyl)-pyridinium	9100	8000	0.88
GUANIDINES				
C8G+	octyl-guanidine	1.9	3.2	1.68
C7G+	heptyl-guanidine	6.7	15	2.24
C5G+	pentyl-guanidine	11	51	4.64
BG+	benzyl-guanidine	21	31	1.48
C4G+	butyl-guanidine	46	89	1.93
gth.	guanethidine	64	360	5.63
BENZIMIDAZOLES				
ABI	2-aminobenzimidazole	900	510	0.57
MNBl	2-mercapto-5-nitrobenzimidazole	1200	490	0.41
ISOQUINOLINES				
dbq.	debrisoquin	67	69	1.03
MIQ+	N-methyl-isoquinolinium	340	560	1.65
OTHER MOLECULES				
BHEMT+	3-benzyl-5-(hydroxyethyl)-4-methylthiazolium	430	370	0.86

		$EC_{50}TH$	$EC_{50}Phaco$	R_{spec}
DMPP+	1,1-dimethyl-4-phenyl-piperazinium	1500	1800	1.20
nic.	nicotine	3600	1400	0.39
DAPT	2,4-diamino-6-phenyl-1,3,5 triazine	>10000	>10000	—
6-OHDA	6-hydroxydopamine	16	15	0.94
DA	dopamine	31	43	1.39
KA	kainic acid	51	130	2.55
GA	glutamtic acid	1200	590	0.49

effective in vivo were toxic in vitro. However, the rank order of potency in vivo is different from that found in vitro. In vivo, 2'-methyl-MPTP is more toxic than MPTP (Sonsalla et al., 1987: Kindt et al., 1987). The observation that 2'-methyl-MPP + is less toxic in vitro than MPP + confirms that the difference of toxicity in vivo results from its more facile oxidation to a pyridinium cation (Sonsalla et al., 1987). In contrast to MPTP, 2'-methyl-MPTP is a substrate not only for MAO-B but also for MAO-A (Sonsalla et al., 1987). Our data indicate that the positive charge of the quarternay nitrogen of the pyridinium is essential for toxicity and shielding it with progressively longer alkyl groups attenuates toxicity.

Paraquat (1,1'-dimethyl-4,4'-dipyridinium), another compound with a structure comparable to that of MPP + failed to selectively destroy dopaminergic neurons (Table 1). This finding provides further evidence against its etiologic role in Parkinson's disease (Perry et al., 1986).

Initial studies with chemicals not directly related to MPTP

So far, studies aimed at elucidating the structural requirements of selective dopaminergic neurotoxins have focussed on close chemical analogs of MPTP (Finnegan et al., 1987: Fuller et al., 1985,1986; Heikkila et al. 1985: Johannessen et al., 1987; Langston et al., 1984; Sonsalla et al., 1987, Youngster et al., 1987). However, structures less closely related to that of MPTP may also exhibit similar toxicity. We started to search for such compounds on the basis of the present knowledge about the mechanism of action of MPTP. According to the current beliefs (Langston, 1985; Singer et al., 1987), the selectivity of MPTP's action is due to the uptake of its metabolite MPP + into the dopaminergic neurons. Following uptake, MPP + is further concentrated in the mitochondria where it inhibits the mitochondrial energy metabolism resulting in cell death. We therefore hypothesized that other compounds satisfying these two criteria (selective uptake and mitochondrial toxicity) might be selective toxins for dopaminergic neurons.

First, we tested a group of guanidines which are known to be accumulated by mitochondria and inhibit their energy metabolism (Pressman, 1963). The addition to the cultures of alkyl-guanidines resulted in unselective degeneration of all cultured cells (Table 1). The effective concentration ($EC50_{Phaco}$) resulting in degeneration of both, dopaminergic neurons and all other cultured cells, decreased with increasing length of the linear side chain and ranged from 3 to 360 µM. Similar to alkylguanidines, benzylguanidine and guanethidine produced only unspecific toxic effects. The EC50 of benzylguanidine was 20.9 µM. Guanethidine, despite the fact that it is known to destroy peripheral catecholaminergic neurons in vivo (Johnson and Manning, 1984), was relatively ineffective (EC50 64.3 µM). It is interesting to note that octylguanidine is considerably more toxic to all cells (ED50 = 3µM) than predicted by its previously reported ED50 for respiration of isolated rat liver mitochondria (ED50 = 12µM) consistent with its being concentrated at the plasma membrane as well as in mitochondria.

Many findings indicate that compounds with a delocalized positive charge such as guanidiniums and pyridiniums have analogous abilities to inhibit mitochondrial function. We therefore tested a limited group of such compounds which, furthermore, could be substrates of the dopamine uptake system (structural relationship to catecholamines or serotonin). A group of compounds with this property are the benzimidazoles. Two such compounds were tested (2-aminobenzimidazole and 2-mercapto-5-nitrobenzimidazole) but were found not to be selectively toxic for dopaminergic neurons (Table 1). Similarly, isoquinolin analogs (debrisoquin and methyl-isoquinolinium) showed no selective toxicity. N-methyl isoquinolinium, recently found to inhibit tyrosine hydroxylation in striatal tissue slices (Hirata et al., 1986) was also not selectively toxic for dopaminergic neurons in culture, indicating that the inhibition of tyrosine hydroxylation is not sufficient for producing dopaminergic toxicity. Other compounds tested in our cultures included a thiazolium, a piperazinium and a substituted triazine (Table 1). The general toxicity varied among the various compounds, however, none of them was selectively toxic for dopaminergic neurons.

Conclusions

Our results obtained on cultures of dissociated neurons from the rat fetal mesencephalon confirm that MPP+ is taken up into and selectively destroys dopaminergic neurons. N-ethyl-pyridinium and 2'-methyl-MPP+ were found to be less toxic than MPP+ itself. Based on the limited number of MPP+ analogs and related compounds tested so far, we tentatively conclude that the structural requirements for the selective toxic effect of MPP+ on dopaminergic neurons are rather strict, but that their relatively non-specific effects on mitochondrial energy production may be mimicked by a variety of cations with delocalized charge. A large number of other compounds will be tested to test this conclusion.

Acknowledgement

These studies were supported by the National Parkinson Foundation.

References

D'Amato R.J.. Alexander G.M.. Schwartzman R.J.. Kitt C.A.. Price D.L., Snyder S.H. (1987) Evidence for neuromelanin involvement in MPTP induced neurotoxicity. Nature 327, 324-326.

DeLaTorre J.C.(1980) An improved approach to histofluorescence using the SPG method for tissue monoamines. J. Neurosci. Methds. 3, 1-5

Finnegan K.T., Irwin I., Delanney L.E., Ricaurte G.A., Langston J.W. (1987) 1,2,3,6-Tetrahydro-1-methyl-4-(methylpyrrol-2-yl)pyridine: studies on the mechanism of action of 1-methyl-4-phenyl-1,2,3,6 tetrahydropyridine. J. Pharmacol. Exp. Ther. 242. 1144-1151.

Fuller R.W., Robertson, D.W., Hemrick-Luecke S.K. (1986) Comparison of the effects of two 1-methyl-1,2,3,6-tetrahydropyridine analogs, 1-methyl-4-(2-thienyl)-1,2,3,6-tetrahydropyridine and 1-methyl-4-(3-thienyl)-1,2,3,6-tetrahydropyridine, on monoamine oxidase in vitro and on dopamine in mouse brain. J. Pharmacol. Exp. Ther. 240. 415-420.

Fuller R.W., Hemrick-Luecke S.K., Robertson D.W. (1985) Comparison of 1-methyl-4-(p-chlorophenyl)-1,2,3,6-tetrahydropyridine, 1-methyl-4-phenyl-1,2,3,6-tetrahydropyridine (MPTP) and p-chloramphetamine as monoamine depletors. Res. Commun. Chem. Pathol. Pharmacol. 50, 57-65.

Harik S.I., Schmidley J.W., Iacofano L.A., Blue P., Arora P.K., and Sayre L.M. (1987) On the mechanisms underlying 1-methyl-4-phenyl-1,2,3,6-tetrahydropyridine (MPTP) neurotoxicity: the effect of perinigral infusion of MPTP, its metabolite and their analogs in the rat. J. Pharmacol. Exp. Ther. 241, 669-676

Heikkila R.E., Hess A., and Duvoisin R.C. (1984a) Dopaminergic neurotoxicity of MPTP in mice. Science 224, 1451-53

Heikkila R.E., Manzino L., Cabbat F.S., Duvoisin R.C. (1984b) Protection against the dopaminergic neurotoxicity of MPTP by monoamine oxidase inhibitors. Nature 311, 67-79

Heikkila R.E., Manzino L., Cabbat F.S., Duvoisin R.C. (1985a) Studies in the oxidation of the dopamine neurotoxin MPTP by monoamine oxidase B. J. Neurochem. 45: 1049-54

Heikkila R.E., Manzino L., Cabbat F.S., Duvoisin R.C. (1985b) Effects of 1-methyl-4-phenyl-1,2,3,6- tetrahydropyridine (MPTP) and several of its analogues on the dopaminergic nigrostriatal pathway in mice. Neurosci. Lett. 58, 133-137

Hirata Y., Sugimura H., Takei H., and Nagatsu T. (1986) The effects of pyridinium salts, structurally related compound of MPP+, on tyrosine hydroxylation in rat striatal tissue slices. Brain Res. 397, 341-844

Javitch J.A. and Snyder S.H. (1984) Uptake of MPP+ by dopaminergic neurons explains selectivity of parkinsonism inducing neurotoxin MPTP. Europ. J. Pharmacol. 106, 455-56

Johannessen J.N., Chiueh C.C., Burns R.S., and Markey S.P. (1985) Differences in metabolism of MPTP in rodent and primate brain parallel differences in sensitivity to its neurotoxic effects. Life Sci.36, 219-224

Johnson E.M., Manning P.T. (1984) Guanethidine-induced destruction of sympathetic neurons. Int. Rev. Neurobiol., 25, 1-37.

Kalaria R.N., Mitchell M.J., and Harik S.l. (1987) MPTP neurotoxicity: correlation with blood-brain-barrier monoamine oxidase activity. Proc. Natl. Acad. Sci. U.S.A. 84, 3521-5

Kindt M.V., Heikkila R.E., Nicklas W.J. (1987) Mitochondrial and metabolic toxicity 1-methyl-4-(2'-methylphenyl)-1,2,3,6-tetrahydropyridine. J. Pharmacol. Exp. Ther. 242, 858-863.

Langston J.W., Ballard P., Tetrud J.W., and Irwin I.J. (1983) Chronic parkinsonism due to a product of a meperidine analog synthesis. Science, 219:979-980

Langston J.W., Irwin I., and Langston E.B. (1984a) Pargyline prevents MPTP induced parkinsonism in primates. Science 225, 1480-82

Langston J.W., Irwin I., Langston E.B. and Forno L.S. (1984b) The importance of the '4-5' double bond for neurotoxicity in primates of the pyridine derivative MPTP. Neurosci. Lett. 50, 289-294

Langston J.E., Irwin, I., and Langston E.B. (1984) Pargyline prevents MPTP-induced parkinsonism in primates. Science 225, 1480.

Markey S.P., Johannessen J.N., Chiueh C.C., Burns, R.S. and Herkenham M.A. (1984)Intraneuronal accumulation of a pyridinium metabolite may produce MPTP-induced parkinsonian syndrome in monkey. Nature 311 464-46

Mytileneou C., and Cohen G.(1984) MPTP destroys DA neurons in explants of rat embryo mesencephalon. Science 225, 529-531

Mytileneou C., and Cohen G. (1985) Deprenyl protects dopamine neurons from the neurotoxic effect of 1-4-phenylpyridinium ion. J. Neurochem 45, 1951-53

Perry T.L., Yong V.W., Wall R.A., Jones K. (1986) Paraquat and two endogenous analogs of the neurotoxic substance N-methyl-4-phenyl-1,2,3,6-tetrahydropyridine do not damage dopaminergic nigrostriatal neurons in the mouse. Neurosci. Lett. 69, 285-289.

Pileblad E., and Carlsson A.(1985) Catecholamine-uptake inhibitors prevent the neurotoxicity of MPTP in mouse brain. Neuropharmacol 24, 689-92

Pressman, B.C. (1963) The effects of alkylguanidines on the energy transfer reactions of mitochondria. J. Biol. Chem. 238, 401-409.

Sanchez-Ramos J.R., Barrett J.N., Goldstein M., Weiner W.J., and Hefti F. (1986) MPP+, but not MPTP is toxic to dopamine neurons in cultures of dissociated rat mesencephalic neurons. Neurosci. Lett. 72, 215-220.

Sanchez Ramos J.R., Michel P., Weiner W.J., and Hefti F. (1988) Selective destruction of cultured dopaminergic neurons from fetal rat mesencephalon by 1-methyl-4-phenylpyridinium (MPP+): cytochemical and morphological evidence. J. Neurochem., 50, 1934-1944.

Singer T.P., Castagnoli N., Ramsay R.R., and Trevor A.J. (1987) Biochemical events in the development of parkinsonism induced by 1-methyl-4-phenyl-1,2,3,6-tetrahydropyridine. J. Neurochem. 49,1-8.

Sonsalla P.K., Youngster S.K., Kindt M.V., Heikkila R.E. (1987) Characteristics of 1-methyl-4-(2'-methylphenyl)-1,2,3,6-tetrahydro-pyridine-induced neurotoxicity in the mouse. J. Pharmacol. Exp. Ther. 242, 850-857.

Youngster S.K., Sonsalla P.K., Heikkila R.E. (1987) Evaluation of the biological activity of several analogs of the dopaminergic neurotoxin 1-methyl-4-phenyl-1,2,3,6-tetrahdyropyridine. J. Neurochem. 48, 929-934.

18

MECHANISMS OF MPP+ NEUROTOXICITY: OXYRADICAL AND MITOCHONDRIAL INHIBITION HYPOTHESES

Juan R. Sanchez-Ramos, Franz Hefti, Gary E. Hollinden,
Thomas J. Sick and Myron Rosenthal

Dept. of Neurology University of Miami School of Medicine, Miami, FL

Recent studies demonstrate that l-methyl-4-phenylpyridinium (MPP+) is selectively toxic to dopaminergic (DA) neurons. Investigations suggest that this selectivity results primarily from the affinity of MPP+ for the dopamine reuptake system which results in preferential accumulation of the toxin within the terminals of the nigrostriatal dopaminergic (DA) system (Javitch and Snyder, 1985;Pileblad and Carlssen, 1985; Mayer et al. 1986).

Two hypotheses predominate as possible explanations of the MPP+ induced degeneration of DA neurons.

Figure 1

A. Mitochondrial Inhibition Hypothesis: 1) MPP + is actively taken up by the high-affinity DA uptake system; 2) MPP + is actively accumulated in mitochondria; 3) MPP + oxidation of NAD-linked substrates, but not succinate; 4) As a consequence of this block, reduced NAD (NADH) increases, O_2 consumption decreases in mitochondria and lactate levels increase in strial slices.

A

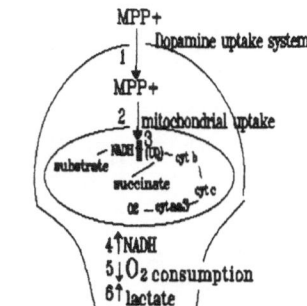

B. Oxyradical Hypothesis: 1) MPP + is accumulated within DA terminals; 2) MPP + is reduced by quinone moieties in neuromelanin or DA storage granules, then oxidized by O_2 in mitochondria generating the superoxide anion in a cyclic manner; 3) superoxide anion is converted to H_2O_2 generating the hydroxyl radical (.OH) which reacts with lipid membranes.

B

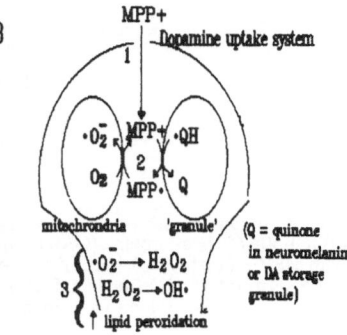

One is that MPP+ produces "energy failure" since it is actively concentrated within mitochondria (Ramsay et al..1986) where it blocks mitochondrial electron transport (Nicklas et al., 1985; Vyas et al., 1986; Singer et al., 1987 (See Figure 1). Consistent with this hypothesis have been findings of MPP+-induced increases in NAD reduction and decreases in O_2 consumption in isolated mitochondria (Nicklas et al., 1985; Ramsay et al., 1986a,b; Vyas et al. 1986), oxidation of cytochrome <u>b</u> in rat striatal slices (Sanchez-Ramos et al., 1988b), and increases in lactate production in mouse striatal slices (Vyas et al., 1986)

An alternative hypothesis for MPP+ neurotoxicity suggests that the toxin enhances oxyradical production within DA neurons (Dimonte et al., 1986; Kopin, 1986) (See Fig 1B). The endogenous protective anti-oxidants (such as glutathione, catalase and superoxide dismutase) which protect the cell from oxyradical damage eventually become depleted and the superoxide anion radical and/or the hydroxyl radical react with polyunsaturated lipids, initiating chain reactions which end in membrane lipid peroxidation (Bannon et al., 1984). Neuromelanin could enhance free radical formation in mitochondria (Kopin, 1986) since it may participate in the redox cycling of MPP+ to generate free radical intermediates (Bannon et al., 1984; Kopin 1986).

Investigations reviewed here were conducted to test these two hypotheses. Among the objectives were: a) to determine whether incubation with anti-oxidants protects against MPP+ toxicity in a DA neuron culture system; b) to determine whether the MPP+-induced putative block in complex I of mitochondria can be bypassed by using succinate as substrate with the effect of promoting survival of DA neurons; c) to determine whether the preincubation of DA neuron cultures with Na pentothal (thiopental) protects against MPP+ neurotoxicity; d)to determine whether effects of MPP+ on the reduction/oxidation status of mitochondrial cytochromes in rat striatal slices is consistent with a putative block of electron transport at complex I; e) to determine whether there are effects on ion transport activity produced by the putative depression of ATP production.

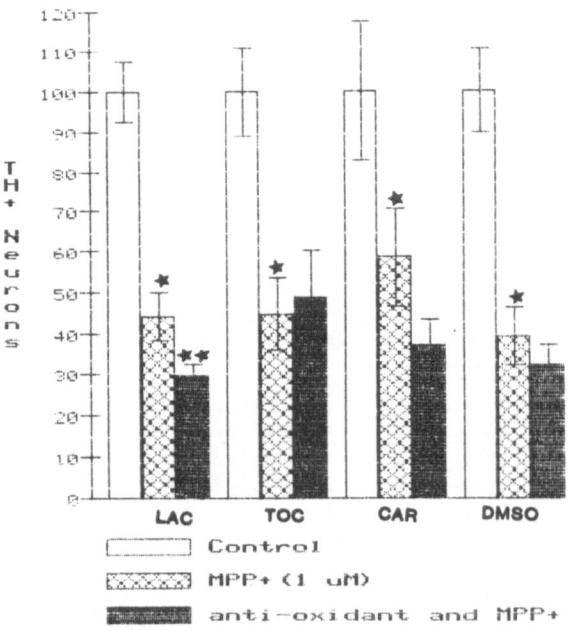

Figure 2

Cultures of DA neurons were incubated with the anti-oxidants (L)-acetylcarnitine (LAC)-10μM, alpha tocopherol (TOC)-30μM, betacarotene (OAR)-30μM or dimethyl sulfoxide (DMS0)-0.5% for 2 hrs prior to co-incubation with MPP+ (1 μM) for 24 hrs. Y-axis represents the percent surviving TH+ neurons per culture dish. (100% corresponds to 1565 ± 136 (mean ± SEM) of THpositive neurons in control cultures. N = 12 for control dishes, N = 6 for MPP+ and N = 6 for Anti-oxidant + MPP+.

* difference between control and MPP+ (p <0.05)
** difference between MPP and LAC + MPP +(p <0.05)

Experiments were conducted using the DA neuron culture system and rat striatal slices (Sanchez-Ramos et al. 1986;1988a:1988b). To assess for selective toxicity, the number of tyrosine hydroxylase (TH) positive neurons were compared to the total number of cells in the cultures surviving MPP+ treatment. In striatal slices the reduction/oxidation states of mitochondrial cytochromes were measured with scanning spectrophotometry and extracellular potassium ion concentration was measured with K+ sensitive microelectrodes (LaManna et al., 1985; Pikarsky et al., 1985; Sick et al., 1982: Sanchez Ramos et al., 1988b).

Anti-oxidants Failed to Protect MPP+ Treated DA Neurons in Culture

MPP+ alone decreased the number of surviving TH+ neurons to 39 - 59 % of the controls (Fig 2). To determine whether anti-oxidants protect against MPP+ toxicity, cultures were incubated with the anti-oxidants (in separate experiments) L-acetylcarnitine (LAC), alpha tocopherol (TOC), beta-carotene (CAR) or dimethyl sulfoxide (DMSO) for 2 hrs prior to coincubation with MPP+ (1 μM) for 24 hrs. Cultures were then processed for TH immunocytochemistry. The antioxidants failed to increase the number of surviving TH+ neurons. Treatment with LAC increased the toxicity of MPP+.

Failure of anti-oxidants to protect against MPP+ toxicity might be due to a)inadequate access of the anti-oxidants to the sites of toxicity or b) to the possibility that oxyradicals do not mediate acute (24 hr) toxicity. The former explanation might be obviated by increasing the concentration or duration of pretreatment with the anti-oxidants. The latter would support the hypothesis that the acute toxicity of MPP+ was a consequence of inhibition of mitochondrial electron transport.

Succinate Did Not Protect MPP+ Treated DA Neurons in Culture

If MPP+ induced energy failure is due to a block of electron transport at complex I in mitochondria (see Fig 1A), then the use of succinate as a substrate might provide a means to

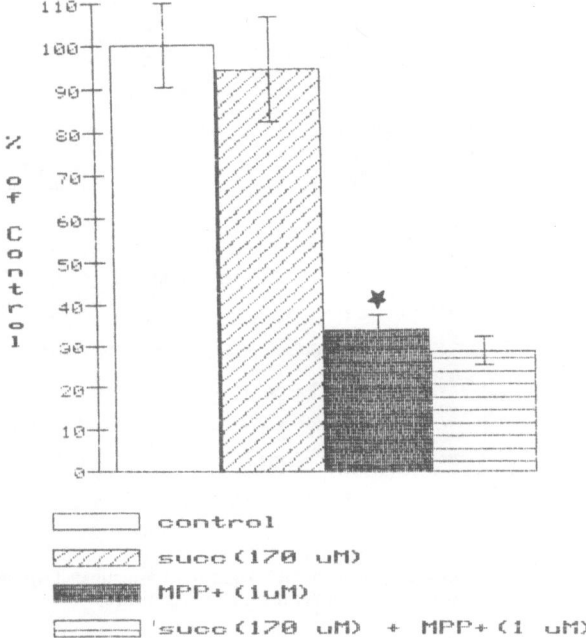

control

//// succ (170 uM)

▓ MPP+ (1uM)

succ (170 uM) + MPP+ (1 uM)

Figure 3

Cultures of DA neurons were incubated for 2 hrs with succinate (170 μM) and then co-incubated with MPP+ for 24 hrs. Y-axis represents the percent surviving TH-positive neurons per dish. (N = 6-12 dishes).

* difference between MPP+ treated cultures and succinate controls (p <0.05)

bypass the block and restore ATP levels as has been suggested in isolated mitochondria (Vyas et al., 1986). Cell cultures were incubated with succinate (170 μM) for 2 hrs followed by coincubation with MPP + (1 μM) for 24 hrs (Fig. 3).

Succinate pretreatment and co-incubation did not protect against the MPP + induced decrease in TH + neurons. The failure of succinate to protect the DA neurons in culture might reflect inadequate cellular uptake of dicarboxylic acids.

Pentothal Increased Survival of MPP + treated DA Neurons in Culture

Cultures were pre-incubated for 2 hours with thiopental (10 or 100 μM) and then co-incubated with MPP + (1 or 10 μM) for 24 hours (Fig 4).

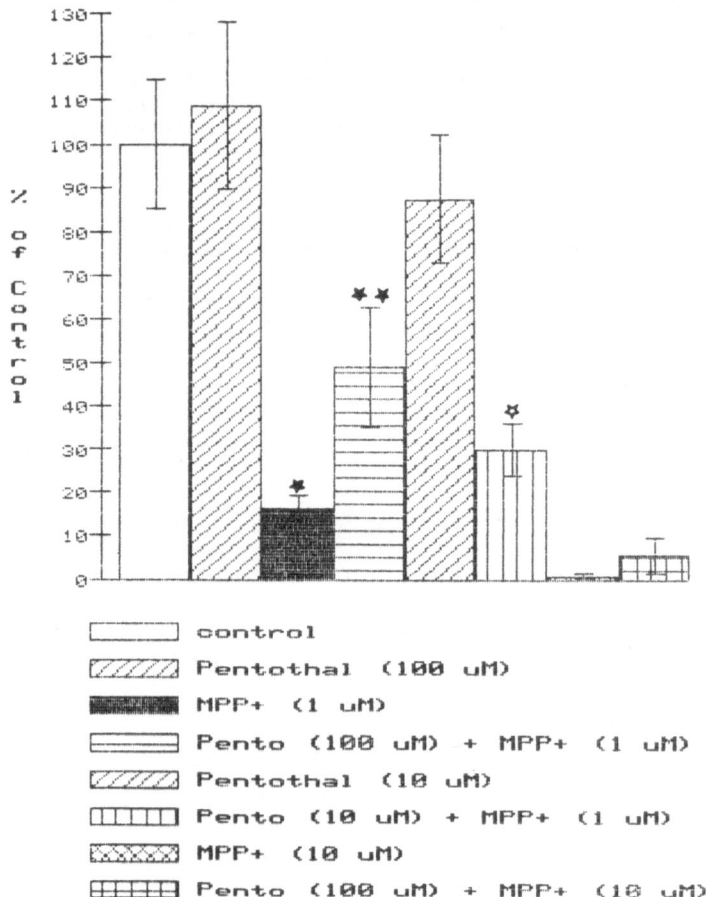

Figure 4

Cultures were incubated for 2 hrs with pentothal (10 or 100 μM) and then co-incubated with MPP + for 24 hrs. Y-axis represents percent surviving TH-positive neurons per dish. (N = 6-9 dishes)

* difference between MPP + -treated cultures and pentothal (100 μM) control cultures (p <0.05).

** difference between pentothal(l00μM) + MPP + (1μM) treated cultures and MPP + -treated cultures(p < 0.05)

open star = difference between pentothal(l0μM) + MPP + (1μM) and MPP + -treated cultures (p <0.05)

Thiopental treatment increased the number of surviving TH+ neurons exposed to low concentrations of MPP+ (1 μM)(p<0.05) but thiopental (100 μM) was not effective in protecting against higher (10 μM) concentrations of the toxin. Protective effects of thiopental may be due to a) competition of the barbiturate with the toxin for the binding site at Complex I (Mayes, 1979), b) barbiturate-induced decreases in mitochondrial respiration, c) slowing of energy dependent activities, or d) decreased energy-dependent uptake of MPP+ by the DA reuptake system.

Effects of MPP+ in Rat Striatal Slices

The experiments with rat striatal slices were undertaken to define the mechanism of MPP+ action on mitochondrial respiration since the reduction of pyridine nucleotide (NAD) and the concomitant effects on O_2 consumption and lactate production described by other investigators (Nicklas et al., 1985; Vyas et al., 1986) could be due to any of 3 conditions: a) hypoxia b) respiratory depression with decreased ADP or c) an actual block in the respiratory

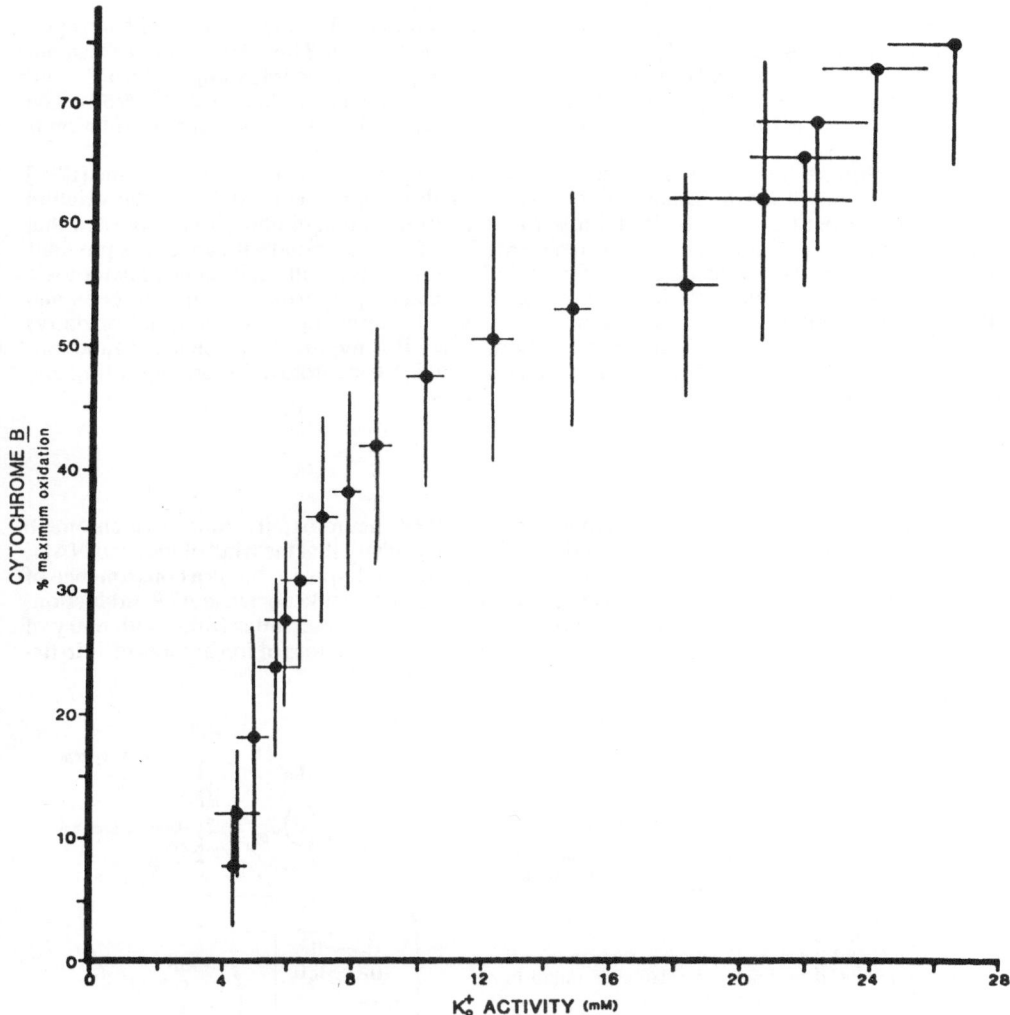

Figure 5

Relationship of cytochrome b oxidation (Y-axis) to extracellular K+ activity (concentration) on X-axis

chain. One would expect that a block in the respiratory chain of mitochondria would produce oxidation of all electron carriers distal to the block and reduction of carriers proximal to the block. Such changes are measurable because oxidation reduction shifts of the electron transport carriers such as NAD and the cytochromes can be recorded by optical techniques (LaManna et al., 1985). Superfusion of rat striatal slices with MPP + resulted in a time-dependent increase in oxidation of cytochrome b which reached approximately 80% of that induced by H_2O_2 (Sanchez-Ramos et al., 1988b). Co-incubation of striatal slices with mazindol (33 μM) and MPP + (1.0 μM) also produced oxidation of cytochrome b but this response was smaller and slower than that produced by MPP + alone (Sanchez-Ramos et al. 1988b). Since cytochrome b is situated at the oxygen side of complex I, the oxidation of cytochrome b is consistent with the concept that MPP + 's site of action is at complex I. The antagonism of MPP + -induced oxidation of cytochrome b by the DA uptake inhibitor, mazindol, indicates that this effect requires the selective uptake of MPP + by DA terminals. This was substantiated by the fact that hippocampal slices, which have a lower density of DA input than striatal slices, were much less affected by equivalent concentrations of MPP +.

Since effects of MPP + on mitochondrial function would likely have effects on the electrophysiological activities of cells, we sought to relate these by recording changes in extracellular potassium ion activity as a signal of ion transport function.

Extracellular [K +] remained stable for 60 to 70 minutes after the onset of MPP + superfusion and then rose very slowly over time in the striatal slices. This MPP + effect was antagonized with mazindol and did not occur in hippocampal slices, confirming that MPP + influences require the selective uptake of MPP + by DA terminals (Hollinden et al., 1987). The relationship of the MPP + -induced increase in extracellular K + to cytochome b oxidation is depicted in Fig 5.

Cytochrome b became oxidized to approximately 50 % of maximum before [K +] started to leak out of DA terminals. Since DA terminals comprise at most 4 % of the striatum (Jellinger, 1987 but extracellular [K+] rose to large (30 μM) magnitudes, it is suggested that cells other than DA terminals were affected by MPP +. The mitochondrial hypothesis provides an explanation for these effects: energy failure in DA terminals results in leakage of neurotoxic substances or metabolites altering membrane conductance properties of adjacent cells and thereby placing additional demand upon ion transport pumps and mitochondrial oxidative phosphorylation. These data indicate that actions of MPP + require the presence of functioning DA terminals but suggest that consequences of MPP + neurotoxicity are not ultimately confined to these terminals.

Conclusions

Figure 6 summarizes the two hypotheses of MPP + neurotoxicity. Since cytochrome b became oxidized in striatal slices exposed to MPP +, it is likely that the triad of reduced NAD, decreased O_2 consumption and increased lactate is not due to hypoxia, but is a consequence of an inhibition of electron transport at complex I. Such a block would decrease ATP production, and in tissues, should result in gradual loss of K + into the extracellular fluid, with entry of Na + and water shifts culminating in edema and death. However, loss of ion homeostasis in tis-

Figure 6

Mechanism of MPP + neurotoxicity: 1) MPP + actively taken up into DA terminals; 2) MPP + accumulated in mitochondria; 3) block of electron transport at complex I results in increase in reduced NADH, decrease in O_2 consumption, and an increase in cytochrome b oxidation; 4) consequence of decreased ATP is gradual leakage of K + to extracellular space; 5) possible damage to cell membranes produced by lipid peroxidation resulting in gradual leakage of K + and other metabolites.

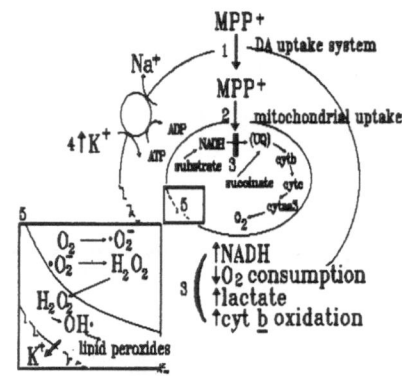

sues may also be a consequence of membrane damage produced by lipid peroxidation. Since MPP + is actively taken up into mitochondria, cytochromes distal to complex I may become oxidized, and oxygen tension at mitochondria may increase. These conditions could promote generation of oxyradicals within mitochondria. Since oxyradicals are generated under normal respiration in isolated mitochondria (Turrens et al., 1985), MPP + addition might potentiate this process and eventually overburden the endogenous oxyradical scavenging capacity of the cell. The failure of oxyradical scavengers to protect against MPP + in cell cultures may be due to the poor accessibility of the antioxidants to the mitochondrial membranes where the initial assault by the toxin occurs.

Alternatively, it may be that failure of antioxidants to protect against acute toxicity of MPP + is because cell death is a direct consequence of energy failure produced by MPP + -induced inhibition of mitochondrial electron transport. In this case, the cell culture system may not be well suited for testing oxyradical mediated damage which occurs over a long time. Attempts to implicate oxyradicals as mediators of toxicity have relied on indirect evidence such as measures of glutathione depletion (Yong et al., 1986). or have been based on demonstrating the protective effects of anti-oxidants (Yong et al., 1986; Perry et al., 1986; Wagner et al., 1985, 1986; Sershen et al., 1985). Most experiments to demonstrate oxyradical mediated toxicity employed MPTP administered in vivo to mice. One source of oxyradicals is from the metabolism of MPTP by MAO to MPP +, so the protective effects of these antioxidants is consistent with scavenging of the radicals generated by the MAO.

Clearly, the mechanism of MPP + neurotoxicity is yet undefined. It is possible that both the mitochodrial inhibition and oxyradical mediated damage may occur but under different dose and temporal dependencies. In acute toxicity, it may be that inhibition of electron transport leads to energy failure, loss of ion homeostasis, edema and death. The damage generated by oxyradicals may require a longer time course of exposure to lower concentrations of MPP +.

The discovery that MPTP produces parkinsonism in man has given great impetus to the search for environmental toxins which may be etiologically linked to idiopathic Parkinson's disease. Hundreds of compounds are being screened for their capacity to selectively destroy dopaminergic neurons. Whether or not a specific compound will be found to be the cause of Parkinson's disease, studies on the mechanism of MPP + -induced degeneration of DA neurons may shed light on the pathophysiology of Parkinson's disease. Understanding the sequence of events which leads to the degeneration of DA neurons may provide a rational approach to preventation of the disease.

Acknowledgements

These studies were supported by grants from the National Parkinson Foundation (to FH and MR) and the American Parkinson Disease Association (to JSR). Dr. Hollinden was a post-doctoral fellow of the American Heart Association, Florida Affiliate.

References

Bannon MJ, Goedert M, Williams B. The possible relation of glutathione, melanin and MPTP to Parkinson's disease. Biochem. Pharmacol. 33: 2697-2698, 1984

DiMonte, D, Sandy MS, Ekstrom G and Smith MT. Comparative studies on the mechanisms of paraquat and MPP + cytotoxicity. Biochem. Biophys. Res. Com. 137: 303-309, 1986

Hollinden GE, Sanchez-Ramos JR, Sick TJ, and Rosenthal M. MPP + increases extracellular potassium in rat striatal slices: preliminary evidence that consequences of MPP + neurotoxicity are spread beyond dopaminergic terminals. Neurosci. Abst. 113; 1502, 1987

Javitch JA and Snyder SH: Uptake of MPP + by dopaminergic neurons explains selectivity of parkinsonism inducing neurotoxin MPTP. Europ. J. Pharmacol. 106: 455-56, 1984

Jellinger K. The pathology of Parkinsonism. In: Movement Disorders. Marsden CD, Fahn S. (eds) Butterworth, Boston 124-150, 1987

Kaufman LM and Barrett JN. Serum factor supporting long term survival of rat central neurons in culture. Science 220:13941396, 1983

Kopin IJ. Toxins and Parkinson's Disease: MPTP Parkinsonism in Humans and Animals. In Advances in Neurology Vol 45: Raven Press, NY. 137-144, 1986.

LaManna LC, Pikarsky SM, Sick TJ, Rosenthal M. A rapid scanning spectrophotometer designed for biological tissues in vitro or in vivo. Anal. Biochem. 144:483-493,1985

Mayer RA, Kindt MV, Heikkila RE, Prevention of the nigrostriatal toxicity of MPTP by inhibitors of 3, 4 dihydroxyphenylethylamine transport. J of Neurochem. 47:1073-1079, 1986

Mayes PA. Biological Oxidation. In, Review of Physiological Chemistry. Eds., Harper HA, Rodwell VW, Mayes PA. Lange, Chicago, 266-284, 1979

Nicklas WJ, Vyas I and Heikkila RE. Inhibition of NADH-linked oxidation in brain mitochondria by MPP+, a metabolite of the neurotoxin MPTP. Life Sci. 36:2503-2508, 1985

Perry TL, Yong VW, Jones K, Wright JM. Manipulation of glutathione contents fails to alter dopaminergic nigrostriatal neurotoxicity of MPTP in the mouse. Neurosci. Lett. 70: 261-265, 1986

Pikarsky SM, LaManna JC, Sick TJ, Rosenthal M. A computer assisted rapid-scanning spectrophotometer with applications to tissues in vitro and in vivo. Comp. Biomed. Res. 18: 408-421, 1985

Pileblad E, and Carlsson A: catecholam!ne-uptake inhibitors prevent the neurotoxicity of MPTP in the mouse brain. Neuropharmacol 24: 689-92, 1985

Ramsay RR, Salach JI, Singer TP. Uptake of the neurotoxin MPP+ by mitochondria and its relation to the inhibition of the mitochondrial oxidatin of NAD+-linkedjsubstrates by MPP+. Biochem. Biophys. Res. Com. 134:743-748, 1986a

Ramsay RR, Salach JI, Dadgar J, Singer TP. Inhibition of mitochondrial NADH dehydrogenase by pyridine derivatives and its possible relation to experimental and idiopathic parkinsonism. Biochem. Biophys. Res. Com. 135:269-275,1986

Sanchez-Ramos JR, Barrett, JN, Goldstein M, Weiner WJ, and Hefti F. MPP+, but not MPTP selectively destroys dopaminergic neurons in cultures of dissociated rat mesencephalic neurons. Neurosci. Lett. 72: 215-220, 1986.

Sanchez-Ramos JR, Michel P, Weiner WJ, and Hefti F. Selective destruction of cultured dopaminergic neurons from embryonic rat mesencephalon: cytochemical and morphological evidence. J. Neurochem. (in press, 1988a)

Sanchez-Ramos JR, Hollinden GE, Sick TJ, Rosenthal M. 1-methyl-4phenylpyridinium (MPP+) increases oxidation of cytochrome b in rat striatal slices. Brain Research 443:183-189, 1988b

Sershen H, Reith MEA, Hashim A and Lajtha A. Protection against MPTP neurotoxicity by the antioxidant ascorbic acid. Neuropharmacol. 24: 1257-1259, 1985

Sick TJ, Rosenthal M, LaManna JC, Lutz PL. Brain potassium ion homeostasis during anoxia and metabolic inhibition in the turtle and rat. Amer. J. Physiol. 243:R281-R288, 1982.

Singer TP, Castagnoli N, Ramsay RR and Trevor AJ. Biochemical events in the development of parkinsonism induced by MPTP. J. Neurochem 49: 1-8, 1987

Turrens JF, Alexandre A, Lehninger AL. Ubisemiquinone is the electron donor for superoxide formation by Complex III of heart mitochondria. Arch. Biochem. Biophys. 237:408-414,1985

Vyas I, Heikkila RE and Nicklas WJ. Studies on the neurotoxicity of MPTP: Inhibition of NAD-linked substrate oxidation by its metabolite MPP+. J. Neurochem. 46: 1501-1507, 1986

Wagner Gc, Carelli RM and Jarvis MF. Ascorbic acid reduces the dopamine depletion induced by methamphetamine and MPP+. Neuropharmacol. 25: 559-61, 1986

Yong VW, Perry TL and Krisman AV. Depletion of Glutathione in brainstem of mice caused by MPTP is prevented by anti-oxidant pretreatment. Neurosci. Lett. 63: 56-60, 1986

19

PLASTICITY OF MESENCEPHALIC DOPAMINERGIC NEURONS

C. B. Jaeger

Department of Physiology and Biophysics
New York University Medical Center, New York, New York 10016

INTRODUCTION

A selective degeneration of mesencephalic dopaminergic (DA) neurons occurs in Parkinson's disease (Parkinson, 1917; Lieberman, 1974). Thus, the study of the DA neurons is very important for gaining insights in the underlying causes of this neurodegenerative disease.

DA neurons comprise a major part of the substantia nigra pars compacta (SNc) and ventral tegmental area (VTA), and their projections have been mapped in great detail (for reviews see Oades and Halliday, 1987; Niewenhuys, 1985; Fallon and Loughlin, 1985; Bjorklund and Lindvall, 1984). However, little is understood about the plasticity of the dopaminergic system and its capacity to conform to alterations of its connections. Plasticity of developing neurons can be determined by experimentally altering the interactions of neuron groups with their targets (Lund, 1978).

In the present study the target areas of the mesencephalic DA neurons were removed in newborn rats and the surviving midbrain DA neurons and their axon terminal regions were mapped in the adult rats. The mapping was done on tissue sections stained immunocytochemically with antibodies to tyrosine hydroxylase (TH), an enzyme necessary for catecholamine synthesis which occurs in DA neurons (Pickel et al., 1975). In addition, the distribution of some efferent projections of substantia nigra neurons were studied by retrograde transport of injected tracers to the reduced target regions. Finally, the question was addressed of whether survival of DA neurons was determined by their projections to alternate targets. This was carried out using roller tube cultures of mesencephalic slices from 5 to 8 day old rats. Such slices were immobilized on glass coverslips and kept in vitro for 12 to 14 days.

MATERIALS AND METHODS

Lesions of the target area

The caudate-putamen was lesioned mechanically in newborn rats during the first (Pl-P4) week of life. The rat pups were anesthetized with ether, put on a lucite platform, and their heads were held in place with an adjustable head holder. A fiberlight was used for trans-illumination which was necessary to aid in positioning a fine glass pipette that was attached to an aspirator. Variable amounts of neural tissue was removed from different parts of the target regions of the mesencephalic DA neurons. This was achieved by altering both the force and duration of the aspiration procedure. All rats were allowed to grow up to adulthood. Surviving mesencephalic DA neurons and their corresponding projection areas were mapped using immunocytochemical evaluation and retrograde tracer injections (see below).

Histological processing

The procedures for histological processing of the brains were described in several previous publications (Jaeger and Lund, 1980; Jaeger, 1985). Perfusion fixed (phosphate buffered paraformaldehyde) brains were cut into 30μm serial sections in the transverse, sagittal, or horizontal planes. Adjacent serial sections were processed for immunocytochemistry and acetylcholinesterase (AChE) histochemistry (Karnovsky and Roots, 1964) .

(a) Immunocytochemistry

The Sternberger (1979) peroxidase anti-peroxidase (PAP) technique was carried out in order to localize DA neurons with antisera generated in rabbits and directed against tyrosine hydroxylase (TH). Specificity of the anti-TH was determined by its ability to precipitate TH from the crude enzyme preparations of bovine adrenal medulla, and by Western blots (Joh and Ross, 1983). For the staining procedure every fourth section of a given series was pretreated with buffer solution containing 5% normal goat serum, to which in some cases triton X-100 (0.1-0.4%) was be added, before transferring the sections to diluted (1: 1000) primary antisera for 12 to 36 hrs. Subsequent incubation with secondary antisera (anti-rabbit immunoglobulin generated in goat) was followed by treatment with the PAP complex and visualization with diaminobenzidine/hydrogen peroxide. Each incubation step in this procedure was followed by rinsing of the tissue in Tris buffer saline (at pH 7.4 to 7.6) containing 1% goat serum.

(b) Tracer injections

Wheatgerm agglutinin conjugated horseradish peroxidase (WHRP) was used in most cases. Additional animals were injected with WHRP and with fast blue (FB), a fluorescent tracer. In these cases the FB was injected three to four days prior to WHRP placement, and the animals were sacrificed 12 to 14 hours later. For the visualization of WHRP the tetramethyl benzidine (TMB) technique (Mesulam and Rosene, 1979) or the glucose oxidase procedures were used. Previously described protocols were followed (Jaeger and Lund, 1980; Jaeger et al., 1983). Fluorescent tracers have the advantage that no specific visualizing procedure is needed in order to follow their transport (Bjorklund and Skagerberg, 1979). Moreover, both types of tracers are also compatible with combined immunocytochemical and histochemical staining procedures. A "Picospritzer" (General valve) was used to apply small quantities of the two tracers unilaterally into different regions of the caudate-putamen (CPU) on either the unlesioned or lesioned side. All the tracers were placed by the use of stereotaxic procedures insuring adequate separation between individual injections.

Section analysis

The TH immunoreactive cells and retrogradely labeled neurons were drawn with a camera lucida in each series of sections. In composite maps the cell charts were superimposed on the remaining areas identified as the DA neuron target regions. Since the dopaminergic pathways are for the most part ipsilateral (Altar et al., 1983; Jaeger et al., 1983; Fallon and Loughlin, 1982), each unilaterally lesioned case also served as its own "internal" control with respect to the TH positive neurons.

Tissue culture

A protocol described by Gahwiler (1981) was used with minor modifications for the in vitro maintenance of mesencephalic slices. In brief, midbrains were dissected aseptically from 5 to 8 day old Sprague Dawley rats. Slices of 450 μm thickness were cut in the sagittal or horizontal plane with a tissue chopper (Brinkman Instr.). The slices were transferred to ice-cold tissue culture medium, and gently separated into individual pieces. Only relatively non-traumatized and intact slices were mounted on glass coverslips within a drop of reconstituted chicken plasma (Gibco) that was clotted by addition of thrombin (Hoffman La Roche; 10 mg/ml). Coverslips with well-attached slices were maintained by the roller tube principle in separate plastic tubes (Falcon #3033) with 1 to 1.5 ml of complete tissue culture medium. The growth medium consisted of 45% Minimum Essential Medium, 40% Hanks Basal Salt Solution, 15% of horse serum, and additional glucose (600mM) and glutamine. Some cultures were treated with cytosine arabinoside (ara C; 5 μg/ml of medium) for 3 to 5 days following slicing in order to inhibit overgrowth by non-neuronal cells.

RESULTS

Absence of Target Region

In previous work (Jaeger et al., 1983) it was demonstrated that large lesions in the forebrain that removed the caudate-putamen, nucleus accumbens, and the olfactory tubercle resulted in severe loss of the mesencephalic DA neurons. DA neurons were lost from all regions of the mesencephalon. This was also found in the present study. Subsequent to a large unilateral forebrain lesion (Fig. 1B) the loss of DA neurons on the lesioned side (Fig. 1C) compared to the

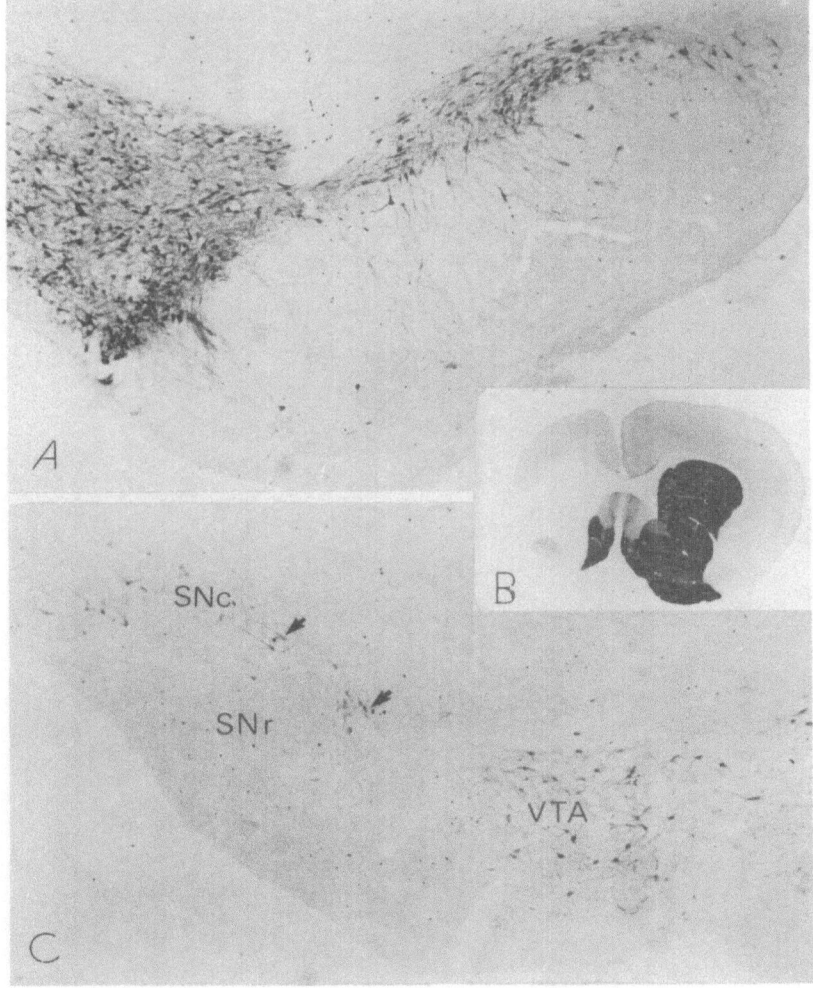

Figure 1. TH immunocytochemistry (A,C) of the substantia nigra in a unilaterally lesioned experimental animal. A) distribution of DA neurons in SNc and VTA that project to the control side of the caudate-putamen. B) Representative profile through the target area stained for AChE histochemistry. Note the complete ablation of the caudate-putamen, nucleus accumbens, and olfactory tubercle on the left (corresponding to the right side of the brain). C) The SN/VTA on the lesiond side of the brain. Several clusters of TH positive neurons (arrows) remain at this level in SNc and within adjacent sections (not illustrated) throughout the substantia nigra compacta.

control side (Fig. 1A) was about 85% to 90% in substantia nigra pars compacta (SNc). Nearly complete disappearance of DA neurons occurred in substantia nigra pars reticulata (SNr, Fig. 1C). A significant loss of DA neurons was also seen in the retrorubral area and within the ventral tegmental area (VTA, Fig. 1C).

Reduced Target Region

Removal of less than one third of the caudate putamen had no observable effect on the DA neuron population of the mesencephalon. However, within the target region itself the density of varicose terminals was much increased. This was observed in all the lesioned cases.

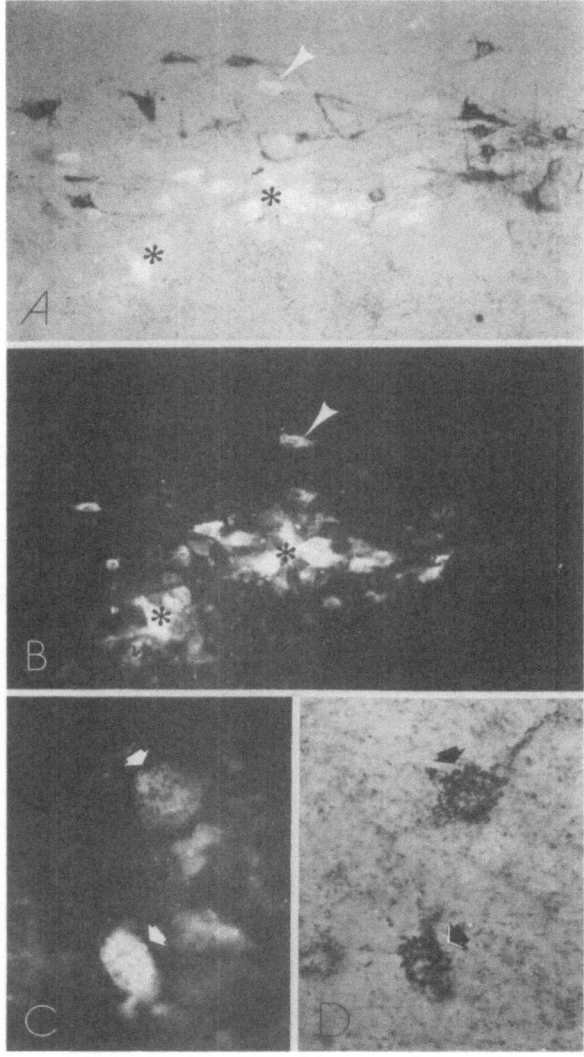

Figure 2: Retrograde transport of two tracers, WHRP and Fast blue, that were placed into an intact caudate-putamen (A,B) and a lesioned one (C,D). A) Double exposure of fluorescence and bright light. FB labeled cell clusters (asterisks) are medially distributed and WHRP labeled cell clusters laterally in SNc. Singely labeled cells intermix. Triangle points to identical neuron in A and B. B) Fluorescence light exposure of same field shown in A. C) Fluorescence photomicrograph of FB labeled neurons in SNc projecting to a reduced CPU. D) Same field as C illustrating WHRP label in identical cells (arrows) as shown in C.

Moreover, lesions that removed more than one third of the caudate putamen were accompanied by a restricted topographic loss of DA neurons in the mesencephalon (Jaeger et al., 1983). In all of the partially reduced target areas the density of TH positive varicosities was much increased. These observations prompted the question as to origin of these dense terminals. Thus, the labeling of SNc neurons was studied following simultaneous retrograde transport of WHRP and FB from the caudate putamen nucleus (CPU). Figure 2 illustrates a case in which two tracers (WHRP and FB) were injected into the same caudate putamen of a control animal (A,B) and into the reduced target region of a lesioned animal (C,D). The tracer injections were well separated and FB was placed medially and caudally and WHRP was injected laterally and rostrally. Groups of singly labeled cells interdigitated in SNc (Fig. 2A, B). The majority of fluorescently labeled cell groups were more medially distributed than the WHRP labeled cells which corresponded with the position of the tracer injections in the CPU. These findings were in agreement with previous work of Van der Kooy (1979) who also rarely found doubly labeled neurons in SNc following injections of two fluorescent tracers into the CPU. In contrast, tracer transport in the lesioned CPU resulted in numerous doubly labeled cells in the SNc (Fig. 2C, D). One explanation for this finding could be that within a spatially reduced target region the terminal arbors of the DA neurons move more closely together and possibly they ramify in overlapping fields (Fig. 3).

Alternate Targets

A small proportion of DA neurons persist following complete ablation of their target (Fig. 1). It was known from previous work that a few of the surviving DA neurons form a contralateral pathway that terminates within the caudate-putamen on the opposite side (Jaeger et al., 1983; Altar et al., 1983). Some interesting new findings were made here that provide further evidence for the possibility that alternative target neurons could support DA neuron survival. Mesencephalic slices, taken from rats at postnatal ages of 5 to 8 days, were maintained in vitro by a novel approach of roller tube culture (Gahwiler, 1981). Surprisingly, a considerable number of the mesencephalic DA neurons survived under conditions of complete removal of the target. Figure 4 shows a representative slice culture stained with antisera to TH following its maintenance for 12 days in culture. Fifteen TH positive neurons were counted on one side of this transverse slice. These cells distributed within the region that corresponded to SNc and SNr of the original slice. A very large DA neuron (35μm; asterisk) projected one of its dendrites into the SNr (open arrow). This neuron also had a very extensive axon that branched locally and formed terminal varicosities. Dense terminal arbors were usually found in the SNc (area a in Fig. 4A) and in the SNr (area b). Additional varicose terminals were also frequently observed in the contalateral substantia nigra by axons that crossed the midline (curved arrow in Fig. 4A). The precise synaptic interactions were not studied in the cultured slice. However, it was found that the mesencephalic slice cultures contained numerous non-dopaminergic neurons that were revealed by AChE histochemistry, Nissl staining, and phase contrast microscopy. The intimate relationships of TH positive boutons and non-dopaminergic neurons (Fig. 4B) were suggestive of local synapses between these neurons.

A summary of the three target manipulations and the respective responses of the DA

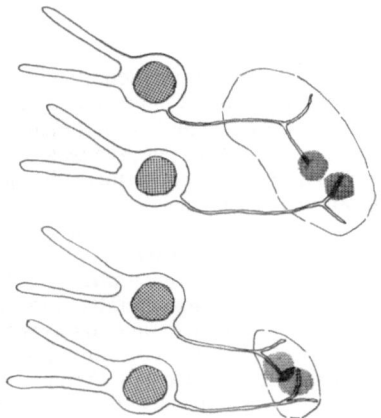

Figure 3: Schematic drawing of hypothetical terminal "crauding" within the remaining target region.

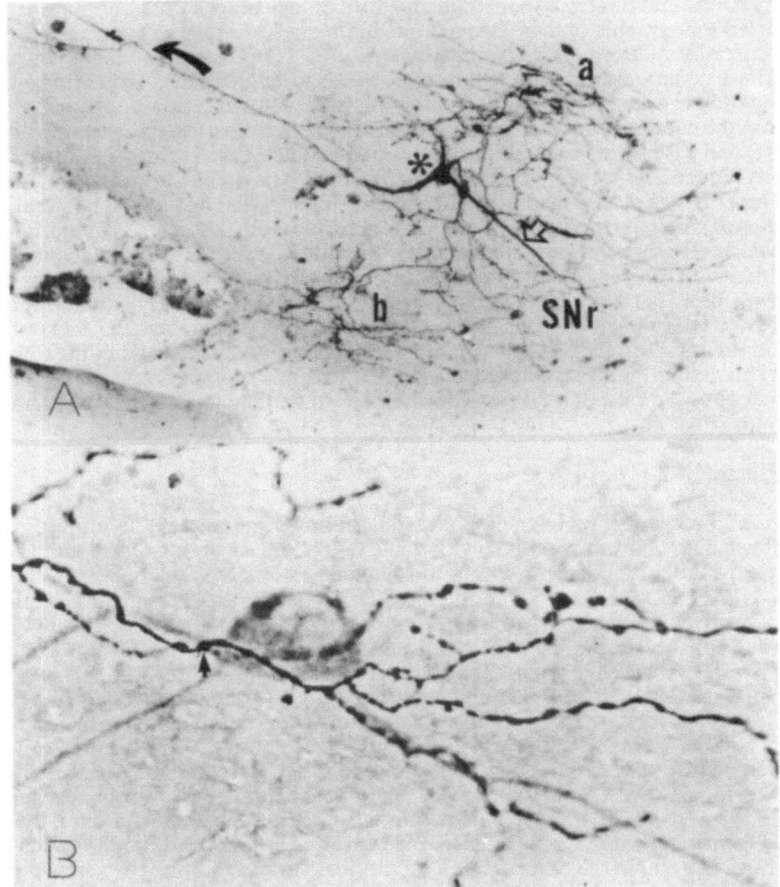

Figure 4: Low power photomicrograph of a roller tube slice culture from the mesencephalon of a 5 day old rat. TH⌐ immunocytochemistry, medial is towards the left. See text for further explanation. B) Phase contrast photomicrograph of TH positive axon in close contact with a non-dopaminergic neuron.

neurons are presented in Figure 5. The plasticity of DA neurons in these cases is discussed below.

DISCUSSION

This work confirmed previous findings by Jaeger and co-workers (1983) that demonstrated the target dependance of mesencephalic DA neurons. Furthermore it was shown that neurons of the caudate-putamen can support a greater density of DA terminals, in cases in which the original target area was substantially reduced, resulting in an increased overlap of terminal fields and consequent double labeling of SNc neurons. Alternative targets of DA neurons were identified in target deprived mesencephalic slice cultures.

The three observed outcomes (A,B,C) of DA neuron plasticity following target manipulation in vivo and in vitro are schematically represented in Figure 5. Removal of the entire caudate-putamen, nucleus accumbens, and olfactory tubercle resulted in the loss of most mesencephalic DA neurons ipsilaterally (Fig. 5A). In this case it could be argued that the DA neurons die because of injury to their axons which project to the target area at the time of target removal (Specht et al., 1981). Alternatively, neurons and (or) glia within the target may provide trophic factors that are necessary for DA neuron survival. The experimental paradigm represented in A does not distinguish between these possibilities. However, DA neuron survival in vitro may

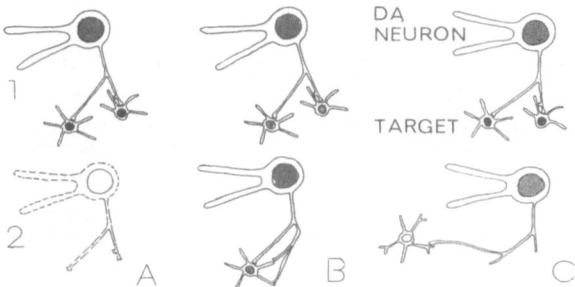

Figure 5: Schematic representations of experimental outcomes (A, B, C) of DA neuron survival following lesions of the target area depicted as region '2'. A) Complete target removal results in degeneration of most DA neurons in vivo. B) Partial target removal causes crowding of DA terminals in remaining target regions. C) In vitro some DA neurons may survive complete target removal by several mechanisms discussed in the text.

also be controlled by the effects of trophic materials. This notion is also supported by findings from previous experiments using co-cultures of embryonic mesencephalon and striatal targets (Hemmendinger et al., 1981; Kotake et al., 1982) and co-transplants (Jaeger, 1986) for the enhancement of DA neuron survival and differentiation.

The outcome depicted in Figure 5 B illustrates that following partial ablation of the caudate-putamen nucleus (CPU, as much as one third) no diminution of the DA neurons occurred. The change that was noticed occurred within the remaining target in which TH positive terminals were more densely packed together (see also Jaeger et al., 1983). This finding is of interest since it suggests a certain flexibility of the terminal arbors of DA neurons projecting to the CPU. The CPU in unlesioned rats receives a non-uniform projection of DA neuron axons. For example, Gerfen and co-workers (1985, 1987) have shown that mesencephalic DA neurons project to two distinct compartments within the caudate putamen, namely the "striatal matrix" and the "striatal patches". The matrix receives DA afferents from the ventral tegmental area (Swanson, 1982), the dorsal tier of the substantia nigra pars compacta, and the retrorubral area (Gerfen et al., 1987). DA neurons of the ventral tier of the substantia nigra pars compacta and the DA neurons of the substantia nigra pars reticulata project to the striatal patches. In addition to a "patch" and "matrix" distribution of terminals the regions abutting fibers of the corpus callosum, also referred to as "rim areas" (Fallon and Laughlin, 1985) receive a denser innervation than the central CPU. The present work suggests that the destruction of such CPU "rim areas" causes a shift of the remaining terminals and possible sprouting of some laterally positioned fibers into the remaining areas of the CPU. It is not known whether under such circumstances the patch and matrix regions are preserved or whether they are altered. The double tracer injections seem to indicate a crowding of the remaining terminals which may suggest a disturbance of the patterned axon distributions. Further studies that utilize combined retrograde transport and immunocytochemical localization are needed to resolve this.

The case drawn in Figure 5 C indicates the observed outcome of DA neuron survival in mesencephalic slice cultures kept in vitro. At least three possibilities exist that could account for the observation that a substantial number of DA neurons survived deafferentation in contrast to the situation shown in Fig. 5 A. New targets within the local neuron sets derived from the midbrain may support the survival of some DA neurons. Alternatively, trophic materials present in the growth medium (e. g. serum) and in newly dividing glia cell populations could serve to support DA neuron survival. Thirdly, surviving DA neurons may represent a population of local interneurons within SNc that remained unaffected by the deafferentation. This latter possibility seems unlikely, because in vivo sparing of substantial numbers of DA neurons failed to occur.

ACKNOWLEDGMENTS

It is a pleasure to thank Dr. T. Joh for his generous gift of antisera directed against the tyrosine hydroxylase enzyme. This work was supported by PHS grant NS-13742 and by The American Parkinson Disease Association.

REFERENCES

Altar, A., Neve, K.A., Loughlin, S.E., Marshall, J.F. and Fallon, J.H. (1983) The crossed mesostriatal projection: neurochemistry and developmental response to lesion. Brain Res. 279: 1-8.

Bjorklund, A. and Lindvall, 0. (1984) Dopamine-containing systems in the CNS. In: A. Bjorklund and T. Hokfelt (eds.) Handbook of chemical Neuroanatomy, Vol. 2, Classical Transmitters in the CNS, part 1, pp. 55-122, Elsevier Sci. Pub., Amsterdam.

Bjorklund, A. and Skagerberg, G. (1979) Simultaneous use of retrograde fluorescent tracers and fluorescence histochemistry for convenient and precise mapping of monoaminergic projections and collateral arrangements in the CNS. J. Neurosci. Meth. 1: 261-277.

Fallon, J.H. and Loughlin, S. E. (1982) Monoamine innervation of the forebrain: Collateralization. Brain Res. Bull. 9: 295-307.

Fallon, J.H. and Loughlin, S. E. (1985) Substantia Nigra. In: G. Paxinos (ed.) The Rat Nervous System. Vol. I, pp. 353- 374. Academic Press, Sydney.

Gahwiler, B.H. (1981) Organotypic monolayer cultures of nervous tissue. J. Neurosci. Meth. 4: 329-342.

Gerfen, C.R. (1985) The neostriatal mosaic. I. Compartmental organization of projections from the striatum to the substantia nigra in the rat. J. Comp. Neurology 236: 454-476.

Gerfen, C. R., Herkenham, M. and Thibault, J. (1987) The neostriatal mosaic: II. Patch- and matrix- directed mesostriatal dopaminergic and nondopaminergic systems. J. Neurosci. 7: 3915-3934.

Hemmendinger, L.M., Garber, B.B., Hoffman, P.C. and Heller, A. (1981) Target neuron-specific process formation by embryonic mesencephalic dopamine neurons in vitro. Proc. Natl. Acad. Sci. U.S.A. 78: 1264-1268.

Jaeger, C.B. (1985) Cytoarchitectonics of substantia nigra grafts: A light and electron microscopic study of immunocytochemically identified dopaminergic neurons and fibrous astrocytes. J. Comp. Neurol. 231: 121-135.

Jaeger, C.B. (1986) Axon terminal clustering in nigro-striatal double grafts. Develop. Brain Res., 24: 309-314.

Jaeger, C.B. and Lund, R.D (1980) Transplantation of embryonic occipital cortex to the brain of newborn rats: A light microscopic study of organization and connectivity of the transplants. J. Comp. Neurol 194: 571-597.

Jaeger, C.B., Joh, T.H. and Reis, D.J. (1983) The effect of forebrain lesions in the neonatal rat: survival of midbrain dopaminergic neurons and the crossed nigrostriatal projection. J. Comp. Neurol. 218: 74-90.

Joh, T.H. and Ross, M.E. (1983) Preparation of catecholamine synthesizing enzymes as immunogen for immunohistochemistry. In: A.C. Cuello (ed.), Immunohistochemistry, Vol 3, pp. 121-138, IBRO Handbook Series: Methods in the Neurosciences, John Wiley & Sons, Chichester.

Karnovsky, M.J. and Roots, L. (1964) A "direct coloring" thiocholine method for cholinesterases. J. Histochem. Cytochem. 12: 219-221.

Kotake, C., Hoffmann, P.C. and Heller, A. (1982) The biochemical and morphological development of differentiating dopamine neurons co-aggregated with their target cells of the corpus striatum in vitro. J. Neurosci. 2: 1307-1313.

Lieberman, A.N. (1974) Parkinson's disease: A clinical review. Amer. J. Med. Sci. 267: 66-80.

Lund, R. D. (1978) Development and Plasticity of the Brain. An Introduction. Oxford University Press, New York.

Mesulam, M.M. and Rosene, D.L. (1979) Sensitivity in horseradish peroxidase neurohistochemistry: a comparative and quantitative study of nine methods. J. Histochem. Cytochem. 27: 763-773.

Nieuwenhuys, R. (1985) Chemoarchitecture of the brain. Springer-Verlag, Berlin.

Oades, R.D. and Halliday, G.M. (1987) Ventral tegmental (A10) system: Neuro-biology. 1. Anatomy and connectivity. Brain Res. Rev. 12: 117-165.

Parkinson, J. (1917) An Essay on the Shaking Palsy. Sherwood, Neely and Jones. London.

Pickel, V.M., Joh, T.H. and Reis, D.J. (1975) Ultrastructural localization of tyrosine hydroxylase in noradrenergic neurons of brain. Proc. Natl. Acad. Sci. U.S.A. 72: 659-667.

Specht, L.A., Pickel, Joh, T.H. and Reis, D.J. (1981) Light microscopic immunocytochemical localization of tyrosine hydroxylase in prenatal rat brain. II. Late ontogeny. J. Comp. Neurol 199: 255-276.

Sternberger, L.A. (1979) Immunocytochemistry, 2nd edition, John Wiley and Sons: New York.

Swanson, L.W. (1982) The projections of the ventral tegmental area and adjacent regions: a combined fluorescence retrograde tracer and immunofluorescence study in the rat. Brain Res. Bull. 9: 321-353.

Van der Kooy, D. (1979) The organization of the thalamic, nigral and raphe cells projecting to the medial vs lateral caudate-putamen in rat. A fluorescent retrograde double labeling study. Brain Res. 169: 381-387.

20

TROPHIC EFFECTS OF STRIATAL PROTEINS ON CENTRAL DOPAMINERGIC NEURONS IN CULTURE

Humberto B. Valdes, Doris Nonner, Dino Rulli,
Leonard Gralnik and John Barrett

Department of Physiology and Biophysics
University of Miami, Miami, FL 33101

The dopaminergic neurons of the substantia nigra (mesencephalic region A9) appear to be influenced by trophic interactions with the tissue they innervate, the striatum. Large lesions in the striatum or the nigrostriatal tract lead to a decrease in tyrosine hydroxylase activity in the substantia nigra, mesencephalic region A9 (Reis et al. 1978). Large lesions in the striatum of newborn rats result in a substantial reduction in the number of dopaminergic neurons in the A9 region (Jaeger et al. 1983). In fact the only A9 dopaminergic neurons surviving after removal of the striatum on one side of the brain may be those which project to contralateral striatum. Jeager et al., point out that the loss of dopaminergic neurons is most likely due to loss of a trophic influence from the striatum rather than a consequence of damage to the dopaminergic axons, since transplanted dopaminergic neurons can survive despite even more extensive damage to their processes. Consistent with this trophic interaction hypothesis, co-culture of the mesencephalic dopaminergic neurons with cells from the striatum enhances both dopamine uptake and the number of neurons with detectable dopamine levels (Prochiantz et al. 1979, Hoffman et al. 1983) and the axonal plexus characteristic of the terminal region in the striatum (Hemmendinger et. al. 1981).

It is not known how these trophic effects are mediated, but by analogy to the effects of NGF on sympathetic ganglion neurons, trophic molecule(s) might be involved. Presumably such molecules would be present in the striatum and so might be obtained from extracts of the striatum. In fact, various extracts made from the striatum do enhance dopamine uptake and/or dopamine synthesis in cultures of ventral mesencephalon cells. A membrane fraction from striatal tissue enhances dopamine uptake by cultures of rat ventral mesencephalon (Prochiantz et al. 1981). A low molecular weight, 1800-2200 dalton, soluble fraction from homogenates of the striatum increases dopamine uptake by cultures of rat mesencephalon (Tomozawa and Appel 1986), and a striatal fraction with apparent molecular weight of 14,000 daltons also enhances dopamine uptake by these cells (Dal Toso et al. 1985, 1986, 1988).

The next part of this chapter summarizes work by Valdes, Nonner, Rulli, Barrett and Barrett (1988). We tested various extraction procedures and molecular weight fractions of striatal extracts in search of striatal components which would enhance dopamine synthesis. These tests were done on cultures of dissociated ventral mesencephalon from day 14 rat embryos. We used only the medial portion of the mesencephalon in a region just caudal to the hypothalmus, thus avoiding the norepinephrine neurons of the more caudal and dorsal locus coeruleus.

A defined culture medium was used (Kawamoto and Barrett 1986) with addition of an acid stable 55,000 dalton serum fraction, NSF55k, which enhances neuron survival (Kaufman and Barrett 1983). The striatal fractions were added one day after plating the cells and readded every 4-5 days. The cultures were assayed for dopamine uptake at 12 days or for dopamine synthesis after three weeks in culture.

Figure 1 shows a photograph of the mesencephalon neurons after 3 weeks in culture.

About 2% of these neurons show a distinct catecholamine fluorescence when stained using a glyoxylic acid method (De la Torre 1976).

Measurement of accumulation of newly synthesized dopamine

A distinctive characteristic of mesencephalic dopaminergic neurons is their ability to synthesize dopamine (but not norepinephrine) from tyrosine. This ability also has a key functional role, since these neurons release dopamine as a neurotransmitter. Thus an assay of dopamine synthesis by these neurons is valuable both because of its specificity for these neurons and its relation to their function.

We assayed the ability of the cultures to synthesize and accumulate labeled dopamine from tritiated tyrosine. The cultures were first incubated 1 hour in medium lacking tyrosine and phenylalanine. The cultures were then incubated 2.5 hours in medium containing H3-tyrosine (10 µM total tyrosine, .01 mCi/ml) to allow time for the accumulation of labeled dopamine to reach an approximately steady-state level. Experiments that terminated the incubation at different times indicated that the accumulation of dopamine was half maximal at about 30 minutes and reached an approximately steady-state level at 2.5 hours. The incubation was terminated by adding a solution containing .1 M HCl, 1 µM unlabeled dopamine, 1 µM norepinephrine, 1 µM DHBA (3,4-dihydroxybenzylamine), .2 mM tropolone, and .2 mM pargyline. This solution halts dopamine synthesis, releases the cell dopamine into the medium and protects dopamine from degradation. Protein was removed from the samples by precipitation with 7% trichloroacetic acid (.1 mM BSA added to samples as carrier protein) or by passing the samples through SP-trisacryl in .1 M HCl buffer. The catecholamines were then isolated by binding to alumina and fractionated using high pressure liquid chromatography on a C-18 column (for details of methods see Woodward et al. 1987).

Elution profiles from this C-18 column show a high level of radiolabel eluting during the dopamine peak. The counts in this dopamine fraction give a measure of the amount of newly synthesized dopamine that accumulated during the incubation period. The amount of newly synthesized dopamine is calculated by correcting the counts measured in the dopamine peak for the counter efficiency, and multiplying by the ratio of total tyrosine to labeled tyrosine in the incubation medium. This calculation assumes that the labeled tyrosine specific activity in the intracellular tyrosine pool is the same as that in the medium. Preincubating the cultures in medium lacking tyrosine and phenylalanine insures this condition by depleting the original intracellular tyrosine pools.

Other catecholaminergic neurons and even some glial cells have high affinity uptake systems for dopamine. Thus the assay of accumulation of newly synthesized dopamine is likely to be more reliable than dopamine uptake as an indicator of the properties of the dopaminergic neurons. However this accumulation assay does not directly measure the dopamine synthesis

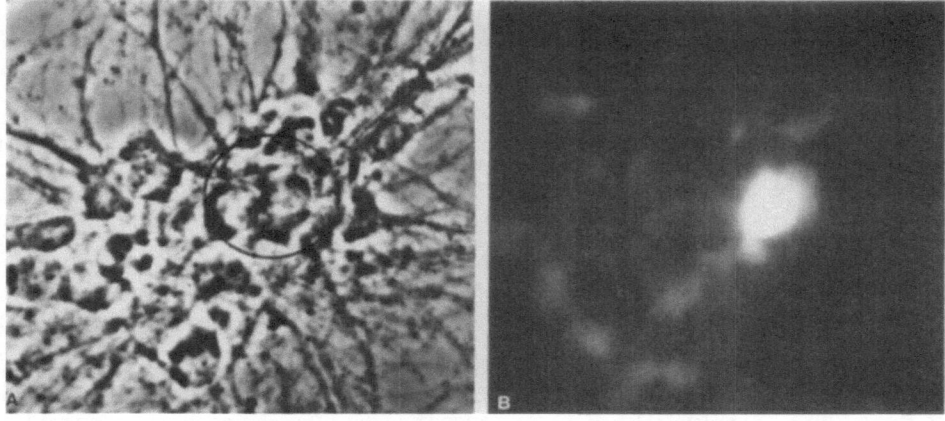

Figure 1 shows a phase contrast photograph (left panel) of a group of mesencephalic neurons after three weeks in culture. The right panel is a photograph of the same field showing catecholamine fluorescence (Nikon V-2 filter block) after treatment with glyoxylic acid. One of the neurons in this field shows distinct fluorescence, indicating that it contained catecholamines.

rate, since the steady-state accumulation of dopamine will be affected by both the rate of synthesis and the rate of dopamine loss. Shorter incubation times (eg. 20 minutes) do give a more direct measure of the dopamine synthesis rate, but these short incubation assays require considerably more isotope to incorporate enough tritium label into dopamine for accurate measurements. Thus most of our assays were done using a 2.5 hour incubation to give an approximately steady-state measurement.

Cultures made from the striatum (without neurons from the mesencephalon) did not synthesize labeled dopamine in this assay, showing the specificity of the assay for dopaminergic neurons. Other catecholaminergic neurons do synthesize dopamine, for example as a precursor to norepinephrine. However, in our ventral mesencephalon cultures the counts in the norepinephrine fraction on C18 chromatography were not above background, indicating the absence of norepinephrine-synthesizing cells in our ventral mesencephalon cultures.

Extracts from striatal tissue

If the striatum does indeed exert a trophic influence on the dopaminergic neurons of the mesencephalon, it is likely that this influence would be mediated by specific molecules synthesized by cells from the striatum. In search of such a molecule, we tested a variety of extracts from the striatum for effects on cultured mesencephalic neurons. We homogenized striatal tissue from 2-3 week old rats in the presence of a protease inhibitor (phenylmethyl sulfonylfluoride) and then centrifuged (15,000 x g for 30 minutes) this homogenate, yielding a soluble supernatant and a pellet. The supernatant was saved for the bioassay and the pellet was extracted using procedures designed to free or solubilize components that might be bound to membrane fragments in the pellet or trapped within vesicular structures in the pellet, eg. secretion vesicles. The solubilization procedures used included treatment with chaotropic agents (0.2M sodium thiocyanate, 0.2M sodium perchlorate), extraction at acidic pH, and hypotonic shock (5 mM solution in distilled water). The thiocyanate and osmotic shock methods gave some activity, but the perchlorate extract did not show biological activity. The acid extract of the striatal pellet had a much greater specific activity than any of the other extracts (i.e. there was much greater activity per milligram of protein in the acid-solubilized fraction). Furthermore the acid extraction seemed to solubilize most of the activity, since further extraction with hypotonic shock released more protein but no dopamine synthesis-enhancing activity. Thus most of our work has concentrated on this acid-solubilized extract from the striatal tissue.

The supernatant obtained after the initial homogenization and centrifugation of striatal tissue also enhanced the accumulation of newly synthesized dopamine. However, the total amount of dopamine synthesis-enhancing activity was less than in the acid striatal extracts, and the specific activity was much lower because of the high concentration of total protein in this initial supernatant.

Molecular size of active components from the striatal extract

The extracts from striatal tissue were fractionated at 4°C using ultrafiltration membranes under nitrogen gas pressure. A serial ultrafiltration protocol was used, starting with a XM100 membrane (apparent molecular weight cutoff of 100 kDa), then using PM30, YM5, YM2 and YCO membranes which have successively lower molecular weight cutoffs. At each stage more buffer was added and the sample was refiltered with the same membrane to enhance the separation. The filtrate from this extra buffer was added to the original filtrate from that membrane and then this total volume was subjected to the next ultrafiltration stage.

Table I

Fraction	Enhancement of dopamine accumulation	μg/ml protein
<0.5 kDa	0.83 + 0.13 (N=12)	<0.1
0.5-1.0 kDa	0.87 + 0.14 (N=12)	5
1-5 kDa	2.38 + 0.31 (N=29)	0.2
5-30 kDa	1.38 + 0.18 (N=39)	14
30-100 kDa	1.75 + 0.35 (N=16)	20

The results of such an experiment are shown in Table I. The highest levels of activity were found in the 1-5 kDa and 30-100 kDa fractions, with the highest activity per mg protein in the 1-5 kDa fraction. We do not yet know whether the activity found in the 1-5 kDa fraction affects the dopaminergic neurons by the same mechanism as the higher molecular weight fractions. It is possible that several different receptors on the dopaminergic neurons might be involved in mediating the effects of these different fractions.

Chemical properties of active material from the striatal extracts

The active components of the striatal extracts were very sensitive to protease activity. No activity could be recovered if protease inhibitors were omitted during the preparation of the extracts. More than 90% of the activity was destroyed by a 2 hour incubation with trypsin at 22 C.

Regional specificity

The acidic extraction procedure followed by fractionation by ultrafiltration was applied to several brain regions including the striatum, frontal cortex and occipital cortex. The 1-5 kDa fraction from occipital cortex produced no enhancement of dopamine synthesis, whereas the extract from frontal cortex produced a small increase (1.33 times control) that did not reach statistical significance at the .05 level. The striatal extract produced a 2.38 fold increase that was statistically significant at the .001 level. Similar regional comparison experiments were also done using the >30K fraction but with dopamine uptake for the assay (see Tomozawa and Appel, 1986, for dopamine uptake assay method). Figure 2 shows that the striatum acid extract prepared from young rats was again most active, with no dopamine uptake-enhancing activity from cerebellum or occipital cortex.

Age and species specificity

Dopamine uptake- and dopamine synthesis-enhancing activity was found in extracts of striata from adult rats and newborn calf, but the yield of biological activity per gram of striatal tissue was less than that from 2-3 week old rats. In the particular experiment shown in Figure 2 the striatal extract (>30k fraction) from adult rats did not produce a significant enhancement of

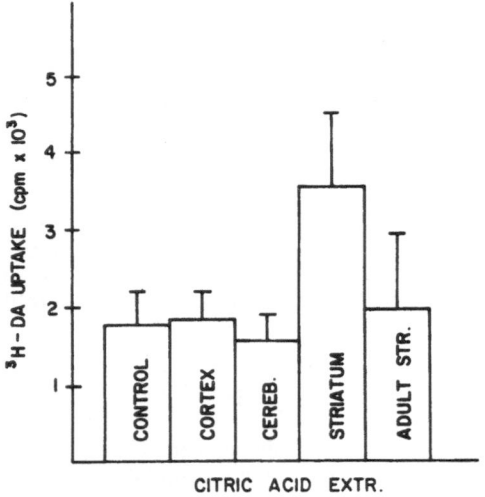

Figure 2 shows measurements of dopamine uptake by cultures of ventral mesencephalon after treatment with acid extracts of various brain tissues. The extracts were prepared by citric acid extraction of the pellet obtained after homogenization and centrifugation of the brain tissue (see text). In this experiment the fraction with apparent molecular weight greater than 30 kDa was used. Although the extracts from all the brain regions were prepared in the same way, only the striatal extract of young (2-3 week old) rats showed statistically significant activity (P<.01, N>5 for each brain region).

dopamine uptake whereas extract prepared the same way, but from young rats (2-3 weeks old) did significantly enhance dopamine uptake at the 0.01 significance level. Extracts from the striatum of a newborn monkey gave approximately the same level of activity as found in the striata from the 2-3 week rats.

Specific binding of striatal extract to ventral mesencephalon

Most of the well-characterized growth factors, such as NGF, bind with high affinity to specific receptors on their target tissues. Thus the active factor(s) in the striatal extract might bind selectively with high affinity to receptors on ventral mesencephalon cells. We radiolabeled the striatal extract by iodination with I-125 (Bolten-Hunter method) and incubated freshly dissociated neurons from various brain regions with the iodinated macromolecules in the striatal extract (>30 kDa fraction). Figure 3 shows that the binding of radiolabeled striatal extract to ventral mesencephalon (per milligram of tissue) is 2.29 times greater than binding to spinal cord, and that similar radiolabeled extracts prepared from cortex do not show this enhanced binding to the ventral mesencephalon. Mixing an excess of unlabeled striatal extract with the labeled fraction greatly reduced the binding, indicating that the sites involved in the binding are saturable. A molecule showing this specific and selective binding might be a trophic factor for some of the neurons of the substantia nigra. Thus analysis of the radiolabeled material showing this binding could give information about the chemical properties of the molecules involved.

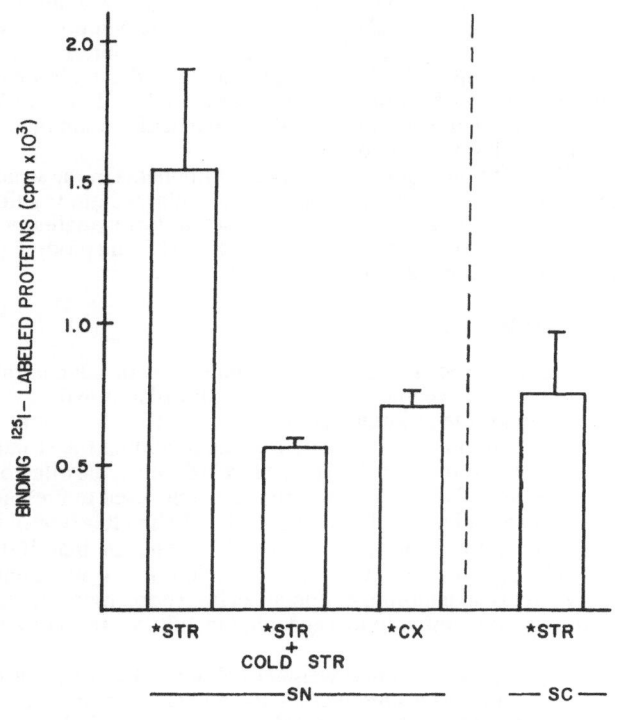

Figure 3 shows the selective and specific binding of radiolabeled striatal extracts to ventral mesencephalic cells. The vertical axis gives the counts of I-125 labeled extract protein that binds to tissue from particular regions of E14 rat embryos. I-125 labeled striatal extracts gave much higher binding to ventral mesencephalon cells than to spinal cord cells. This binding was greatly reduced by adding an excess of cold striatal extract, demonstrating the saturable nature of the binding sites. Similarly I-125 labeled extracts from cortex did not give the high binding to ventral mesencephalon.

Effect of identified peptides present in the striatum

The striatum contains a considerable array of small peptides that might account for the dopamine synthesis-enhancing activity found in the 1-5 kDa range. These include the enkephalins, endorphins, tachykinins and several other peptides. Black et al. (1985,1986) found that two tachykinins, substance P and substance K, produce an increase in tyrosine hydroxylase activity (the rate limiting enzyme in dopamine synthesis) in cultures of rat ventral mesencephalon. This increase was not found when tetrodotoxin was present in the medium in sufficient concentration to block sodium-dependent action potentials. Thus the tachykinins may influence tyrosine hydroxylase activity by altering action potential firing rates. In our assay substance P and substance K did not enhance the measured accumulation of newly synthesized dopamine. The tachykinins were probably rapidly degraded in our cultures, and were not added frequently enough in our assay to be effective. Since the striatum extract was effective at protein concentrations overlapping the concentrations tested for the tachykinins, it is unlikely that the activity of the striatum extract in our assay was due to tachykinins. Thus there may be other small striatal peptides in addition to the tachykinins which influence dopamine metabolism. other tested peptides include somatostatin, secretin, galanin, met-enkephalin, leuenkephalin, beta-endorphin and bombesin. Some of these peptides did produce a small enhancement: namely secretin (1.22 + .13 N=14), neurotensin (1.24 + 0.12 N=11) and somatostatin (1.31 +.27, N=8), but none were sufficient to account for the activity in the striatal extract. Similarly Tomozawa and Appel (1986) found that somatostatin produced a small enhancement of dopamine uptake, but that their striatal extract produced a much greater increase.

High molecular weight striatal fraction and the 55,000 dalton serum fraction

The striatal extract fraction in the 30-100k apparent molecular weight range also enhanced dopamine synthesis. It is possible that this activity might be related to the 55kd fraction from serum which enhances the survival of the mesencephalic neurons. A similar 55kd fraction of cerebral spinal fluid also supports the survival of these neurons. Thus cells within the brain might produce the molecules identical or similar to those that underlie the activity of this 55 kDa serum fraction. It is also possible that molecules produced by another tissue, such as the liver, might enter the blood and be transported into particular brain regions by specialized transport systems across the blood-brain barrier.

Some of the bioactive higher molecular weight components of the striatal extracts might also break down to yield the activity found in lower molecular weight fractions. Several large peptides found in the striatum are known to be processed to form smaller peptides. These include tachykinin, the precursor of substance P and substance K, and prodynorphin, a precursor for the enkephalins and dynorphins (Zamir et al. 1984).

Insulin-like growth factors (IGF's)

There are high concentrations of IGF-II-like molecules in the central nervous system (Haselbacher et al., 1985), and receptors for these insulin-like growth factors are present in many brain regions including the striatum (Baskin et al. 1987). Messenger RNA-s for both IGF-I and II have been found in nearly all brain regions including the striatum, although the striatal messenger levels are less than one fourth the levels found in the olfactory cortex or cerebellum (Rotwein et al., 1988). The acid extraction conditions used to prepare our most active striatal extract fractions would be expected to extract both IGFI and IGFII from the brain tissue.

A preliminary report by Fellows et al. (1987) indicates that IGF-I increases the dopamine content of cells cultured from the mesencephalon of day 19 rat embryos. We found that both IGF-I and IGF-II (1- 10 nanomolar) enhanced both dopamine uptake and the accumulation of newly synthesized dopamine in mesencephalon cultures from day 13-14 rat embryos (Figure 4).

The IGF monomers have a molecular weight of about 8 kDa, and thus may contribute to the activity of striatal extract fractions retained by the 5 kDa cutoff ultrafiltration filter. Since some IGF-like activity is found in molecular weight fractions up to 38 kDa from brain extracts (Haselbacher 1985), IGF-like activity might also contribute to activity in the 30-100 kDa fraction.

The possibility of a role for IGF-like molecules in contributing to the regulation of the dopaminergic neurons is also consistent with the finding of messenger RNA for the IGFs in the striatum. Perhaps some forms of Parkinson,s disease might be correlated with inadequate amounts the IGFs or other trophic molecules from the striatum.

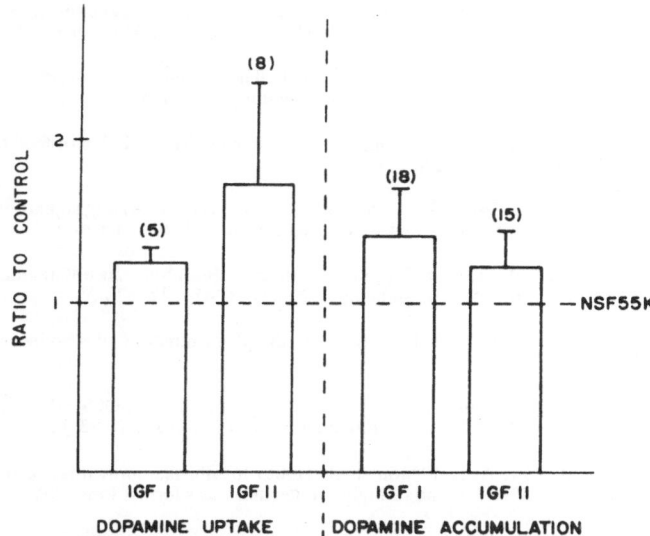

Figure 4 shows enhancement of dopamine uptake, and accumulation of newly synthesized dopamine by the insulin like growth factors, IGF-I and IGF-II. Cultures were exposed to 1 nM IGF-I or IGF-II for two weeks before measuring dopamine uptake, and three weeks before measurements of dopamine accumulation. All the cultures were fed with medium containing the NSF55k serum fraction and the measurements are plotted as a ratio of the activity in the IGF treated cultures to controls receiving only the serum fraction.

Summary and Conclusions

Dopaminergic neurons from the ventral mesencephalon of embryonic rats were maintained in culture for many weeks in the presence of an acid-stable fraction prepared from serum. Extracts prepared from the striatum more than double both dopamine uptake and the accumulation of newly synthesized dopamine in these cultures. Several small peptides (secretin, somatostatin, and insulin like growth factors I and II) enhance dopamine synthesis by 20 to 40% in the presence of the serum fraction, but do not account for all the biological activity found in the striatal extracts. Thus dopaminergic neurons of the mesencephalon may be regulated by a variety of peptides and proteins, the most potent of which may not yet be identified.

References

Ahnert-Hilger, Engele, Reisert, I., and Pilgrim, C., (1986) Different developmental schedules for dopaminergic and noradrenergic neurons in dissociation culture of fetal rat midbrain and hindbrain. Neuroscience 17: 157-165

Daguet, M.C., DiPorizo, Prochiantz, A., Kato, A., and J. Glowinski (1980) Release of dopamine from dissociated mesencephalic dopaminergic neurons in primary cultures in absence or presence of striatal target cells. Brain Res. 191: 564-568

Dal Toso, R., Giorgi, O., Presti, D., Favaron, M., Leon, A., and G. Toffano (1985) Purification and characterization of neuronotrophic activity from bovine caudate nuclei: possible modulation by GMl ganglioside. Neurosci. Abst. 11: 949

Dal Toso, Benvegnu, D., Ferrari, G., Soranzo, C., Doherty, P., Walsh, F.S., Toffano, A., and A. Leon (1986) Striatal derived neuronotrophic factor: biological characterization and production of monoclonal antibodies. Neurosci. Abst. 12: 1100

Dal Toso, Giorgi, Soranzo, C., Kirschner, G., Ferrari, G., Favaron, Benvegnu, D., Presti, D., Vicini, S., Toffano, G., Assone, and A. Leon (1988) Development and survival of neurons in dissociated fetal mesencephalic serum-free culture: I. Effects of cell density and of an adult mammalian striatal derived neuronotrophic factor (SDNF). J. Neurosci. 8: 733-745

De la Torre, J.C. and Surgeon, J.W. (1976) A methodological approach to rapid and sensitive monoamine histofluorescence using a modified glyoxylic acid technique: the SPG method. Histochem. 49: 81-93

Denis-Donini, S., Glowinski, J., and A. Prochiantz (1984) Glial heterogeneity may define the three-dimensional shape of mouse mesencephalic dopaminergic neurons. Nature 307: 641-643

Fellows, Al-Hader, and R. Kadle (1987) IGF-I supports survival and differentiation of fetal rat brain neurons in serum-free defined medium. Neurosci. Abs. 13: 1615

Friedman, W.J., Dreyfus, C.F., McEwen, B.S., and I.B. Black (1985) Depolarizing signals increase tyrosine hydroxylase development in cultured mouse substantia nigra. Neurosci. Abs. 11: 1142

Friedman, W.F., Dreyfus, C.F., McEwen, B.S., and I.B. Black. (1986) Substance K regulates tyrosine hydroxylase in cultured embryonic mouse substantia nigra. Neurosci. Abs. 12: 378

Fonnum, F. (1975) A rapid radiochemical method for the determination of cholineacetyltransferase, J. Neurochem. 24: 407409

Gammeltoft, S., Haselbacher, G., Humbel, R., Fehlmann, M., and E. Van Obberghen (1986) Two types of receptor for insulin-like growth factors in mammalian brain. EMBO J. 3407-3412

Haselbacher, G. Schwab, M. Pasi, A. and Humbel, R. (1985) Insulin-like growth factor II (IGF II) in human brain: Regional distribution of IGF II and of higher molecular mass forms. Proc. Natl. Acad. Sci. USA 82: 2153-2157

Hemmendinger, L.M., Barber, B.B., Hoffmann, P.C. and A. Heller (1981) Target neuron-specific process formation by embryonic mesencephalic dopamine neurons in vitro. Proc. Natl. Acad. Sci. USA 78: 1264-1268

Horowitz, P.M. (1985) Rapid fluorescamine based protein assay usable in the presence of interfering substances. J. Chromatogr. 319: 446-449

Hoffman, PC., Hemmendinger, LM., Kotake C., and Heller A.,(1983) Enhanced dopamine cell survival in reaggregates containing telencephalic target cells., Brain Res 274: 275-281

Jaeger, C.B., Joh, T.H., and D.J. Reis (1983) The effect of forebrain lesions in the neonatal rat: survival of midbrain dopaminergic neurons and the crossed nigrostriatal projection. J. Comp. Neurology 218: 74-90

Kaufman, L.M. and Barrett, J.N. (1983) Serum factor supporting long-term survival of rat central neurons in culture. Science 220: 1394-1396

Kawamoto, J.C. and J.N. Barrett (1986) Cryopreservation of primary neurons for tissue culture. Brain Res. 384: 84-93

Lenoir, D., and Honegger, P., (1983) Insulin-like growth factor I (IGF-I) stimulates DNA synthesis in fetal rat brain cell cultures. Dev. Brain Res. 7: 205-213

Murphy, L.J., Bell, G.I. and H.G. Friesen (1987) Tissue distribution of insulin-like growth factors I and II messenger ribonucleic acid in the adult rat. Endocrinology 120: 1279-1282

Prochiantz, A., Di Porzio, Kato, A., Berger, B. and J. Glowinski (1979) In vitro maturation of mesencephalic dopaminergic neurons from mouse embryos is enhanced in the presence of their striatal target cells. Proc. Natl. Acad. Sci. 76: 5387-5391

Prochiantz, A., Daguet, M.C., Herbert, A., and Glowinski, J. (1981) Specific stimulation of in vitro maturation of mesencephalic dopaminergic neurons by striatal membranes. Nature 293: 570-572

Puymirat, J., Faivre-Bauman, A., Barret, A., Loudes, C., and Tixier-Vidal, A., (1985) Does triiodothyronine influence the morphogenesis of fetal mouse mesencephalic dopaminergic neruons cultured in chemically defined medium? Brain Res 355: 315-317

Recio-Pinto, E., and Ishii, D., (1984) Effects of insulin, insulin-like growth factor-II and Nerve Growth Factor on neurite outgrowth in cultured human neuroblastoma cells. Brain Res. 302: 323-334

Reis, D.J., Gilad, G., Pickel, V.M., and T.H. Joh (1978) Reversible changes in the activities and amounts of tyrosine hydroxylase in dopamine neurons of the substantia nigra in response to axonal injury as studied by immunochemical and immunocytochemical methods. Brain Res. 144: 325-342

Rotwein, P., Burgess, S.K., Milbrandt, J.D., and J.E. Krause (1988) Differential experession of insulin-like growth factor genes in rat central nervous system. Proc. Natl. Acad. Sci. USA 85: 265-269

Shalaby, I., Kotake, C., Hoffman, P., and Heller, A., (1983) Release of dopamine from coaggregate cultures of mesencephalic tegmentum and corpus striatum. J. Neurosci. 3;1565-1571

Shalaby, I., Hoffman, P., and Heller, A. (1984) Release of dopamine from mesencephalic neurons in aggregate cultures: influence of target and non-target cells. Brain Res. 307: 347-350

Tomozawa, Y. and Appel, S.H.(1986) Soluble stiatal extracts enhance development of mesencephalic dopaminergic neurons in vitro. Brain Res. 399: 111-124

Valdes, H.B., Nonner, D., Rulli, D., Barrett, E., and J. Barrett. Effects of striatal extracts and the insulin-like growth factors on rat dopaminergic central neurons in culture (in preparation)

Woodward, W.R., Seil, F.J.,and J.P. Hammerstad (1987) Cerebellum plus locus coeruleus in tissue culture. II: Development and metabolism of catecholamines. J. Neurosci. Res. 17: 184-188

Zamir, N., Palkovits, M., Weber, E., Mezey, E., and M. Brownstein (1984) A dynorphinergic pathway of leu-enkephalin production in rat substantia nigra. Nature 307: 643-645

21

AGE-RELATED CHANGES OF DOPAMINERGIC FUNCTIONS

Caleb E. Finch

Andrus Gerontology Center and Dept. of Neurobiology
University of Southern California Los Angeles, CA 90089-0191

INTRODUCTION

This summarizes recent findings on several aspects of Parkinson's disease (PD), which are part of our studies on questions of gene expression in the brain during aging and neurological diseases. Although there is no evidence that age per se determines the onset of PD or that PD is simply accelerated senescence, the strong age-related increase of PD after midlife (Mutch et al., 1986; Ayd, 1961) can not be ignored, and draws attention to the types of age changes and their individual variations that occur in these and other neuronal systems (Finch, 1976, 1987). A particular issue concerns the long-standing assumptions that non-dividing cells such as neurons inevitably suffer senescent involution and age-related loss. This question bears on the often presumed inevitability of PD as a feature of later life, with the presumed generality of nigrostriatal degeneration. While true that few of central neuron types in mammals are formed de novo after puberty or can regenerate after injury (Jacobson, 1978; Rakic, 1985), there is ample evidence to question whether senescent involution generally occurs in most neurons. For example, two neurosecretory systems show no evidence of general failure up through the average lifespan. The LHRH neurons that maintain sustained elevations of pituitary gonadotropins long beyond menopause (Scaglia et al., 1976) and no loss of these cells occurs in female mice showing major agerelated disturbances of pituitary functions (Hoffman and Finch, 1986; Finch et al., 1984). At the molecular level, an analysis of whole brain RNA from male rats across their lifespan did not detect changes in the levels of messenger RNA (polyA + mRNA isolated from polyribosomes) or in its nucleotide sequence complexity, which assays the number of different types of mRNA species (Colman et al., 1980). On the other hand, many types of neurons show statistically significant trends for degenerative changes during aging, particularly in the hippocampus, cerebral cortex, and in monoaminergic projection systems (Coleman and Flood, 1987; Morgan et al., 1987b). In regard to the basal ganglia, the striatum is the only major brain region to show decreases of bulk (ribosomal) RNA during aging in rodents (Chaconas and Finch, 1973; Shaskan, 1977). At a cell level, the changes include alterations in dendritic structure, loss of receptors, decreases in cell body RNA, and to an unknown extent, death of neurons. It is interesting to compare the nucleolar shrinkage, an index of ribosomal RNA transcription, in the substantia nigra and locus ceruleus of neurologically normal humans, since both cell types accumulate neuromelanin with age: the greater nucleolar atrophy of nigra neurons (Mann and Yates, 1979) could indicate selective vulnerability of these cells to age-related damage. Future studies may reveal how many of these changes occur in all individuals. The extent of age-related neuron loss is highly controversial (Coleman and Flood, 1987; Terry et al., 1987), and may be far less general than once thought. Its extent in the basal ganglia is based on a relatively limited number of specimens, even though a large number of neurons were counted in these demanding studies (McGeer et al., 1977).

Age changes in relation to Huntington's disease (HD) and PD

The basal ganglia are intensively studied for age changes because of its involvement in several diseases with strong age-related incidence, particularly PD and HD. A major question concerns causes of the onset in characteristic decades of life: in most cases HD is first manifested between 25 and 45 years (Brackenridge, 1973), while PD is usually first manifested after 45 (Mutch et al., 1986). We and others have found distinctly different effects of age on specific aspects of basal gangliar function (Finch, 1980). The age-related loss of striatal D2 receptors may be the most generally observed of age changes in the mammalian brain. Humans, monkeys, rabbits, and rodents show decreased concentrations in the neostriatum long before old age (Severson and Finch, 1980; Severson et al., 1982; Lai et al., 1987; Thal et al., 1980; Morgan et al., 1987a,b; Seeman et al., 1987). In humans, decreases of D2 receptors may be significant by 30 years according to PET scanning with positron labeled DA antagonists (Wong et al., 1984). The decrease of striatal D2 receptors is progressive across the adult lifespan and may reach a 20 to 40% deficit in the neostriatal concentration of D2 receptors in rodents and humans. In situ binding studies show that the loss of D2 sites is relatively greater in ventro-lateral zones of the rat neostriatum; changes in this receptor field may be correlated with the sensorimotor deficits of old rats (Joyce et al., 1986). Assays for the regulation of these receptors include pharmacologic blockade by haloperidol, which in young animals induces extensive increases of DA receptors and supersensitization by behavioral indices; however, older mice have smaller increases (Randall et al., 1981). Impairments of compensatory super-sensitization may predispose older patients to Parkinsonian side effects from neuroleptics (Ayd, 1961).

In contrast to D2 receptors, the D1 (adenylate cyclase-linked) DA receptors do not decrease as sharply and may even increase during aging in humans (Morgan et al., 1987a). These assays used (^{3}H)fluphenazine with discrimination of D1 and D2 sites by spiperone and apomorphine (Morgan and Finch, 1986). The relative complement of receptors thus shifts the D1/D2 ratio by 2-fold over the lifespans of neurologically normal humans according to this assay. However, many more samples and other ligands must be analyzed to establish these trends in the D1 receptor, and several other studies report age-related decreases or no change (O'Boyle and Waddington, 1984; reviewed in Morgan et al., 1987a; Seeman et al., 1987).

The presynaptic levels of dopamine in the nigrostriatal path also decrease, but more gradually than the loss of D2 receptors. In the rodent striatum, the loss of DA is about half that of D2 receptors; humans at the average lifespan have greater DA loss (Carlsson, 1981; Hornykiewicz, 1985; Osterberg et al., 1981; Rogers and Bloom, 1985). The loss of striatal DA may accelerate at later ages in both rodents and humans. We recently compared the levels of DA catabolites in PD and in Alzheimer disease (AD). As expected, the loss of DA and the catabolite DOPAC in caudate and putamen was far greater in PD than in AD. However, the HVA loss was equivalent, about 40% (Morgan et al., 1987b). Because levels of HVA reflect extraneuronal metabolism, while DOPAC reflects the intraneuronal compartment, we suggest that there is a reduced presynaptic release of DA during AD.

In regard to the typically post midlife age of PD onset, it seems noteworthy that the onset of DA loss in neurologically normal humans is also modest until after midlife. In PD, the loss of nigro-striatal neurons is usually much greater than in normal individuals of the same ages (McGeer et al., 1977). On the other hand, HD has an earlier age-related distribution that coincides with the earlier onset of D2 receptor loss. In HD, the aggregate loss of striatal cells may overlap with the age-related trend for neostriatal cell loss (Bugiani et al., 1978; Finch, 1979), but it likely that not all types are lost in common. The selectivity of cell changes is important in regard to the goal of relating degenerative changes to selective gene regulation.

Genotypic variations in receptor regulation

Inbred mouse strains which differ in their monoaminergic mechanisms are useful models for age-related receptor changes, in particular the CBA, C57BL/6J, and the BALBc mouse. These strains have been inbred for more than 100 generations, and are available through the Jackson Laboratory in Bar Harbour, Maine. The CBA/J and BALBc/J are distantly related, but share no ancestry with the C57BL/6J. The CBA mouse has fewer nigra neurons and a lower reduced number of DA receptors (Ross et al., 1976; Severson et al., 1981), and is a model for accelerated aging; it also shows less supersensitization to haloperidol, as described above for aging C57BL/6J mice (Randall et al., 1981). The CBA strain has similar impairments in the regulation of cortical beta-1 receptors as probed by supersensitization in response to chronic propranolol (Severson et al., 1986). We speculate that CBA mice may differ in some protein component of supersensitization shared by striatal dopaminergic and cortical beta-adrenergic systems. These and other strains also have different rates of receptor loss during aging (Lephron-Greenwood and Cinander, 1987). There is thus a potential that genetic polymorphisms can predetermine patterns of aging in the basal ganglia and may be a determinant in the risk for a variety of neurological diseases and in the age of onset and their time course. The concepts of multiple and pleiotropic gene interactions are well accepted in developmental genetics but their role in aging processes is only beginning to be appreciated. Although the contribution of heredity to PD remains controversial, influences of genotype on receptor regulation suggest that diverse polymorphisms could nonetheless influence the disease and its age of onset, whatever the primary cause.

Studies on the regulation of tyrosine hydroxylase in a rat model of PD

We are investigating the genomic mechanisms of age-related change which predispose to neuronal dysfunctions and loss during aging in the basal ganglia. Microspectrophotometric observations of postmortem brain from elderly normal individuals and from those with age-related neurological diseases often show nucleolar shrinkage or loss of cell body RNA in large neurons (Doebler et al., 1987; Mann and Yates, 1983; Mann et al., 1984), e.g. in the substantia nigra of PD or basal cholinergic nuclei and hippocampus of AD. The nucleolar shrinkage in the remaining neurons of the substantia nigra in PD was particularly puzzling, because lesions of this pathway induce hyperactivity in the remaining neurons in young rats, as judged by increased synthesis and release of DA, and increased TH (tyrosine hydroxylase) (Agid et al., 1973; Hefti et al., 1980; Stachowiak et al., 1987).

We recently developed a lesion model which demonstrates the PD-like atrophy of nigra neurons, yet which also shows the enhanced synthesis at the remaining terminals (Pasinetti et al., 1987). Adult male rats were given unilateral 6-OHDA injection into the substantia nigra, and sacrificed 9 months later. Measurements of striatal catecholamines showed major depletion of DA on the lesioned side, but increased DOPAC/DA ratios; this indicates increased release of DA at the remaining striatal terminals, as expected from the compensatory responses of the nigrostriatal terminals (see above). To our knowledge, the maintenance of this compensatory increase for such a long portion (35%) of the rodent lifespan had not been indicated before. The cell bodies of these lesioned neurons, however, showed an opposite response to 6-OHDA lesions; they were atrophied just as seen in PD. In these measurements, dopaminergic neurons were identified by immunocytochemistry with antisera to rat TH. The cell bodies, nuclei, and nucleoli of TH-immunoreactive neurons were about 30% smaller than in the contralateral (non-lesioned) side. The smaller nucleoli imply decreased synthesis of ribosomes, which would be consistent with neuron RNA loss in the nigra during PD (Mann and Yates, 1983).

To establish the extent of change in neuron RNA we examined two messenger RNA populations using in situ hybridization: TH-mRNA and beta-tubulin-mRNA. For these measurements, brain sections were prepared for imunocytochemistry to TH, followed by hybridization to cRNA (complementary) antisense strand probes made in a transcription vector. The loss of TH-mRNA was larger than for beta-tubulin. It is of interest to calculate the yield of DA per unit TH mRNA on the lesioned side. With a loss of TH mRNA of >50% per unit area, a 50% shrinkage of cell volume, and an 80% loss of DA cells, the lesioned side has 20% x 50% x 50% = 5% of the TH mRNA content found on the nonlesioned side. The striatal concentration of DA, however is reduced much less, to 35% of the nonlesioned side. Thus, the remaining nigra neurons are 35%/5% or 7-fold more efficient in utilizing the available TH mRNA in producing presynaptic DA, the final product. This calculation most likely underestimates the relative increase of DA synthetic activity per neuron, because there is at least a 30% increase in DA turnover and release.

These results pose some interesting questions. In regard to increased DA metabolism, the opposite changes of the TH-mRNA in its cell body and of DA synthesis and release at its striatal terminals imply a dichotomous regulation. Is the efficiency of TH mRNA translation increased several-fold to compensate for reduced mRNA at a time when more DA is synthesized and released per neuron? The greater deficit of TH mRNA than tubulin mRNA is also interesting, and illustrates that even during neuron cell body atrophy with nucleolar shrinkage, some mRNA's are relatively unaffected. We plan to include additional mRNA in these studies to ascertain the extent of coordinate regulation of mRNA for cytoskeletal proteins, housekeeping enzymes, and specialized cell functions.

Several mechanisms for these changes can be considered. It is possible that 6-OHDA has long-lasting toxic effects that cause damage to even the remaining neuron cell bodies, while allowing the terminals to manage some compensation. Alternatively, there may be transynaptic effects through the striato-nigra pathways containing GABA and substance P, which influence the non-dopaminergic neurons of the substantia nigra pars reticulata (Saji and Reis, 1987). We hypothesize that the net reduction of DA release after 6-OHDA lesions causes changes in the striato-nigra afferents. This perspective may apply to age-related loss of striatal RNA in aging rodents (Chaconas and Finch, 1973; Shaskan, 1977) and the age-related slowing of striatal DA turnover (Finch, 1973; Osterburg et al., 1981). We hypothesize that the slowed release of DA would cause decreased RNA synthesis in striatal neurons, particularly in the medium spiny I neurons that atrophy in PD (McNeil et al., 1987). According to this view, some neurotransmitters may be linked to neurotrophic factors. The reversal of cholinergic neuron atrophy in the striatum and basal forebrain of 2 yr old rats by infusion of nerve growth factor (Fischer et al., 1987) indicates an extensive role of neurotrophic factors, but need not result from primary deficiencies of such factors. We suggest that many other examples of age-related neuron atrophy could be interpreted as the result of deafferentation syndromes. Apparently similar atrophy of basal forebrain cholinergic neurons also can be induced by decortication (Pearson et al., 1983). Moreover, lesions that kill 30% of basal forebrain cholinergic nuclei in rats cause slowly evolving transynaptic cascades that lead to neuronal degeneration in the entorhinal cortex and hippocampus (Arendash et al., 1987); these changes give a striking model for neuronal atrophy in these same pathways during AD (Hyman et al., 1984). We need more details about changes in specific RNA and protein species to establish if neuron atrophy has the same final common regulatory pathway in aging, diseases like PD with major neuron loss, or after experimental deafferentation. Finally, these results bear on the choice of drugs used to treat PD: it will be important to learn how various drug treatments influence the levels and translation of TH mRNA: it may be possible to increase the translatable pool of TH mRNA using assays such as described here.

Molecular approaches are also being developed to study the genomic basis for these phenomena in the human postmortem brain. The strategy involves in situ hybridization for TH mRNA and other mRNA in TH-immunoreactive neurons, as well as hybridization analyses on RNA extracted from the brain. The results, though preliminary, are encouraging. First, we were able to obtain high molecular weight poly(A)RNA from postmortem human hippocampus and cortex (May et al., 1987; Johnson et al., 1986). After processing more than 50 samples, we concluded that the postmortem interval is not crucial (0-24 hrs) but that other factors such as premorbid hypoxia or wasting conditions may be more important. In situ hybridization studies with conventionally fixed, paraffin embedded human brain specimens show reasonable signals in cortex with cRNA to coding sequences for glial fibrillary acidic protein (GFAP). This abundant mRNA was localized over glial cells that reacted with GFAP antisera. While less abundant sequences are not evaluated through this approach, we are encouraged to pursue these studies with TH mRNA, to evaluate if the PD brain shows deficits of TH mRNA as observed in rodents after 6-OHDA lesions.

A separate project on AD is also mentioned here, because its results indicate the possibility of obtaining clones for sequences from the substantia nigra that could be used to analyze processes of neurodegeneration and compensatory responses. Poly(A)RNA from the AD brain was found to be relatively intact in favorable specimens and can be cloned by conventional procedures that yielded double stranded cDNA of average size 1.5 kb (range 0.5-5.0 kb)(May et al., 1987). A library of recombinants in the cloning vector lambda gt10 yielded 61 plaques that showed ≥ 2 fold change in Alzheimer's disease through differential plaque hybridization screening procedures (AD + clones). The clones isolated so far fall into two classes: a substantial number (80%) that cross-hybridized to each other, demonstrating extensive homology. On the basis of the partial sequence, these AD + clones contain coding sequences for glial fibrillary acidic protein (GFAP). By reference to mouse GFAP data in GENBANK, this human GFAP clone has an 83% identity of base sequence, and contains a nearly complete coding sequence for GFAP. Blot hybridization indicates 3-fold increases of mRNA for GFAP in hippocampus and cortical regions that degenerate in AD, but no changes in the cerebellum (Morgan et al.,in prep.; May et al., 1987). Another clone showed select increases in the hippocampus, but not degenerating regions of neocortex; the fragments sequenced so far have no homologue in GENBANK (May et al., 1987). These results indicate that the PD brain should also be amenable to analysis at the level of gene regulation. For example, it should be possible to obtain clones of sequences differentially regulated in the remaining nigra neurons.

Summary

The extensive literature on age-related neurochemical and anatomical changes in the basal ganglia gives a strong basis for future studies at the level of gene regulation. Besides identifying the intracellular (transacting) factors that regulate transcription of specific genes, this new field will also identify new therapeutic approaches to ameliorating neurotransmitter deficiencies thorough drugs that may increase transcription of genes that encode relevant enzymes.

References

Agid, Y., Javoy, F., Glowinski, J., (1973). Hyperactivity of remaining dopaminergic neurones after partial destruction of the nigro-striatal dopaminergic system in the rat. Nature. 245: 150-151.

Arendash, G.W., William, J.M., Dunn, A.J., Meyer, E.M., (1987). Long-term neuropathological and neurochemical effects of nucleus basalis lesions in the rat. Science. 238: 952-955.

Ayd, R.J. (1961). A survey of drug-induced extra pyramidal reactions. J. Am. Med. Assoc. 175: 1054-1060.

Brackenridge, C.J., (1973). The relation of sex of affected parent to the age at onset of Huntington's disease. J. Med. Genetics 10: 333-336.

Bugiani, O., Perdelli, F., Salvarini, S., Leonardi, A., Mancardi, G.L., (1980). Loss of striatal neurons in Parkinson's disease: a cytometric study. Eur. Neurol. 19: 339-344.

Carlsson, A., (1981). Aging and brain neurotransmitters. In Pfaff D (ed.): Funktionsstorungen des Gehirns im Alter. F.K. Schatlauer, Verlag. Stuttgart. 67-87.

Chaconas, G., Finch, C.E., (1973). The effect of ageing on RNA/DNA ratios in brain regions of the C57BL/6J male mouse. J. Neurochem. 21: 1469-1473.

Colman, P.D., Kaplan, B.B., Osterburg, H.H., Finch, C.E., (1980). Brain poly(A)RNA during aging: Stability of yield and sequence of complexity in two rat strains. J. Neurochem. 34: 335-345.

Coleman, P.D., Flood, D.G., (1987). Neuron numbers and dendritic extent in normal aging and Alzheimer's disease. Neurobiol. Aging. 8: 521-546.

Doebler, J.A., Markesbery, W.R., Anthony, A., Rhoads, R.E., (1987). Neuronal RNA in relation to neuronal loss and neurofibrillary pathology in the hippocampus in Alzheimer's disease. J. Neuropathol. Exp. Neurol. 46: 28-39

Finch, C.E., (1976). The regulation of physiological changes during mammalian aging. Q. Rev. Biol. 51: 49-83.

Finch, C.E., (1980). The relationships of aging changes in the basal ganglia to manifestations of Huntington's chorea. Ann. Neurol. 7: 406-411.

Finch, C.E., Felicio, L.S., Mobbs, C.V., Nelson, J.F., (1984). Ovarian and steroidal influences on neuroendocrine aging processes in female rodents. Endocrine Rev. 5: 467-494

Finch, C.E., (1987). Neural and endocrine determinants of senescence: investigation of causality and reversibility of laboratory and clinical interventions. In, H. Warner (ed.) Modern Biological Theories of Aging. Raven Press, New York, pp. 261-306.

Fischer, W., Wictorin, K., Bjorklund, A., Williams, L.R., Varon, S., Gage., F.H., (1987). Amelioration of cholinergic neuron atrophy and spatial memory impairment in aged rats by nerve growth factor. Nature 329: 65-68.

Hefti, F., Melamed, E., Wurtman, R., (1980). Partial lesion of the dopaminergic nigrostriatal system in rat brain. Brain Res. 195: 123-137.

Hoffman, G.E., Finch,C.E., (1986). LHRH neurons in the female C57BL/6J mouse brain during reproductive aging: No loss up to middle-age. Neurobiol. Aging 7: 45-48.

Hornykiewicz, O., (1985). Brain dopamine and aging. Interdiscipl. Topics Geront. 19: 143-155.

Hyman, B.T., Van Hoesen, G.W., Damasco, A.R., Barnes, C.L., (1984). Alzheimer's disease: cell-specific pathology isolates the hippocampal formation. Science. 225: 613-617.

Jacobson, M., (1978). Developmental Neurobiology, 2nd (ed.), Holt, Rinehart, Winston. New York. Plenum.

Johnson, S.A., Morgan, D.G., Finch, C.E., (1986). Extensive postmortem stability of RNA from rat and human brain. J. Neurosci. Res. 16: 267-280.

Joyce J.N., Loeschen S.K., Sapp D.W., and Marshall J.F. (1986). Age-related regional loss of caudate-putamen dopamine receptors revealed by quantitative autoradiography. Brain Res. 378: 158-163.

Lai, H., Bowden, D.M., Horita, A., (1987). Age-related decreases in dopamine receptors in the caudate nucleus and putamen of the rhesus monkey (Macaca mulatta). Neurobiol. Aging 8: 45-49.

Lephron, C.E., Greenwood, Cinader, B., (1987). Variations in age-related decline in striatal D2 dopamine receptors in a variety of mouse strains. Mech. Ageing Develop. 38: 199-206.

Mann, D.M.A. and Yates, P.O. (1979) The effects of aging on the pigmented neurons of the human locus ceruleus and substantia nigra. Acta Neuropathol. 47: 93-97.

Mann, D.M.A., Yates, P.O., (1983). Possible role of neuromelanin in the pathogenesis of Parkinson's disease. Mech. Ageing Devel. 21: 193-203.

Mann, D.M.A., Yates P.O., and Marcyniuk, B. (1984). Alzheimer's presenile dementia, senile dementia of Alzheimer type, and Down's syndrome in middle age form an age-related contunuum of pathological changes. Neuropathol. Appl. Neurobiol. 10: 185-207.

May, P.C., Johnson, S.A., Masters J.N., Lampert-Etchells M., Finch, C.E., (1987). Cloning of poly(A)RNA sequences differentially expressed in Alzheimer's disease hippocampus. Soc. Neurosci. Abstracts. 13: 1325.

McGeer, P.L., McGeer, E.G., Suzuki, J.S., (1977). Aging and extrapyramidal function. Arch. Neurol. 34: 33-35.

McNeil, T., Brown, S.A., Shoulson, I., Lapham, L., Eskin, T., Rafols, J., (1987). Regression of striatal dendrites in Parkinson's Disease. In, The Basal Ganglia II. Carpenter M. and Jayaraman, A. (eds.), pp. 475-482.

Morgan, D.G., Finch, C.E., (1986). (^{3}H)Fluphenazine binding to brain membranes: Simultaneous measurement of D-1 and D-2 receptor sites. Neurochem. 46: 1623-1631.

Morgan, D.G., Marcusson, M.N., Gordon, Ljung, L., Yang, E., Gorgy, M., Lerner, S.P., Winblad, B., Gottfries, C., Bird, E.D., Riederer, P., Tourtellotte, W., Finch, C.E., (1987a). Stability of dopamine and DOPAC, but not HVA and 3-MT in Alzheimer's Disease. Contrast with Parkinson's Disease. Soc Neurosci. Abstracts. 13: 564

Morgan, D.G., Marcusson, J. 0., Nyberg, P., Wester P., Winblad, B., Gordon, M.N., and Finch, C.E. (1987b). Divergent changes in D-1 and D-2 dopamine binding sites in the human brain during aging. Neurobiol. Ageing 8: 195-201.

Mutch, W.J., Dingwall-Fordyce I.,Downie A.W., Paterson J.G., and Roy S.R. (1986). Parkinson's disease in a Scottish city. Brit. Med. J. 292: 534-535.

O'Boyle, K.M. and Waddington, J.L. (1984) Loss of rat striatal dopamine receptors with aging is selective for D-2 but not D-1 sites: association with increased nonspecific binding of the D-1 ligand [3H]piflutixol. Eur. J. Pharmacol. 105: 171-174.

Osterberg, H.H., Donahue, H.G., Severson, J.A., Finch, C.E., (1981). Catecholamine levels and turnover during aging in brain regions of male C57BL/6J mice. Brain Res. 224: 337-352.

Pasinetti, G.M., Lerner, S.P., Johnson, S.A., Morgan, D.G., Telford, N.A., Myers, M.M., Finch, C.E., (1987). Chronic lesions differentially decrease messenger RNA in dopaminergic neurons of rat substantia nigra. Soc. Neurosci. Abstracts. 13: 378

Pearson, R.C.A., Gatter, K.C., Powell, P.S., (1983). Retrograde cell degeneration in the basal nucleus in monkey and man. Brain Res. 261: 321-326.

Rakic, P., (1985) Limits of neurogenesis in primates. Science. 227: 1054-1056.

Randall, P.K., Severson, J.A., Finch, C.E., (1981). Aging and the regulation of striatal dopaminergic mechanisms in mice. J. Pharm. Exp. Ther. 219: 695-700.

Robertson, G.L. and Rowe, J., (1981). The effect of aging on neurohypophysed function. Peptides. 1 (Suppl.1): 158-162.

Rogers, J. G. and F.Bloom (1985). Neurotransmitter metabolism and function in the aging central nervous system. In, Handbook of the Biology of Aging. 2nd edition, C.E. Finch and E.L. Schneider, (eds.), Van Nostrand Rheinhold, New York, pp. 645-691.

Ross, R.A., Judd, A.B., Pickel, V.M., Joh, T.H., Reis, D.J., (1976). Strain-dependent variations in number of midbrain dopaminergic neurones. Nature. 264: 654-656.

Saji, M., Reis, D.J., (1987). Delayed transneuronal death of substantia nigra neurons prevented by gamma-aminobutyric acid agonist. Science. 235: 66-69.

Scaglia, H., Medina, M., Pinto-Ferreira, A.L., Vasques, G., Gual, C., Perez-Palacios, G., (1976). Pituitary LH and FSH secretion and responsiveness in women of old age. Acta Endocrinol. 81: 673-679.

Seeman, P., Bzowej, N.H., Guan, H.C., Bergeron, C., Becker, L.E., Reynolds, G.P., Bird, E.D., Reiderer, P., Jellinger, K., Watanabe, S., (1987). Human Brain dopamine receptors in children and aging adults. Synapse. 1: 399-404.

Severson, J.A., Finch, C.E., (1980). Reduced dopaminergic binding during aging in the rodent striatum. Brain Res. 192: 147-162.

Severson, J.A., Randall, P.K., Finch, C.E., (1981). Genotypic influences on striatal dopaminergic regulation in mice. Brain Res. 210: 201-215.

Severson, J.A., Marcusson, J., Winblad, B., Finch, C.E., (1982). Age-correlated loss of dopaminergic binding sites in human basal ganglia. J. Neurochem. 39: 1623-1631.

Severson, J.A., Pittman, R.N., Gal, J., Molinoff, P.B., Finch, C.E., (1986). Genetic influence on the regulation of beta-adrenergic receptors in mice. J. Pharm. Exp. Therap. 236: 24-29.

Shaskan, E.G., (1977). Brain regional spermidine and spermine levels in relation to RNA and DNA in aging rat brain. J. Neurochem. 28: 509-516.

Stachowiak, M.K., Keller, R.W., Striker, E.M., Zigmond, M.J., (1987). Increased dopamine efflux from striatal slices during development and after nigrostriatal bundle damage. J. Neurosc. 7: 1648-1654.

Terry, R.D., DeTeresa, R., Hansen, L.A., (1987). Neocortical cell counts in normal human adult aging. Ann. Neurol. 21: 530-539.

Thal, L., Horowitz, S. G., Dvorkin, B., and Makman, M.H. (1980). Evidence for loss of brain 3H-spiperidol and 3H-ADTN binding sites in rabbit brain with aging. Brain Res. 192: 185- 194.

Wong, D.F., Wagner, H.N., Dannals, R.f., Links, J.M., Frost, J.J., Ravert, H.T., Wilson, A.A., Rosenbaum, A.E., Gjedde, A., Douglass, K.H., Petronis, J.D., Folstein, M.F., Tuong, J.K., Burns, H.D. Kuhar, M.J., (1984). Effects of age on dopamine and serotonin receptors measured by positron tomography in the living human brain. Science. 226: 1393-1394.

22

ADRENAL MEDULLA TRANSPLANTS IN RODENTS AND NONHUMAN PRIMATES: REGENERATIVE RESPONSES IN THE HOST BRAIN

Don Marshall Gash, Massimo S. Fiandaca, Jeffrey H. Kordower and John T. Hansen

University of Rochester Medical Center
Rochester, New York 14642

INTRODUCTION

The first clinical trials using intrastriatal adrenal autografts for the treatment of parkinsonism are now underway. From the early published reports it is clear that the procedures being used and the reported benefits from the grafts vary considerably from study to study. There are two general procedures being employed for the implantation of tissue. The first clinical experiments in Sweden (1,2) entailed the stereotactic placement of adrenal medullary tissue into the caudate or putamen using, in some instances, a steel spiral tissue carrier and, in other instances, the direct injection of tissue into the graft sites. In the first series involving two patients, the stereotactic implants were into the caudate nucleus (1). Backlund and his colleagues reported a moderate transient improvement during the first ten days postsurgery in the first patient and a more lasting improvement in the second patient which they suggested extended out to approximately six months. The criteria used were relatively subjective, making it difficult to quantitate the degree of recovery.

The second series of transplants in Sweden involved the stereotactic placement of adrenal medulla into the putamen of two patients (2). Of all the studies conducted to date, this series has provided the most extensive objective evaluations. Clinical assessment of the patients, CT scans, sensory and movement evoked potentials, quantitative electroencephalograms, regional cerebral blood flow measurements, neurochemical measurements of catecholamine and catecholamine metabolites in the cerebrospinal fluid and positron emission tomography scans were used to evaluate the patients for up to six months post-transplantation. One patient had longer "on" periods for about two months postsurgery and the other patient exhibited some improvement in balance and gait for about two months. Lindvall and his colleagues concluded from their study that the adrenal medulla autografts resulted in only a transient beneficial effects in the patients.

Using a homologous surgical approach in which adrenal autografts were stereotactically placed in the right caudate nucleus with a silver tissue carrier, Jiao and his associates at the Capital Institute of Medicine in Beijing, China, have reported (3) more positive clinical benefits in their study involving four patients. In their study, an acute phase of recovery, lasting for the first two to three days postsurgery, was followed by a return to the preoperative state of impairment. At three to four weeks post-grafting, a gradual, continual improvement was seen in all four patients which was maintained for six to eight months (the longest time period studied). The criteria for evaluating recovery was similar to that used in the first Swedish study. From the subjective criteria used, it is difficult to quantitate the actual degree and extent of behavioral recovery. The use of drugs, such as Amantadine and traditional Chinese herbal medications, as well as the involvement of the patients in an extensive in-hospital physical therapy program following surgery adds to the difficulty in assessing the effects of the transplant versus other treatments on the reported alleviation of symptoms.

A quite different neurosurgical approach has been used by Madrazo, Drucker-Colín and their associates in Mexico, for intrastriatal adrenal autogafts (4,5). They have used an open microsurgical technique in which the head of the caudate nucleus is visualized through a transcortical approach. Caudate tissue is removed to create a transplantation cavity exposed to the lateral ventricle into which adrenal medullary pieces are implanted and held in place by surgical staples. They suggest that this approach allows cerebrospinal fluid to bathe the adrenal medulla implant, which may be important for medullay tissue survival. Using several criteria, including neuropsychological tests, electromyograms and somatosensory evoked potentials, recovery was reported (5) in a series of 11 patients to begin from periods ranging from immediately following surgery to 30 days postsurgery. The degree of recovery varied and was estimated to be up to 90 percent for the younger patients. No diminution of recovery was noted over time. Two of the patients died, one 45 days postsurgery and the second five months after surgery. The investigators felt that these deaths were not related to the neural implants (5).

With only these few published reports which differ significantly in their methodologies and results, it is quite difficult to assess the safety and efficacy of intrastriatal adrenal autografts. In order to better evaluate the clinical trails which are now underway, our group has been conducting a series of intrastriatal adrenal autografts into rodents and nonhuman primates using procedures similar to those being practiced clinically. Neurotoxins, 1-methyl-4-phenyl-1,2,5,6-tetrahydropyridine (MPTP) and 6-hydrosydopamine (6-OHDA), which damage and destroy nigral dopamine neurons in primates and some other species have been employed to create nigrostriatal lesions. The results from these studies are discussed below.

Intrastriatal Adrenal Medulla Grafts into MPTP Treated Mice

Nigrostriatal lesions closely resembling the neuropathological features of Parkinson's disease can be created in nonhuman primates using the neurotoxin MPTP. MPTP administered to mice also creates a biochemical lesion in the nigrostriatal system and provides a useful paradigm in which to examine the properties of intrastriatal adrenal medullary transplants. The dose level of MPTP and the time course of administration is very important for creating effective lesions in these studies. In our experiments (6), two injections of 50 mg MPTP/kg body weight, spaced 16 hours apart, were effective in reducing the dopamine level in the striatum by 86% for time periods up to 5 weeks post-treatment. Lower dose levels administered over greater time periods can produce severe initial depletion of striatal dopamine levels, but the treated animals often exhibit total recovery within five weeks. This illustrates that, in young mice, the striatal dopaminergic system possesses a significant potentials for recovery from the neurotoxic effects of MPTP.

Using the higher dosing rate that resulted in a prolonged depletion of striatal dopamine levels, the fate of adult adrenal medullary isografts was studied (6) in young adult C57BL/6 mice. One week after MPTP administration overlying cortical and callosal tissue was removed to expose the striatum in which a transplantation cavity was constructed. Adrenal medullary fragments were placed directly into the transplantation cavity and held in place using gelfoam. The control groups were unoperated MPTP treated mice and MPTP treated mice undergoing the surgical procedures (including gelfoam placement) but not receiving tissue implants. The viability of grafted cells and expression of enzymes involved in the catecholamine biosynthesis pathway (tyrosine hydroxylase, TH: dopamine-B-hydroxylase, DBH; and phenylethanolamine N-nethyltrasferase, PNMT) were assesed by immunocytochemical staining procedures at 2,4 and 6 weeks after transplantation.

While clusters of surviving chromaffin cells could be identified in many of the hosts, the total number of surviving implanted cells was relatively low. In no animal could more than 100 surviving chromaffin cells be identified. Many of the surviving cells expressed TH and occasional cells staining for PNMT were also seen. Most of the grafted cells remained in the endocrine phenotype; occasional cells did begin to elaborate short neuritic processes, but in no instances could fibers from these cells be traced from the graft into the host brain. In the striatum ipsilateral to the graft, a marked hyperinnervation of TH immunoreactive fibers was observed in each of the transplant recipients. This recovery could have been due to 1) a sprouting or regeneration of the lesioned host dopaminergic fibers or 2) a biochemical recovery or increase in TH expression in fibers which were already present. Our data, at present, do not allow us to distinguish between these two possibilities.

The recovery of host dopaminergic systems innervating the striatum is quite possibly due to a neurotrophic effect from the experimental procedures. There are at least four possible sources for these putative trophic factors. 1. Cells from the adrenal implant, either chromaffin cells or other cell types, such as Schwann cells, could be elaborating a factor(s) inducing the observed response. 2. The damaged induced by the grafting procedures could have produced a gliosis which resulted in the production of a neurotrophic factor(s). 3. The inflamatory cells, which were predominately macrophages, may have produced a trophic factor(s). 4. Alterations in the blood-brain barrier from the implantation of a peripheral paraneural tissue could result in the influx of blood-borne factors accounting for the response. While it is not possible at this time to distinguish between these alternatives it is important to note that the surgical procedures alone, without a tissue implant, did not produce a marked increase in host TH immunoreactive fiber staining. Clearly it is important to identify the specificity of the response and the cell types responsible for the observed neuroanatomical recovery.

Adrenal Medullary Grafts in 6-OHDA Lesioned Rats

Many groups have productively employed, as a rodent model of Parkinson's disease, the unilateral 6-OHDA nigrostriatal-lesioned rat model developed by Ungerstedt (7). Some of the pioneering work using adrenal medulla tissue implants in this model was conducted by Freed and his colleagues and this work is reviewed elsewhere in this volume. Our group has also studied (8) the survival and functional properties of dispersed adult adrenal medullary cell implants grafted into the striatum of adult Long-Evans rats. The host animals, all with unilateral 6-OHDA nigrostriatal lesions, received implants of dispersed adrenal medullary cells. Donor tissue came from juvenile Long-Evans rats and was dispersed using a trypsin-collagenase treatment. Each rat recipient received a total of 40,000 dispersed medullary cells suspended at a density of 10,000 cells/ul. Control animals were treated identically to the graft recipients except that only culture media was injected. All animals were sacrificed one month after transplantation and the viability of the implanted tissue assessed by immunocytochemical staining for TH. Some surviving adrenal chromaffin cells could be identified in every graft recipient, however, the number of surviving cells was small and never more than 100 TH immunoreactive grafted cells could be identified in any one host. The grafted cells were predominately of the endocrine phenotype. Few cells possessed processes and the processes which were present were short. TH immunoreactive fibers were present adjacent to the transplants but the fibers appeared to be of host origin rather than from the grafts. In comparison to the control group, the adrenal medulla graft recipients demonstrated a significant reduction in amphetamine-induced rotational behavior. Given the limited number of surviving grafted chromaffin cells, this behavioral response may have been due to the recovery of host TH immunoreactive fiber systems rather than from catecholamines produced and released by the grafted cells.

Adrenal Autografts in Nonhuman Primates

Our initial studies have focused on evaluating the adrenal autograft approach of using procedures in nonhuman primates virtually identical to those being currently used in the clinical treatment of Parkinson's disease. We have recently analyzed data from an experimental series designed to assess the fate of grafted adrenal tissue and the response of the host brain to tissue implants. In our studies, 12 Cebus apells monkeys received either sham surgery or intrastriatal adrenal medullary autografts. The open microsurgical approach (which is being widely used clinically in the United States) was used on three of these animals. A stereotactic transplant approach (used in some institutions in the United States and in Sweden and China) was utilized for the other nine animals. In the stereotactic group, seven animals received adrenal implants and the other two served as surgical controls by undergoing tissue carrier implantation without adrenal medullary tissue being implanted. One month following transplantation the animals brains were evaluated by TH immunocytochemistry, electron microscopy, and routine histological methods.

A preliminary analysis of the fate of the grafted tissue and the response of the host brain to the implant has been completed (9,10). The overall results are quite similar to those seen with adrenal isografts in MPTP treated mice. Each of the nonhuman primate graft recipients received grafts containing between 250,000 to 400,000 adrenal medullary cells. One month later only 20-40 surviving adrenal cells could be identified in two of the animals receiving implants via a stereotactic approach. No surviving chromaffin cells were found in animals receiving grafts using the open microsurgical approach. The primary constituents of the implants sites were inflammatory cells, especially macrophages. Despite the minimal survival of adrenal medullary cells, a marked potentiation of TH immunoreactive fiber staining was noted in the nigrostriatal pathway (see Figures 1 and 2) and the striatum (see Figures 3 and 4) on the side ipsilateral to the transplant.

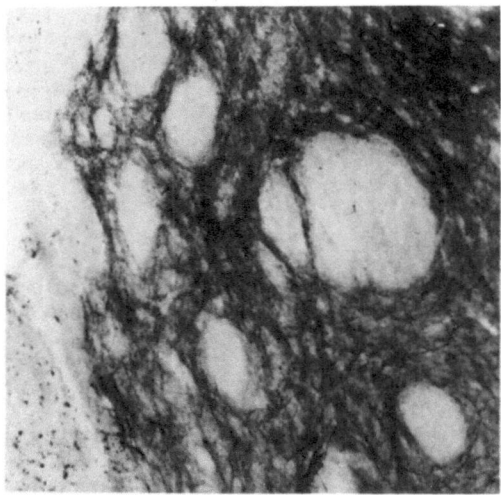

Figure 1. The caudate nucleus near the site of a tissue carrier implant shows heavy TH immunoreactivity. 130X, TH immunocytochemistry.

Figure 2. In contrast, the contralateral striatum in the same animal shown in figure one, possesses few TH immunoreactive fibers. 130X, TH immunocytochemistry.

Our results to date suggest that few adrenal chromaffin cells are surviving either the open microsurgical or the stereotactic approaches being currently employed for the clinical treatment of parkinsonism. The number of surviving cells may vary from patient to patient and it is possible that modifications in the neurosurgical technique might result in better tissue survival. Our study does provide evidence that the implantation procedure can induce some neuroanatomical recovery by stimulation of the remaining host catecholaminergic systems. This neuroanatomical recovery may be the basis of any behavioral recovery observed in subjects undergoing adrenal autograft procedures.

Conclusions

The original rationale for using adrenal medullary tissue for transplants was that the chromaffin cells within the medulla might provide sufficient titers of catecholamines, especially dopamine and norepinephrine, to compensate for deficient dopamine levels in the striatum. However in studies conducted using mouse, rat and nonhuman primate models of Parkinson's disease, the survival of grafted chromaffin cells appears to be quite variable with too few cells surviving to account for the recovery that is observed. Our studies would suggest that alternate hypotheses need to be considered. Following implants in mice, rats, and nonhuman primates, we have seen an increase in host TH immunoreactive fibers, an observation which is consistant with the hypothesis that the implantation procedures result in an increase in neurotrophic activities effecting the host dopaminergic systems. Further elucidation of the mechanism leading to behavioral recovery can be crucial in developing better treatments for parkinsonism and other neurological disorders.

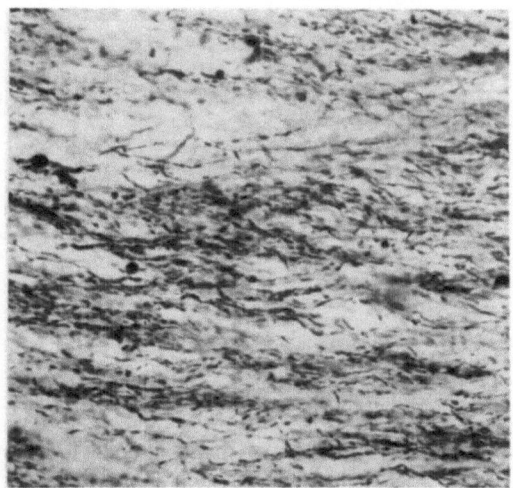

Figure 3. A dramatically enhanced TH immunoreactive fiber staining is clear in the nigrostriatal pathway of this Cebus monkey on the side ispsilateral to the tissue carrier implant. Note the thick varicose fibers running in the pathway. 130X, TH immunocytochemistry.

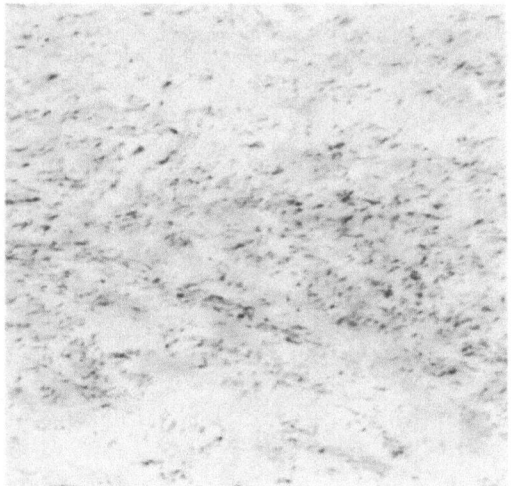

Figure 4. The contralateral nigrostriatal pathway on the same section in the same animal as shown in figure 3. TH immunoreactive staining allows the boundaries of the pathway to be distinguished but staining is less robust and the size of this fiber bundle is clearly diminished as compared to the implanted side. 130X, TH immunocytochemistry.

Acknowledgements

The studies reported in this review were conducted in collaboration with Drs. Martha Bohn, Gouying Bing, Frederick Marciano, Shiga-Hisa Okawara, and Mary F.D. Notter and their participation is gratefully acknowledged. Funding for this research came in part from grants from the American Health Foundation and NIH Grants NS15109 and NS25778. I thank Kim Gesell for her secretairal assistance and Leja Allyn for assistance with the manuscript preparation.

References

1. Backlund, E.-O., P.-O. Granberg, B. Hamberger, E. Knutsson, A. Martensson, G. Sedvall, A. Seiger, and L. Olson. Transplantation of adrenal medullary tissue to striatum in parkinsonism. First clinical trials. J. Neurosurg 62:169-173, 1985.

2. Lindvall, O., E.-O. Backlund, L. Farde, G. Sedvall, R. Freedman, B. Hoffer, A. Nobin, A. Seriger, and L. Olson. Transplantation in Parkinson's disease: Two cases of adrenal medullary grafts to the putamen. Ann. of Neurology 22:457-468, 1987.

3. Jiao, S., W. Zhang, J. Cao, Z. Zhang, H. Wang, M. Ding, Z. Zhang, J. Sun, Y. Sun and M. Shi. Study of adrenal medullar tissue transplantation to striatum in parkinsonism. Prog. Brian Res. (in press).

4. Madrazo, I. R. Drucker-Colín, V. Díaz, J. Martínez-Mata, C. Torres and J.J. Becerril. Open microsurgical autograft of adrenal medulla to the right caudate nucleus in two patients with intractable Parkinson's disease. N. Eng. J. Med. 14:831-873, 1987.

5. Drucker-Colín, R.,I. Madrazo, F. Ostrosky-Solís, M. Shkurovich, R. Franco and Caesar Torres. Adrenal medullary tissue transplants in the caudate nucleus of Parkinson patients. Prog. Brian Res. (in press).

6. Bohn, M.C., L. Cupit, F. Marciano and D.M. Gash. Adrenal medulla grafts enhance recovery of striatal dopaminergic fibers. Science 237:913-916, 1987.

7. Ungerstedt, U. and G.W. Arbuthnot. Quantitative recording of rotational behavior in rats after 6-hydroxy-dopamine lesions of the nigrostriatal dopamine system. Brain Res. 24:483-493, 1970.

8. Bing, G., M.F.D. Notter, J.T. Hansen and D.M. Gash. Comparison of adrenal medullary, carotid body and PC12 cell grafts in 6-OHDA lesioned rats. Brain Res. Bull. (in press).

9. Gash, D.M., J. Kordower, M. Fiandaca, S.H. Okawara, M.F.D. Notter and J.H. Hansen. Adrenal medulla autografts in nonhuman primates: Effects on the host brain, (manuscript in preparation).

10. Hansen, J.H., J. Kordower, M. Fiandaca, S.H. Okawara and D.M. Gash. Adrenal medulla autografts in nonhuman primates: Fine structural analysis. (manuscript in preparation).

23

PRELIMINARY EVALUATION AND REPORT ON HUMAN ADRENAL MEDULLARY GRAFTING

Roy A. E. Bakay

Chief of Neurosurgery, V.A. Hospital, Decatur, Georgia
and Assistant Professor of Medicine Section of Neurosurgery
Department of Surgery Emory
University School of Medicine Atlanta, Georgia

"Brain Transplantation" is an area of high interest and heightened activity by neuroscientists, neurosurgeons, and neurologists. It has been called the "ultimate transplant" which reflects its perceived importance and utility. This is, in fact, a misnomer since the brain is not actually being transplanted but rather small fragments of neural tissue are being grafted into the brain. A specific tissue is being utilized to correct a specific deficit. Thus, for Parkinson's disease, dopamine producing tissue is placed in the striatum. While the procedure is still capable of producing front page news, the technical aspects of the surgery are embarrassingly simple. Most of the sophistication resides in the knowing of who should have surgery, when in the course of the disease surgery should be performed, what tissue should be used, where the tissue should be placed, and how best to place the tissue. Incredibly, none of these issues have been totally resolved.

WHO SHOULD HAVE THE SURGERY - YOUNG VS. OLD?

Age is frequently used as a criterion to restrict surgery. Certainly, younger subjects, whether they be human or lower mammals, demonstrate variability both in response to injury and in potential for recovery simply based on age. Early studies demonstrated that adrenal medullary tissue grafted into young rats reversed behavioral deficits whereas similar grafts were ineffective in older rats (1). Subsequent studies have demonstrated that it is possible for older rats to make a recovery through refinements of technique (2,3). Nonhuman primate studies similarly suggest that younger Parkinson-like monkeys recover to a greater extent than aged monkeys. If the age of human subjects which have undergone grafting procedures for Parkinson's disease is considered in relation to degree of improvement, there appears to be a suggestion of greater benefit for younger patients (4-12). But does this mean that only younger patients should be investigated at this time? I think not. Parkinson's is basically a disease of older persons and is uncommon in people less than fifty. Furthermore, a modest improvement in an older patient may be as important or more important than the marked improvements seen in a younger patient, for neither are going to be cured and neither are going to be able to enjoy the quality of life they had prior to their disease. At this point age alone as a factor should not be a criterion for exclusion from surgery. A careful analysis of the relative degree of improvement related to age is still needed. Secondly, the age of the patient does not necessarily speak to the duration or severity of the disease.

WHEN SHOULD SURGERY BE PERFORMED - MODERATE VS. SEVERE DISABILITY?

Severity of the disease is frequently used as a restrictive criterion for surgery. Certainly, from animal studies in both rats and nonhuman primates, modestly affected individuals do far better than those that have severe end-stage symptomatology. Careful scrutiny of the human Parkinson grafting reports to date would suggest the same. I think that severity of the disease should be used as part of the restrictive criteria for surgery since end-stage disease of any type is least responsive to any treatment. A Hoehn and Yahr stage "5" and a Schwab and England activity scale of 20% or less suggest a degree of disability which will be unresponsive to any therapy. Information to date also suggest this class of patients is subject to the greatest postoperative cognitive and psychiatric problems. Furthermore, with end-stage disease comes a variety of other systemic failures. The mortality and morbidity which this operation generates have been most notable in severely affected patients, especially those with multiple system disease. In our investigations we eliminate those patients who are so severely affected as to be unable to care for themselves at all. Those with swallowing difficulties represent an additional exclusion category, since these are most likely to require intubation postoperatively and most likely to develop aspiration pneumonia. In addition, no patients with major systemic disease (i.e., previous history of myocardial infarction, severe COPD, etc.) are included in our initial study.

WHAT TISSUE SHOULD BE USED - FETAL VS. ADRENAL?

The most important question is which tissue to use. For after this question is resolved, the others tend to follow. As a substitute for the missing dopaminergic neurons of the substantia nigra, the obvious first and best choice is fetal tissue of the substantia nigra. There is an intrinsic beauty in replacing damaged neurons with functional neurons of the exact same type. In Parkinsonism, fetal mesencephalic transplantation has the theoretical advantage that the precise neurons which are required for reversal of the neurological deficit can be integrated into the host. Mesencephalic grafts have demonstrated neurons with spontaneous firing activity at rates which are indistinguishable from normal nigral neurons, and which are capable of sustaining intracranial self-stimulation, suggesting that the transplanted neurons, under certain circumstances, axonally convey specific, temporally-organized information to the reinnervated striatum (3,5). Fetal mesencephalic tissue from humans is readily obtainable and has been used to reverse neurological deficits in rats (13). Because this tissue would come from genetically dissimilar individuals of the same species, they would be allografts and therefore subject to potential rejection and worsening of neurological deficits. Although there are clinical means of decreasing the risk of such rejection, the transplanted patients would be at risk for the remainder of their lives and in Parkinsonian patients this may be quite a number of years. In addition, at least in the United States, there are a number of ethical and legal problems to be resolved before such tissue could be utilized.

Adrenal medullary transplantation has the advantage of utilizing an autograft, with a much lower risk of rejection and adverse immunological reactions. The disadvantage is that this tissue although neuronal in origin does not contain functional neurons. Although differentiation to a "neuronal" phenotype with extension of fiber process is occasionally observed, these tissues are not integrated into the brain and simply serve as a "dopamine infusion pump" (1,3,5). There is some suggestion that neurotrophic factors may also be released by the adrenal medullary grafts (14). It may be that this factor causes stimulation of the host brain thus promoting recovery by enhanced dopaminergic neuron re-innervation of the striatum. Whatever the mechanism of action adrenal medullary grafts do not work as well as fetal mesencephalic grafts in correcting turning behavior in the rat, especially in older animals (1,3,5). Adrenal medullary grafts in nonhuman primates have a poor survival rate and very little differentiation compared to the rats (15,16).

Erik-Olof Backlund, working with Lars Olson, began human brain grafting experiments at the Karolinska Hospital in 1982. The first patient was a 55 year old man with severe Parkinsonism, who no longer responded adequately to drug therapy (4-6). Using a stereotactic technique, small fragments of adrenal medullary tissue were placed in the head of the caudate nucleus. Minor improvements were observed during the first week after the operation. However the patient subsequently returned to his debilitated state and remained so since that time. The second patient, a 46 year old woman was operated on five months later in a similar fashion. She also was severely affected and demonstrated some degree of improvement lasting only four months before returning to her previous debilitated state. These two patients have progressively worsened over the years suggesting that the grafts failed to alter the progression of the course of the disease.

Two years passed before performance of a second group of operations. The third and fourth patients, 2 men age 65 and 47, differed significantly from the first two patients in that they continued to have some control of their Parkinsonian symptoms by drug therapy. In addition, the target was different with two separate grafts made into the patient's putamen. Immediately postoperatively, both patients experienced a decreased dependency on medication, as well as a decrease in the symptomatology. Unfortunately the improvement was only for a matter of a few months.

These results coupled with additional negative animal laboratory findings appeared to end the human adrenal medullary transplantation studies. It is against this background that the neuroscientific community was surprised by Iagnacio Madrazo and colleagues in Mexico with report of dramatic and persistent bilateral improvement in patients using adrenal medullary transplantation to the right caudate nucleus (7-9). There were, however, a number of significant differences between the Swedish study and the study from Mexico. Madrazo's patients were much younger, very early in the course of their Parkinsonism, and the grafted tissue was placed through a ventricular approach rather than being placed in an intraparenchymal position. The intraventricular placement allows cerebrospinal fluid to bathe the graft. This intraventricular graft site appears to be far superior to intraparenchymal graft sites based on a number of studies in the rat and a few studies in the monkey. These results have provided stimulus to return to the lab and reinvestigate adrenal medullary tissue grafting and to expand the number of clinical programs investigating both intraventricular and intraparenchymal medullary transplantation techniques. It will probably be several years before the efficacy of this particular strategy is known.

Incredibly, the question of fetal versus adrenal tissue for transplantation has already reached human investigational levels (10). I can remember in 1985, after presenting work on fetal transplantation in Parkinsonian monkeys, being asked by a reporter when the same technique would be used in humans, and replying "in 3-5 years, but it will be done in this country first". I did not realize at that time how prophetic that statement would be. In every animal model studied, fetal tissue is superior to adrenal tissue in improving behavioral performance. The major questions, however, remain to be resolved in terms of the implications of immunological rejection and potential for immunological damage which could extend far beyond the graft area itself. Our investigations of allografts in nonhuman primates failed to demonstrate any immunological rejection (17), but we are aware of at least one report where immunological rejection was demonstrated and resulted in loss of behavioral improvement. Use of immunological suppression therapy will fail to prevent such functional damage. I strongly believe that fetal transplantation should not be employed until an adequate and thorough investigation of its immunological potential has been performed in nonhuman primates. Moral and legal questions also need time to be resolved. Clinically, all of our investigations are with adrenal medullary tissue. There is reason to believe that adrenal tissue does provide some benefit to Parkinson patients.

WHERE SHOULD THE TISSUE BE PLACED - CAUDATE VS. PUTAMEN?

There is still a raging debate as to whether the optimal target is actually the caudate or the putamen. Biochemical and electrophysiologic studies would suggest on at least a theoretical basis that the putamen should be the target. Animal transplantation studies have been predominantly grafting to what would be the anatomic equivalent of the caudate and these have proved quite successful in improving a number of behaviors (1-3). The use of multiple targets may be the most efficacious (18). Both the caudate and putamen have been used as separate targets for adrenal medullary grafts in man with some success (4, 8, 19). The target selection, however, is in large part restricted by the technique utilized for implantation which brings us to our next question.

HOW BEST TO PLACE THE TISSUE - OPEN VS. STEREOTACTIC SURGERY?

This is really a question of whether intraventricular techniques or intraparenchymal techniques should be used. Initial animal studies were all performed with open techniques, especially those using delayed cortical windows or intraventricular placement of grafts. Stereotactic techniques were developed later and were less satisfactory until the technology improved significantly (3,5). Stereotactic techniques, of course, have the advantage of being able to place tissue in multiple targets. The human investigations initiated in Sweden using stereotactic techniques failed to produce long-term significant behavioral changes. This is in marked contrast to the reports from Mexico. Reportedly, the Chinese have been quite successful in stereotactically placing tissue in young Parkinsonian patients and demonstrating improvement, but the details are quite sketchy and assessment at this time is most difficult (12). There are other institutions (including some in this country) which are exploring stereotactic techniques in Parkinson patients. I think it is important to follow-up on the observations from Mexico and define the efficacy and safety of this procedure before proceeding with alternative surgical techniques. Stereotaxis is the future of neural transplantation. We are investigating techniques for improving survival of grafts in nonhuman primates using stereotactic techniques and in the future are quite hopeful that the efficacy of such technology will be proven.

CONCLUSION

In summary, we are attempting to evaluate the open surgical technique of transcortical transventricular placement of adrenal medullary tissue into the right caudate of patients with moderately severe Parkinson's disease who have no other major systemic disease. It is too early to discuss specific details of our results or those of other investigators in this country. I think, however, there are some clear messages that can be given at this point. The first is, this is major surgery with a potential for mortality and major morbidity. Disasters have occurred in this country as well as internationally. Proper patient selection may help prevent many of these complications in the future.

Early results of our investigation and others in this country are encouraging, but the improvement appears to be modest at best. It is not known if this modest increase in activity and reduced requirement for replacement therapy medication is of lasting value. Follow-up on these patients is entirely too short to determine whether or not a diminution of improvement over time occurs as has been reported in nonhuman primates (16).

The future for neural tissue grafting is unquestionably bright. One can envision in the future the stereotactic placement of one's own cells which have been genetically engineered to produce dopamine or neurotrophic factors implanted into the brain to produce desired behavioral improvements. What we must guard against at this point is over-optimism and unrealistic expectations. There are many years of research ahead of us and while it may be easy to be the first, it is much more difficult to develop the techniques which are the best.

REFERENCES

1. Freed, W.J.: Functional Brain Tissue Transplantation: reversal of lesion-induced rotation by intraventricular substantia nigra and adrenal medulla grafts, with a note on intracranial retinal grafts. Biological Psychiatry, 18: 1205-1267, 1983.

2. Stromberg, I., Herrera-Marchitz, M., Ungerstedt, U., Ebendal, T., Olson, L.: Chronic implants of chromaffin tissue into the dopamine-denervated striatum. Effects of NGF on graft survival, fiber growth and rotational behavior. Exp Brain Res. 60: 335-349, 1985.

3. Perlow, M.J.: Brain grafting, in Koller, W.C. (Ed): Handbook of Parkinson's Disease, Marcel Dekker, Inc., 1987, pp 437-454.

4. Backlund, E-O., Granberg, P.O., Hamberger, B., Knutsson, E., Martenson, A., Sedvall, G., Seiger, A., Olson, L.: Transplantation of adrenal medullary tissue to striatum in parkinsonism. First clinical trials. J Neurosurg. 62: 169-173, 1985.

5. Olson L., Backlund, E-O., Gerhardt, G., Hoffer, B., Lindvall, O., Rose, G., Seiger, A., Stromberg, I.: Nigral and adrenal grafts in parkinsonism: recent basic and clinical studies, in Yahr, M.D., Bargmann, K.J. (Eds): Advances in Neurology, 1986, 45: 85-94.

6. Lindvall, O., Olson, L., Seiger, A., Backlund, E-O.: Towards a transplantation therapy in parkinson's disease: a progress report from ongoing clinical experiments. II. Neurological Assessment, in Azmitia, E., Bjorklund, A. (Eds): Cell and Tissue Transplantation into the Adult Brain. New York, New York Academy of Sciences, 1986, pp 568-673.

7. Madrazo, I., Drucker-Colin, R., Diaz, V., Martinez-Mata, J., Torres, C., Becerril, J.J.: Open microsurgical autograft of adrenal medulla to the right caudate nucleus in two patients with intractable parkinson's disease. N. Engl. J. Med. 316: 831-834, 1987.

8. Madrazo, I., Drucker-Colin, R., Leon, V., Torres, C.: Adrenal medulla transplanted to caudate nucleus for treatment of parkinson's disease: disease report of 10 cases. Surg. Forum, 38: 510-511, 1987.

9. Drucker-Colin, R., Madrazo, I., Shkurovich, M., Ostrosky-Solis, F., Torres, C.: Open microsurgical autograft of adrenal medulla to caudate nucleus of patients with parkinson's disease, in Schmitt Neurological Sciences Symposium, Transplantation into the Mammalian CNS, University of Rochester, Rochester, New York, June 30-July 3, 1987.

10. Madrazo, I., Leon, V., Torres, C., Aguilera, C., Varela, G., Alvarez, F., Fraga, A., Drucker-Colin, R., Ostrosky, F., Skurovich, M., Franco, R.: Transplantation of fetal substantia nigra and adrenal medulla to the caudate nucleus in two patients with parkinson's disease. N. Engl. J. Med. 318: 51, 1988.

11. Allen, G.S., Burns, R.S., Tilipan, N.B.: Human adrenal autografts as a potential therapy for parkinson's disease, in Schmitt Neurological Sciences Symposium, Transplantation into the Mammalian CNS, University of Rochester, Rochester, New York, June 30-July 3, 1987.

12. Jiao, S-S., Zhang, W.C., Ding, M.C., Sun, J.B.: The clinical study of adrenal medullary tissue transplantation to striatum in parkinsonism, in Schmitt Neurological Sciences Symposium, Transplantation into the Mammalian CNS, University of Rochester, Rochester, New York, June 30-July 3, 1987.

13. Brundin, P., Clarke, D.J., Strecker, R.E., Widner, H., Nilsson, O.G., Astedt, B., Lindvall, O., Bjorklund, A.: Experimental basis for clinical trials with dopamine neuron grafting in patients with parkinson's disease, in Schmitt Neurological Sciences Symposium, Transplantation into the Mammalian CNS. University of Rochester, Rochester, New York, June 30-July 3, 1987.

14. Bohn, M.C., Cupit, L., Marciano, F., Gash, D.M.: Adrenal medulla grafts enhance recovery of striatal dopaminergic fibers. Science 237: 913-916, 1987.

15. Morihisa, J.M., Nakamura, R.K., Freed, W.J., Mishkin, M., Wyatt, R.J.: Adrenal medulla grafts survive and exhibit catecholamine-specific fluorescence in the primate brain. Exp. Neurol. 84: 643-653, 1984.

16. Bankiewicz, K.S., Plunkett, R.J., Oldfield, E.H., Jacobowitz, D.M., Porrino, L.J., Vaidya, U., DiPorzio, U., Schuette, WH., Markowitz, A., London, W.T., Kopin, I.J., in Schmitt Neurological Sciences Symposium, Transplantation into the Mammalian CNS. University of Rochester, Rochester, New York, June 30-July 3, 1987.

17. Fiandaca, M.S., Bakay, R.A.E., Sweeney, K.M., Chan, W.C. (Emory University, Atlanta, Georgia, USA): Immunological response to intracerebral fetal neural allografts in the Rhesus monkey, in Schmitt Neurological Sciences Symposium, Transplantation into the Mammalian CNS, University of Rochester, Rochester, New York, June 30-July 3, 1987.

18. Lindvall, O., Backlund, E-O., Farde, L., Sedvall, G., Freedman, R., Hoffer, B., Nobin, A., Seiger, A., Olson, L.: Transplantation in parkinson's disease: two cases of adrenal medullary grafts to the putamen. Ann. of Neurol. 22: 457-468, 1987.

19. Dunnett, S.B., Bjorklund, A., Gage, F.H., Stenevi, U.: Transplantation of mesencephalic dopamine neurons to the striatum of adult rats, in Bjorklund A, Stenevi U (Eds): Neural Grafting in the Mammalian CNS, Elsevier Science Publishers, Amsterdam, 1985, pp 451-469.

24

BEHAVIORIAL PERFORMANCE IMPROVES AFTER FETAL SUBSTANTIA
NIGRA TRANSPLANT IN BONNET MONKEYS WITH
MPTP INDUCED PARKINSONISM

C. R. Freed*, J. B. Richards, and M. L. Reite

Departments of Medicine, Pharmacology, and Psychiatry
University of Colorado Health Sciences Center
Denver, Colorado 80262   U.S.A.

INTRODUCTION

Nearly 10 years have elapsed since the first reports in rats that transplantation of fetal dopamine neurons have an influence on motor behavior in animals with a unilateral lesion of the nigrostriatal tract (Bjorklund and Stenevi, 1979; Perlow et al., 1979). These experiments were important in establishing several basic principles for brain transplantation. First it is a good experimental model. The unilaterally lesioned rat which circles in response to amphetamine or apomorphine as described by Anden et al. (1966) and Ungerstedt (1969) has been useful for studying transplant effects. Second was the discovery that only fetal tissue of a narrow developmental range was useful for transplant. Only the dopamine cells from ventral mesencephalon of fetal rat embryonic day 13 to 15 could successfully reinnervate dopamine denervated striatum. In this animal model, the transplants produced changes in circling behavior within three to six weeks of transplant (Bjorklund and Stenevi, 1979; Perlow et al., 1979; Dunnett et al, 1984). With amphetamine treatment, animals which formerly circled ipsilateral to the side of the dopamine denervation lesion may show a reversal in their circling direction to circle contralateral to the initial lesion. Successful transplants have dopamine cells which survive along the transplant track. In addition, dopamine concentration increases from near total depletion to about 10% of the normal intrinsic dopamine concentration (Schmidt et al., 1984).

While these experiments in the rat have been useful for showing that tissue transplantation can have behavioral effects, the rat model for brain dopamine deficiency is quite different from human Parkinsonism and therefore may not predict the value of transplantation for treating human Parkinson's disease. The recent discovery of the neurotoxin n-methylphenyl-tetrahydropyridine (MPTP) by Davis et al. (1979) and the demonstration that this neurotoxin will produce a Parkinsonian syndrome in non-human primates by Burns et al. (1983) has led several investigators to attempt to treat this syndrome with transplantation of adrenal medulla and fetal substantia nigra.

After Bakay and coworkers (1985) first described fetal transplants into Parkinsonian monkeys, other laboratories have performed transplants of fetal tissue into MPTP lesioned monkeys (Redmond et al., 1986; Freed et al., 1986; Bankiewicz et al., 1987). There is still considerable controversy about the correct fetal age for transplant, the nature of controls, and the evaluation of improvement. A major problem with the MPTP model is that animals spontaneously recover from the behavioral deficit over weeks to months despite persistent dopamine depletion (Eidelbert et al., 1986). Interpretation of histology is complicated by the fact that the monkey has tryosine hydroxylase immunoreactive cells in caudate/putamen even in the absence of nigral transplants (Dubach et al. 1986). For these reasons there has been no report of behavioral improvement following transplant that can be definitely related to the transplant itself.

Rejection of transplanted tissue has been largely ignored in the experiments on monkeys primarily because rejection has not been a problem in rat transplants (Low et al., 1985). The brain is a relatively privileged site immunologically. When transplantation has been done between species, immunosuppression with cyclosporine A has been shown to be of value (Brundin et al., 1985). Despite the success of previous transplants within species without immunosuppression, we believe we have seen transplant rejection in one monkey following a second tissue transplant seven months after a successful first transplant.

Methods

Adult male Bonnet monkeys (Macaca radiata) were food deprived to 90% of initial weight and were chair trained to pull levers with the left and right arms for a banana pellet reward. Training was under the control of a Commodore 64 computer which also monitored the pull rate of the animals. Following several months of training, animals were lesioned with injections of MPTP hydrochloride 0.5 mg/kg i.m. for five doses given every other day. After lesioning with MPTP, most animals were followed for at least six months to measure the rate of spontaneous improvement.

Animals were transplanted with substantia nigra from embryonic day 15 rat fetus or from monkey fetus 18, 45, or 150 mm crown rump length. We found that ultrasound was the most satisfactory method for accurately sizing monkey fetus. Fetuses as small as 15 mm can be detected by this technique. Tissue was obtained by caesarean section and the region of the ventral mesencephalon was dissected and disrupted by passage through a 21 gauge needle. Transplants were performed unilaterally into caudate, putamen, and globus pallidus. The side was randomly chosen. Five cannula passes were made with four droplets of tissue along each track. Each transplant droplet was two microliters of substantia nigra suspension. Recovery of animals was measured by behavioral testing performed three times per week and by monthly administration of amphetamine 0.5 mg/kg to test for circling behavior induced by the transplant.

Results

Treatment with MPTP stopped lever pulling in all animals. Performance resumed in three of five animals after one to two months but at a much reduced rate. A sixth animal was treated with this regimen but had to be sacrificed because of a persistent downhill course. Another severely affected animal was supported with intensive nursing care for two months after receiving MPTP injections. This animal was transplanted and subsequently improved. Hand feeding was required for the sickest animals and an air mattress was used to reduce the severity of decubitus ulcers.

Because some animals did not resume the pull task after MPTP, we devised a simpler task in which animals were timed while reaching for pellets presented by the experimenter. Timing this task gave a measure of the severity of the Parkinson syndrome.

All animals which were not killed by the first course of MPTP showed spontaneous recovery. Recovery to normal occurred in three of the five animals so that additional courses of MPTP were required. One animal required four courses of MPTP to establish a stable Parkinsonian syndrome. Recovery took place over months.

Because of the many uncertainties associated with this experiment, animals were followed for long periods and each received a different kind of tissue transplant. The slope of behavioral improvement was measured for each animal for months before and after transplant. Significant improvement was defined as a change in the slope of spontaneous improvement following the transplant. Behavioral results are individually summarized below.

One animal required two courses of MPTP over a two month period to become Parkinsonian. The animal's clinical state was then followed for six months to establish a rate of spontaneous improvement. Following transplantation of fetal substantia nigra from a 45 mm crown rump length monkey embryo into left caudate, there was a change in the slope of the behavioral improvement curve with greater improvement shown on the right than on the left although both sides showed some improvement. The changes in the slope were significant bilaterally. Paired comparisons of right and left arm performance before and after transplant showed significantly better right arm performance than left arm prformance only in the post transplant period.

The animal was sacrificed six months after transplant and a few tyrosine hydroxylase positive neurons were seen in transplant tracks in left caudate/putamen. Dopamine assays were done which showed up to 99.7% depletion in non transplanted caudat/putamen. On the transplanted side, one micropunched track showed a seven-fold increase in dopamine concentration, although this increase restored levels of dopamine to only 2% of normal.

Another animal required four courses of MPTP before establishing a stable Parkinsonian syndrome. After the final course, the animal could not maintain a posture and had loss of normal hand function with a flexed wrist with extended fingers as occasionally seen in Parkinson patients. Because of his downhill course, this animal was transplanted earlier than other animals in this series. He received nigral tissue from an 18 mm crown rump length monkey fetus six weeks after his last course of MPTP. Following transplant, improvement was gradual first appearing at six weeks and reaching maximum about 12 to 16 weeks. Left hand performance improved with return of normal grasp. The left hand became the preferred hand.

Despite the improvement in left hand function and in maintenance of posture and ambulation, the aminal still remained severely Parkinsonian. Therefore, the animal underwent a seond trasplant on the other side of brain seven months after the first. Left caudate/putamen was transplanted with tissue from an 18 mm crown-rump length fetus. The second fetus was related to the first, having the same father but a different mother. The procedure was uncomplicated and the animal was in his preoperative condition on the morning after surgery. Six days after surgery, the animal showed a dramatic loss of function of the previously improved left hand and left leg. He lost normal grasp with the left hand. Right hand performance was unchanged despite the recent transplant into left brain.

We assumed that the second transplant was being rejected and so attempted to abort the rejection with prednisone and cyclosporine A therapy. This treatment was continued for 10 days without clinical change. The drug was then discontinued for five days to look for further clinical deterioration. There was no further change in behavior. The animal was sacrificed and substantia nigra tissue processed for tyrosine hydroxylase immunoreactivity using Eugene Tech antibody. Results showed that very large numbers of tyrosine hydroxylase positive cells were seen along transplant tracks in caudate, putamen, and globus pallidus on the right side of brain. After hemotoxylin staining, lymphocytes and oligodendrogliocytes were also seen in the transplant tracks. The new transplant tracks on the left side of brain showed fresh debris with rare tyrosine hydroxylase positive cells seen. Serum from this animal before and after the second transplant was tested for antibodies to lymphocytes from the parents of the fetal donor. No antibodies were detected. These data suggest that cell mediated immunity may have destroyed the previously functioning transplant in right caudate/putamen.

Transplants were also performed on three other animals. One received nigra from a 46 mm crown rump length monkey embryo and also showed behavioral improvement although the improvement rate post transplant was not much greater than the spontaneous improvement rate pretransplant. Histologic analysis of the transplant showed fewer cells than seen with the 18 mm embryo. An animal receiving fetal tissue from a late gestational fetus measuring 150 mm crown rump length had essentially no nigral cell survival. As with the previously described animal, the rate of improvement post transplant could not be distinguished from the improvement rate occurring pretransplant. A monkey receiving a transplant of pooled nigra from six rat embryos from embryonic day 15 was sacrificed one month after transplant. The monkey had been treated with cyclosporine A and prednisone, however at autopsy the brain showed few tyrosine hydroxylase positive cells.

Discussion

Our results show that transplants of cell suspensions of fetal substantia nigra tissue from first trimester monkey fetus (particularly 18 mm crown rump length) can produce improvement in behavioral performance which is associated with surviving tyrosine hydroxy-lase positive cells in caudate and putamen on the transplanted side. Transplanted cells were always found in close association with the transplant track. Recovery was delayed but progressive starting about two months after transplant and continuing for up to eight months.

Survival of cells from 18 mm fetus was much better than from 45 mm fetus. Tissues from older, near term monkey fetus (150 mm crown-rump length) or from rats showed very little limited cell survival and were ineffective at producing significant behavioral improvement. Because all animals with a moderate Parkinsonian syndrome showed behavioral improvement over months, it was only possible to assess a positive transplant effect by comparing the rate of spontaneous improvement from months of baseline recording to the rate of improvement following transplantation.

Rejection of a transplant may have occurred in one animal in this series. The animal received its first transplant from ventral mesencephalon of an 18 mm crown-rump length embryo. That transplant led to excellent behavioral improvement in the limb contralateral to the transplant and behavior was stable for seven months. Six days following transplantation of a second graft on the opposite side of brain, the animal lost function of the previously improved arm and leg. While loss of function is an unfortunate complication of a transplant, in this experimental circumstance it offered dramatic evidence that the first transplant had a definite effect on motor performance. Deliberate induction of rejection may be a useful tool for proving transplant function in this difficult research area.

References

Anden, N.E., Dahlstrom, A., Fuxe, K., and Larsson, K. (1966) Functional role of the nigro-neostriatal dopamine neurons. Acta Pharmacol. et Toxicol. 24:263-274.

Bakay, R.A.E., Barrow, D.L., Schiff, A., and Fiandanca, M.S. (1985) Biochemical and behavioral corrections of MPTP Parkinson-like syndrome by fetal cell transplantation. Appl. Neurophysiology 48:358-361, 1985.

Bankiweicz, K.S., Jacobowitz, D.M., Plunkett, R.J., Oldfield, E.H., and Kopin, I.J. (1987) Injury induced sprouting into the caudate nucleus after solid tissue implantation in MPTP-induced Parkinsonian monkeys. Society for Neuroscience Abstracts, Abstract 46.16.

Björklund, A. and Stenevi, U. (1979) Reconstruction of the nigrostriatal dopamine pathway by intracerebral nigral transplants. Brain Res. 177:555-560.

Brundin, P., Nilsson, O.G., Gage, F.H., and Björklund, A. (1985) Cyclosporin A increases survival of cross-species intrastriatal grafts of embryonic dopamine-containing neurons. Exp. Brain Res. 60:204-208.

Burns, R.S., Chiueh, C.C., Markey, S.P., Ebert, M.H., Jacobowitz, D.M., and Kopin, I.J. (1983) A primate model of parkinsonism. Selective destruction of dopaminergic neurons in the pars compacta of the substantia nigra by N-methyl-4 phenyl-1,2,3,6-tetrahydropyridine. Proc. Natl. Acad. Sci. USA 80:4546-4550.

Davis, G.C., Williams, A.C., Markey, S.P., Ebert, M.H., Caine, E.D., Reichert, C.M., and Kopin, I.J. (1979) Chronic Parkinsonism secondary to intravenous injection of meperidine analogues. Psychiatry Res. 1:249-254.

Dubach, M., Schmidt, R., Bowden, D.M., Kunkel, D., Martin, R., and German, D. (1986) Neurons containing tyrosine hydroxylase-like immunoreactivity in the caudate and putamen of a nonhuman primate. Soc. for Neurosci. Abstracts 12:1327, abstract 362.6.

Dunnett, S.B., Björklund, A., Schmidt, R.H., Stenevi, U., and Iversen, S.D (1984) Intracerebral grafting of neronal cell suspensions, V. Behavioral recovery in rats with unilateral 6-OHDA lesions following implantation of nigral cell suspensions in different forebrain sites. Acta Physiologica Scandinavica, Suppl. 522:29-37

Eidelberg, E., Brook, B.A., Morgan, W.W., Walden, J.G., and Kokemoor, R.H. (1986) Variability and functional recovery in the N-methyl-1-4-phenyl-1,2,3,6-tetrahydropyridine model of Parkinsonism in monkeys. Neurosci. 18:817-822.

Freed, C.R., Richards, J.B., Alianiello, E., Peterson, R., Ruppe, L., Singh, S. and Reite, M. (1986) Fetal dopamine cell transplantation as a treatment for Parkinson's syndrome in bonnet monkeys. Soc. Neurosci. Abstracts 12:1476, abstract 397.6.

Low, W.C., Daniloff, J.K., Bodony, R.P., and Wells, J. (1985) Cross-species transplants of cholinergic neurons and the recovery of function. In Neural Grafting in the Mannalian CNS. (eds. A. Björklund and U. Stenevi). Elsevier, Amsterdam.

Perlow, M.J., Freed, W.J., Hoffer, B.J., Seiger, A., Olson, L., and Wyatt, R.J. (1979) Brain grafts reduce motor abnormalities produced by destruction of nigrostriatal dopamine system. Science 204:643-647.

Redmond, D.E., Jr., Sladek, J.R., Jr., Roth, R.H., Collier, T.J., Elworth, J.D., Deutch, A.Y., and Haber, S. (1986) Fetal neuronal grafts in monkeys given methylphenyltetrahydropyridine. The Lancet, May 17, No. 8490; 11-25-1127.

Schmidt, R.H., Björklund, A., Stenevi, U., Dunnett, S.B., and Gage, F.H. (1984) Intracerebral grafting of neuronal cell suspensions, III. Activity of intrastriatal nigral suspension implants as assessed by measurements of dopamine synthesis and metabolism. Acta Physiologica Scandinavica, Suppl. 522:29-38.

Ungerstedt U., Butcher, L.L., Butcher, S.G., Anden, N.-E., and Fuxe, K. (1969) Direct chemical stimulation of dopaminergic mechanisms in the neostriatum of the rat. Brain Res. 14:461-471.

25

MECHANISMS OF ACTION OF SUBSTANTIA NIGRA
AND ADRENAL MEDULLA GRAFTS

William J. Freed[1] Jill B. Becker[2]

[1]National Institute of Mental Health,
Saint Elizabeths Hospital, Washington, D.C. 20032
[2]The University of Michigan, Department of Psychology and
Neuroscience Program Ann Arbor, Michigan 48104-1687

Loss of the dopaminergic neurons of the substantia nigra (SN) and concomitant degeneration of the nigrostriatal dopamine system has been linked to the clinical syndrome of Parkinson's disease (Bernheimer et al., 1973). Although brain tissue transplantation techniques have now been applied to several experimental models, it is the nigrostriatal system that has been the most extensively studied. Two brain grafting techniques have been found to produce a long-term alleviation of consequences of SN lesions in experimental models of Parkinson's disease: intracerebral adrenal medulla grafts, and embryonic SN grafts. The two techniques appear to operate through different mechanisms. SN grafts produce dopamine-containing neurites which reinnervate the striatum. Adrenal medulla grafts, on the other hand, do not significantly reinnervate the striatum.

Both of these brain transplantation procedures can reduce the behavioral consequences of SN lesions, but neither procedure is completely effective. The efficacy of adrenal medulla grafts is limited, and the mechanism of action of these grafts is, at present, unknown. It has been suggested that adrenal medulla grafts act by secreting catecholamines and possibly other substances which enter into the host brain through diffusion. Evidence for this and other possible mechanisms of adrenal medulla graft function are discussed below. The limited efficacy of SN grafts is probably related to the limited ability of neurites from SN grafts to penetrate into the host corpus striatum. Axons and dendrites from SN grafts enter the striatum (Perlow et al., 1979; Freed et al., 1980; Bjorklund et al., 1980; Dunnett et al., 1983a,b), reduce denervation supersensitivity (Freed et al., 1983a) and form synaptic contact with host neurons in parts of the striatum immediately adjacent to the grafts (Freund et al., 1985; Mahalik et al., 1985). Even nearly two years after transplantation, however, the penetration of graft-derived neurites into the host striatum is limited to about 1.5 mm (Freed, 1983; Freed et al., 1983a). A "hyperinnervation" of SN grafts in the lateral ventricle by catecholamine-containing neurites derived from the grafts themselves has also been noted (Perlow et al., 1979; Freed et al., 1980), suggesting that graft derived neurites partially fail to enter the host brain. In order to improve the performance of SN grafts we have, therefore, undertaken a series of experiments aimed at elucidating the factors which control reinnervation of the dopamine-denervated striatum by intraventricular SN grafts.

MATURITY OF THE TARGET TISSUE

Several studies suggest that there are factors in the embryonic or immature corpus striatum which can promote the survival and extension of neurites by developing SN neurons (Hemmendinger et al., 1981; Denis-Donini et al., 1983). It is quite possible, therefore, that the absence of these substances in the mature brain accounts for the limitations of the ability of SN grafts to innervate the striatum. In order to investigate the ability of intraventricular SN grafts to innervate immature striatal target tissue, dual grafts of SN and striatum from 17-day gestational rat embryos were transplanted together into the lateral ventricles of animals with unilateral SN lesions (de Beaurepaire and Freed, 1987). The SN grafts were found to completely reinnervate the embryonic striatal grafts. On the other hand, little or no reinnervation of the host striatum was observed in animals with combined intraventricular SN and striatal grafts, and in these animals the graft was not hyperinnervated by catecholaminergic neurites. In control animals that received intraventricular SN grafts without a concomitant striatal graft, a typical limited reinnervation of the host brain and hyperinnervation of the graft were invariably observed.

Thus the capacity of embryonic brain grafts to innervate embryonic striatum is not limited. When presented with a "choice" between an embryonic and a fully mature but denervated striatal target, grafted SN dopaminergic neurons exclusively innervate the grafted embryonic target, and do not produce the typical innervation of the dorso-medial striatum.

A means of exploiting the favorable neurite-promoting properties of embryonic target tissue in a behavioral model was therefore sought. It is difficult to create a situation in a developmentally immature animal where a brain graft would be employed to ameliorate the effects of a lesion. Instead, therefore, the usual experimental paradigm was reversed so that normal rats received embryonic SN grafts or sciatic nerve control grafts on the first day after birth, and subsequently received bilateral SN lesions (Schwarz and Freed, 1987). Reinnervation of the striatum as well as catecholaminergic innervation of the septum was observed. Behaviorally, the neonatal grafts were very effective. The incidence of aphagia and adipsia was markedly reduced in the animals with SN grafts, and the akinesia and rigidity were reduced as well. Animals with SN grafts were 3.7 times as active as control animals. Most of the animals with SN grafts did not become aphagic and adipsic following bilateral SN lesions, and maintained relatively normal eating, drinking, and body weight for at least one month after lesioning. Rigidity was also noticeably reduced.

Thus it appears that when SN grafts are implanted into neonatal animals, they can successfully compete with the endogenous striatal dopaminergic innervation. Whether some of the reinnervation of the striatum produced by these neonatal grafts occurs subsequent to the SN lesions has not been entirely ruled out. Increased eating was, however, observed as soon as two days after lesioning in the animals with SN grafts, suggesting that a major part of the reinnervation of the striatum produced by these grafts was already in place at the time of lesioning. It is clear, however, that the behavioral efficacy of SN grafts is markedly increased when implanted neonatally. Neonatally-implanted SN grafts protect animals from the aphagia, adipsia, and akinesia following subsequent SN lesions which produce severe debilitation in control rats.

CORTICAL LESIONS

Several studies suggest that competition between more than one system of neuronal fibers for a single terminal region can restrict the growth of either system, as compared to a comparable situation where only one fiber system is present (Schneider, 1973; Leong and Lund, 1973; Olson et al., 1978). When an embryonic SN graft is transplanted to the lateral ventricle, dopamine-containing neurites penetrate for short distances into the host striatum. In these experiments, the SN lesions are usually made long before the SN grafts are implanted. During this interval, synapses which had been occupied by dopaminergic terminals would either have become filled by other competing axonal systems derived from interneurons or other inputs to the striatum, or have been blocked by glia. One of the largest afferents to the striatum is the glutamate containing corticostriatal pathway (McGeer et al., 1977; Webster, 1961). It was therefore hypothesized that lesioning of the overlying cerebral cortex would remove a competing innervation of the striatum and allow for an increased penetration of graft-derived neurites into the host brain. On the other hand, substances which promote the survival of neurons in tissue culture, as well as after transplantation to the brain, are secreted from injured brain tissue (Nieto-Sampedro et al., 1982, 1983). In any study involving brain injury, the possibility of stimulatory effects of the brain injury due to secretion of trophic substances should also be considered.

In order to evaluate the effects of cortical lesions on SN grafts, SN grafts were implanted ten days after the production of cortical lesions or sham operations (Freed and Cannon-Spoor, 1988). Because of the localized nature of the brain injury model that was employed, reinnervation of the striatum was evaluated in four separate regions, corresponding to the most dorsal one-fourth of the lateral ventricle (region "a"), the second and third most dorsal regions ("b" and "c") and the most ventral one-fourth of the ventricle ("d"). In the most dorsal sector ("a") mean dopaminergic neurite ingrowth was increased in animals with cortical lesions from $0.52 + 0.14$ mm to $0.78 + 0.14$ mm (p<0.01). There were no significant differences in sectors "b", "c", or "d".

In several of the animals with cortical lesions the reinnervation of the caudate-putamen was found to closely follow the roof of the caudate putamen, directly underneath the cortical lesion. In some cases, the lateral extent of the reinnervation corresponded closely to the lateral extent of the cortical lesion. In other animals, clumps of catecholamine-containing neurites were observed in areas of the caudate-putamen close to the cortical lesions.

Injury of the cerebral cortex therefore stimulates the growth of neurites into the denervated striatum from intraventricular SN grafts. Because of the localized nature of the increased neurite growth, it does not appear that this effect is due to the removal of competing corticostriatal pathway. The topography of this pathway has been examined in rats in detail (Webster, 1961) and the lesions which were employed would be expected to remove the corticostriatal innervation from the medial striatum extending at least through the "a", "b", and "c" sectors. It appeared that cortical lesions influenced the dopaminergic neurites, rather than cell parenchyma, because the differences were seen only in the parts of the striatum physically closest to the lesion. An effect on the cell bodies of the grafted neurons would be expected to be manifest equally in all four sectors.

Examination of published photographs of growth of dopaminergic neurites from cortical cavities suggests that reinnervation of the striatum using this procedure somewhat exceeds that which occurs when similar grafts are implanted into the lateral ventricle (Bjorklund et al., 1980). The present study suggests that some of this difference could be due to direct effects of the lesions on dopaminergic neurites.

ADRENAL MEDULLA GRAFTS

The chromaffin cells of the adrenal medulla have several unusual properties which suggest their possible use for transplantation to the brain as a substitute for embryonic SN. First, these cells produce catecholamines, including dopamine (Snyder et al., 1977), and can release dopamine (Lishajko, 1970). Depending upon their environment, the morphological and biochemical properties of these cells can change: when chromaffin cells are removed from proximity to the adrenal cortex and are no longer in contact with high corticosteroid concentrations, synthesis of epinephrine from norepinephrine is decreased (Wurtman et al., 1972), and these cells develop neurite-like processes (Olson et al., 1980; Unsicker et al., 1978). Although the properties of the adrenal chromaffin cell have been extensively studied, the mechanisms through which adrenal medulla grafts produce behavioral effects are not well understood.

Intraventricular grafts of adrenal medulla from young adult donors have been found to reduce rotational behavior of animals with unilateral SN lesions, similar to embryonic SN grafts (Freed et al., 1981). These grafts were found to contain substantial but very variable concentrations of dopamine (Freed et al., 1983b). Adrenal medulla grafts are behaviorally effective even when taken from mature donors; however, when one- or two-year old aging donors were employed, the grafts were ineffective (Freed, 1983). Even though surviving catecholaminergic cells were found in grafts from aging donors, these grafts did not produce a behavioral effect.

Intraventricular adrenal medulla grafts are also more effective than intraparenchymal adrenal medulla grafts. This appears to be due to the relatively poor survival of adrenal chromaffin cells in the parenchyma of the corpus striatum (Stromberg et al., 1984; Freed et al., 1986). Concentrations of nerve growth factor in the striatum are very low (Korsching et al., 1985) and infusions of nerve growth factor into the striatum have been reported to increase the survival of adrenal chromaffin cell grafts (Stromberg et al., 1985). In primates, limited survival of adrenal medulla grafts in the striatal parenchyma has also been reported (Morihisa et al., 1984). In this latter study, adrenal medulla grafts were not found to survive in the ventricles of primates although it was suspected that this was due to poor adhesion of the grafts to the ventricular ependyma.

There are several proposed or conceivable mechanisms through which adrenal medulla grafts could exert their behavioral effects. Initial clinical trials of adrenal medulla grafts in patients with Parkinson's disease reported only a transient improvement in clinical symptoms

(Backlund et al., 1985; Lindvall et al., 1987). Recent clinical trials employing adrenal medulla grafts in contact with the ventricular system report, however, a substantial and sustained clinical improvement (Madrazo et al., 1987). It has therefore become increasingly important to understand the mechanisms through which adrenal medulla grafts produce behavioral effects.

The first and most obvious possibility is that the grafted adrenal chromaffin cells may reinnervate the corpus striatum, in the same way as embryonic SN grafts. Although grafted adrenal chromaffin cells do extend processes (Freed et al., 1981, 1983b, 1986), these processes generally remain within the graft and do not reinnervate the host brain. The formation of neurite-lite process by intraparenchymal adrenal chromaffin cell grafts is greatly increased following nerve growth factor treatment (Stromberg et al., 1985). Under these circumstances, therefore, it is possible that reinnervation of the host brain plays a role in the effects of adrenal medulla grafts. It is also conceivable that adrenal medulla grafts produce a reinnervation of the striatum that cannot be detected by catecholamine fluorescence techniques. At the present time, however, it appears that adrenal medulla grafts do not appreciably innervate the host striatum.

A recent study by Bohn and her colleagues (1987) reported that adrenal medulla grafts promote tyrosine hydroxylase immunoreactivity in the striatum of mice with SN damage produced by the neurotoxin MPTP. These findings suggest that some property of adrenal medulla can either promote tyrosine hydroxylase activity in residual striatal dopaminergic neurites, or possibly induce reinnervation of the striatum from host systems. It should be noted that very few of the transplanted chromaffin cells survived, so this effect may be induced either by degenerating tissue or by some component of adrenal medulla other than chromaffin cells (such as ganglion cells or connective tissue). Since most of the grafted cells died, it is possible that the tyrosine hydroxylase enzyme or fragments of the enzyme with immunoreactive epitopes released from the degenerating cells might have been taken up into endogenous striatal terminals. Finally, MPTP produces only partial lesions of the SN in mice, so the same phenomena might not occur in animals with the more complete SN lesions that are employed for behavioral studies. Indeed, regrowth of endogenous catecholaminergic fibers following adrenal medulla grafts has not been observed in animals with 60HDA- induced SN lesions (Freed, 1983; Freed et al., 1981, 1983b, 1986; Stromberg et al., 1985). Nevertheless, it is certainly possible that, under certain circumstances, adrenal medulla grafts exert trophic effects upon host brain systems. Many substances other than catecholamines, including dopamine beta-hydroxylase, ATP, enkephalins, and other peptides are produced and released by adrenal chromaffin cells (Carmichael, 1986) and could exert a variety of effects upon host brain systems.

It has often been noted that adrenal medulla grafts, when studied by catecholamine fluorescence histochemistry, are surrounded by a "halo" of diffuse fluorescence. In addition, within the striatum there is a gradient of decreasing dopamine concentration away from the graft (Freed et al., 1983; Becher and Freed, 1988). These results have led to the hypothesis that grafted chromaffin cells simply release catecholamines non-specifically into the surrounding extracellular fluid, and that these released catecholamines enter into the striatum and the CSF by simple diffusion (Freed et al., 1981). There are three points which argue against this hypothesis. First, the diffusion of dopamine in the brain parenchyma can occur over only very limited distances (Hargraves and Freed, 1987; Sendelbeck and Urquhart, 1985). Secondly, no detectable dopamine was found in the ventricular CSF of animals with intraventricular adrenal medulla grafts, even though dopamine metabolites were elevated in the CSF (Becker and Freed, 1988). Finally, even though resting state concentrations of dopamine in the striatal parenchyma were not elevated, striatal dopamine release could be induced by amphetamine in animals with behaviorally effective adrenal medulla grafts (Becker and Freed, 1988). If dopamine released from intraventricular adrenal medulla grafts is reaching the striatum by diffusion through extra-cellular fluid, dopamine should also have been found in the CSF. Thus it appears unlikely that this mechanism explains the behavioral effects of adrenal medulla grafts. It is still a possibility that release and diffusion of one of the other substances released by adrenal chromaffin cells is responsible for the behavioral effects.

A final possibility (Becker and Freed, 1988; Rosenstein and Brightman, 1986), is that dopamine is released from adrenal chromaffin cells into blood vessels, and carried into the corpus striatum via the circulation. Blood vessels within and adjacent to intraventricular grafts of brain tissue have been found to be permeable to blood-borne proteins. Thus, it is conceivable that catecholamines are released from adrenal medulla grafts into blood vessels, and are delivered into adjacent striatal tissue by leaking from permeable blood vessels in the striatum adjacent to the grafts (Becker and Freed, 1988).

DISCUSSION

Intraventricular SN grafts consistently extend dopaminergic neurites which almost invariably produce a limited reinnervation of the host striatum. Reinnervation of the septum is occasionally observed, but only when the grafted dopaminergic neurons are located immediately adjacent to the septum but separated from the striatum by other parts of the graft (unpublished observations). Reinnervation of other structures which sometimes adjoin the grafted SN, such as the corpus callosum or hippocampus, has never been observed. Therefore, the dopamine-denervated striatum is a relatively favorable target for graft-derived dopaminergic neurites. On the other hand, the fully mature host striatum is not an ideal target. Penetration of dopaminergic neurites into the host striatum is very limited, and complete reinnervation of the striatum is never observed.

It appears that the state of maturity of the host striatum is very important in controlling reinnervation by SN grafts. Embryonic striatum transplanted to the lateral ventricle in combination with SN grafts is innervated in preference to the host striatum, and SN grafts produce very pronounced behavioral effects when transplanted into immature brains. Thorough investigation of the molecular mechanisms that control innervation of immature striatal neurons by dopaminergic neurites will therefore become of importance for the ultimate success of SN grafts.

It also appears to be possible to enhance the growth of SN graft-derived dopaminergic neurites by local injury to the brain. Of course, it is unlikely that brain injury per se could be employed to enhance the function of SN grafts if and when these grafts are applied clinically. Nevertheless, this provides a clue which could eventually result in the development of techniques for enhancing the performance of SN grafts. For example, perhaps the local administration of brain-injury induced neuronotrophic factor(s) would enhance the performance of SN grafts. Or, means of inducing the brain to produce such factors without actual injury to the brain may be found. Although the mechanisms through which brain injury and target cell immaturity enhance the growth of dopaminergic neurites are unclear, the possibility that the two effects are related is intriguing. For example, brain injury might induce nearby neurons to adopt some of the properties of immature neurons which are responsible for the attraction of dopaminergic neurites. Alternatively, brain-injury induced neuronotrophic factor(s) may be similar or identical to substances produced by immature striatal neurons which attract a dopaminergic innervation. Further study of these possibilities is likely to yield data that will not only be useful in enhancing the function of SN grafts, but which can also be useful in developing an understanding of mechanisms of growth and plasticity of dopaminergic systems.

SN grafts appear to influence behavior by extending axonal processes which form synapses with host neurons. There is no evidence, on the other hand, that synaptic connections are formed between grafted adrenal chromaffin cells and host neurons. Recent clinical studies have reported that adrenal medulla grafts in contact with the lateral ventricle produce substantial improvements in patients with Parkinson's disease (Madrazo et al., 1987). On the other hand, no lasting benefits were obtained in earlier studies of grafts which were entirely intraparenchymal (Backlund et al., 1985; Lindvall et al., 1987). It has therefore become extremely important to understand the mechanisms through which adrenal medulla grafts produce behavioral effects. At the time of this writing, several possible mechanisms of action have been proposed. One possibility is that adrenal medulla grafts promote regrowth of catecholaminergic neurons (Bohn et al., 1987). At the present time, there is no evidence that this putative growth-stimulatory effect occurs under the same circumstances that result in a behavioral effects of adrenal medulla grafts. A second possible mechanism of action is secretion of dopamine into the extracellular fluid, followed by diffusion away from the graft, into the CSF and the host brain. Again, there is no evidence that this phenomenon takes place with adrenal medulla grafts. A final possibility is that adrenal chromaffin cells secrete dopamine into local blood vessels, and that the secreted dopamine enters the host brain through permeable blood vessels adjacent to the graft (Becker and Freed, 1988; Rosenstein, 1987). Further studies will also be required to determine whether this mechanism is responsible for the behavioral effects of adrenal medulla grafts. Investigation of these and other possible mechanisms has become of primary importance. In order to demonstrate that any proposed mechanism is responsible for the behavioral effect, it will be essential not only to determine that the suggested process takes place, but it will also be necessary to show that the suggested mechanism and the behavioral effects of these grafts are directly related.

Only through an understanding of the mechanisms of action of these grafts will it become possible to accurately predict when these grafts will be clinically effective, and to find ways of increasing their efficacy.

REFERENCES

Backlund, E.O., Granberg, P.O., Hamberger, B., Knutsson, E., Martensson, A., Sedvall, G., Seiger, A. and Olson, L.: Transplantation of adrenal medullary tissue to striatum in parkinsonism. J. Neurosurg. 62: 169-173, 1985.

de Beaurepaire, R. and Freed, W.J.: Embryonic substantia nigra grafts innervate embryonic striatal co-grafts in preference to mature host striatum. Exp. Neurol. 95: 448-454, 1987.

Becker, J.B. and Freed, W.J.: Neurochemical correlates of behavioral changes following intraventricular adrenal medulla grafts: Intraventricular microdialysis in freely moving rats. Prog. Brain Res. (in press).

Bernheimer, H., Birkmayer, W., Hornykiewicz, 0., Jellinger, K. and Seitelberger, K.: Brain dopamine and the syndrome of Parkinson and Huntington. Clinical, morphological, and neurochemical correlations. J. Neurol. Sci. 20: 415-455, 1973.

Bjorklund, A., Dunnett, S.B., Stenevi, U., Lewis, M.E. and Iversen, S.D.: Reinnervation of the denervated striatum by substantia nigra transplants: Functional consequences as revealed by pharmacological and sensorimotor testing. Brain Res. 199: 307-333, 1980.

Bohn, M.D., Marciano, F., Cupit, L. and Gash, D.M.: Adrenal medulla grafts enhance recovery of striatal dopaminergic fibers. Science 237: 913-916, 1987.

Carmichael, S.W.: The Adrenal Medulla, Volume 4. Cambridge University Press, Cambridge, 1986.

Denis-Donini, S., Glowinski, J. and Prochiantz, A.: Specific influences of striatal target neurons on the in vitro outgrowth of mesencephalic dopaminergic neurites: A morphological quantitative study. J. Neurosci. 3: 2292-2299, 1983.

Dunnett, S.B., Bjorklund, A., Schmidt, R.H., Stenevi, U. and Iversen, S.D.: Intracerebral grafting of neuronal cell suspensions. IV. Behavioral recovery in rats with unilateral 6-OHDA lesions following implantation of nigral cell suspensions in different forebrain sites. Acta Physiol. Scand. Suppl. 522: 29-37, 1983a.

Dunnett, S.B., Bjorklund, A. and Stenevi, U.: Dopamine-rich transplants in experimental parkinsonism. Trends Neurosci. 6: 266-270, 1983b.

Freed, W.J., Perlow, M.J., Karoum, F., Seiger, A., Olson, L., Hoffer, B.J. and Wyatt, R.J.: Restoration of dopaminergic function by grafting of fetal rat substantia nigra to the caudate nucleus: Long-term behavioral, biochemical, and histochemical studies. Ann. Neurol. 8: 510-519, 1980.

Freed, W.J., Morihisa, J.M., Spoor, E., Hoffer, B.J., Olson, L., Seiger, R. and Wyatt, R.J.: Transplanted adrenal chromaffin cells in rat brain reduce lesion-induced rotational behavior. Nature 292: 351-352, 1981.

Freed, W.J.: Functional brain tissue transplantation: Reversal of lesion-induced rotation by intraventricular substantia nigra and adrenal medulla grafts, with a note on intracranial retinal grafts. Biol. Psychiatry 18: 1205-1267, 1983.

Freed, W.J., Ko, G.N., Niehoff, D.L., Kuhar, M.J., Hoffer, B.J., Olson, L., Cannon-Spoor, H.E., Morihisa, J. M. and Wyatt, R.J.: Normalization of spiroperidol binding in the denervated rat striatum by homologous grafts of substantia nigra. Science 222: 937-939, 1983a.

Freed, W.J., Karoum, F., Spoor, H.E., Olson, L., Morihisa, J. and Wyatt, R.J.: Catecholamine content of intracerebral adrenal medulla grafts. Brain Res. 269: 184-189, 1983b.

Freed, W.J., Cannon-Spoor, H.E. and Krauthamer, E.: Intrastriatal adrenal medulla grafts in rats: Long-term survival and behavioral effects. J. Neurosurg. 65: 664-670, 1986.

Freed, W.J. and Cannon-Spoor, H.E.: Cortical lesions increase reinnervation of the dorsal striatum by substantia nigra grafts. Brain Res. (in press).

Freund, T.F., Bolan, J.P., Bjorklund, A., Stenevi, U., Dunnett, S.B., Powell, J.F. and Smith, A.D.: Efferent synaptic connections of grafted dopaminergic neurons reinnervating the host neostriatum: A tyrosine hydroxylase immunocytochemical study. J. Neurosci. 5: 603-616, 1985.

Hargraves, R. and Freed, W.J.: Chronic intrastriatal dopamine infusions in rats with unilateral lesions of the substantia nigra. Life Sci. 40: 959-966, 1987.

Hemmendinger, L.M., Garber, B.B., Hoffman, P.C. and Heller, A.: Target neuron-specific process formation by embryonic mesencephalic dopamine neurons in vitro. Proc. Natl. Acad. Sci. (USA) 78: 1264-1268, 1981.

Korsching, S., Auburger, G., Heumann, R., Scott, J. and Thoenen, H.: Levels of nerve growth factor and its mRNA in the central nervous system of the rat correlate with cholinetgic innervation. EMBO Journal 4: 1389-1393, 1985.

Leong, S.K. and Lund, R.D.: Anomalous bilateral cortiofugal pathways in albino rats after neonatal lesions. Brain Res. 62: 218-221, 1973.

Lindvall, O., Backlund, E.O., Farde, L., Sedvall, G., Freedman, R., Hoffer, B., Nobin, A., Seiger, A. and Olson, L.: Transplantation in Parkinson's disease: Two cases of adrenal medullary grafts to putamen. Ann. Neurol. 22: 457-468, 1987.

Lishajko, F.: Dopamine secretion from the isolated perfused sheep adrenal. Acta Physiologica Scand. 79: 405-410, 1970.

Madrazo, I., Drucker-Colin, R., Diaz, V., Martinez-Mata, J., Torres, C. and Becerril, J.J.: Open microsurgical autograft of adrenal medulla to the right caudate nucleus in two patients with intractable Parkinson's disease. N. Engl. J. Med. 316: 831-834, 1987.

Mahalik, T.J., Finger, T.E., Stromberg, I. and Olson, L.: Substantia nigra transplants into denervated striatum of the rat: Ultrastructure of graft and host connections. J. Comp. Neurol. 240: 60-70, 1985.

McGeer, P.L., McGeer, E.G., Scherer, U. and Singh, K.: A glutamatergic coricostriatal path? Brain Res. 128: 369-373, 1977.

Morihisa, J.M., Nakamura, R.K., Freed, W.J., Mishkin, M. and Wyatt, R.J.: Adrenal medulla grafts survive and exhibit catecholaminespecific fluorescence in the primate brain. Exp. Neurol. 84: 643-653, 1984.

Nieto-Sampedro, M., Lewis, E., Cotman, C., Manthorpe, M., Skaper, S., Barbin, G., Longo, F. and Varon. S.: Brain injury causes time-dependent increase in neuronotrophic activity at the lesion site. Science 217: 860-861, 1982.

Nieto-Sampedro, M., Manthrope, M., Barbin, G., Varon, S. and Cotman, C.W.: Injury-induced neuronotrophic activity in adult rat brain: Correlation with survival of delayed implants in the wound cavity. J. Neurosci. 3: 2219-2229, 1983.

Olson, L., Seiger, A. and Alund, M.: Locus coeruleus fiber growth in oculo induced by trigeminotomy. Med. Biol. 56: 23 27, 1978.

Olson, L., Seiger, A., Freedman, R. and Hoffer, B.J.: Chromaffin cells can innervate brain tissue: Evidence from intraocular double grafts. Exp. Neurol. 70: 414-426, 1980.

Perlow, M.J., Freed, W.J., Hoffer, B.J., Seiger, A., Olson, L. and Wyatt, R.J.: Brain grafts reduce motor abnormalities produced by destruction of nigrostriatal dopamine system. Science 204: 643-647, 1979.

Rosenstein, J.M.: Adrenal medulla grafts produce blood-brain barrier dysfunction. Brain Res. 414: 192-196, 1987.

Rosenstein, J.M. and Brightman, M.W.: Alterations of the blood-brain barrier after transplantation of autonomic ganglia into the mammalian central nervous system. J. Comp. Neurol. 250: 339-351, 1986.

Schwarz, S.S. and Freed, W.J.: Brain tissue transplantation in neonatal rats prevents a lesion-induced syndrome of adipsia, aphagia, and akinesia. Exp. Brain Res. 65: 449-454, 1987.

Schneider, G.E.: Early lesions of superior colliculus: Factors affecting the formation of abnormal retinal projections. Brain Behav. Evol. 8: 73-109, 1973.

Sendelbeck, S.L. and Urquhart, J.: Spatial distribution of dopamine, methotrexate, and antipyrine during continuous intracerebral microperfusion. Brain Res. 328: 251-258, 1985.

Snyder, S.R., Sahar, D., Prasad, A.L.N. and Fahn, S.: Changes in adrenal dopamine concentration after metyrapone or ACTH administration: Implications for the in vivo regulation of dopamine beta-hydroxylase by glucocorticoids. Life Sci. 20: 1077-1086, 1977.

Stromberg, I., Herrera-Marschitz, M., Hultgren, L., Ungerstedt, U. and Olson, L.: Adrenal medullary implants in the dopamine-denervated rat striatum. I. Acute catecholamine levels in grafts and host caudate as determined by HPLC-electrochemistry and fluorescence histochemical image analysis. Brain Res. 297: 41-51, 1984.

Stromberg, I., Herrera-Marschitz, M., Ungerstedt, U., Ebendal, T. and Olson, L.: Chronic implants of chromaffin tissue into the dopamine-denervated striatum. Effects of NGF on graft survival, fiber growth, and rotational behavior. Exp. Brain Res. 60: 335-349, 1985.

Ungerstedt, U.: Postsynaptic supersensitivity after 6-hydroxydopamine induced degeneration of the nigro-striatal dopamine system. Acta Physiol. Scand. Suppl. 367: 69-93, 1971.

Unsicker, K., Krisch, B., Otten, U. and Thoenen, H.: Nerve growth factor-induced fiber outgrowth from isolated rat adrenal chromaffin cells: Impairment by glucocorticoids. Proc. Natl. Acad. Sci. (USA) 75: 3498-3502, 1978.

Webster, K.E.: Cortico-striatal interrelations in the albino rat. J. Anatomy (Lond.) 95: 532-545, 1961.

Wurtman, R.J., Pohorecky, L.A. and Baliga, B.S.: Adrenocortical control of the biosynthesis of epinephrine and proteins in the adrenal medulla. Pharmacol. Rev. 24: 411-426, 1972.

26

RECENT ADVANCES IN DOPAMINERGIC IMPLANTS

I. J. Kopin, K.S. Bankiewicz, R. J. Plunkett, L. Porrino,
D. M. Jacobowitz*, W.T. London and E. H. Oldfield

National Institutes of Health
National Institute of Neurological and Communicative Disorders
and Stroke
*National Institute of Mental Health
Bethesda, Maryland

INTRODUCTION

The first attempts at transplanting neural tissue into brains of experimental animals began nearly 100 years ago. Initially progress was slow and halting, but during the last two decades the tremendous advances in the neurosciences have provided the necessary biochemical methods, histological techniques, and specific toxins with which to produce animal models of neurological disorders and to examine the consequences of attempts at replacing damaged neuronal systems using tissue implants. Perhaps the most spectacular and most promising advances have been related to the dopaminergic systems and Parkinson's disease treatments.

The discoveries that dopamine is present in highest concentrations in the basal ganglia and that the degeneration of the dopaminergic neurons in the substantia nigra was responsible for dopamine depletion in the caudate-putamen of patients with Parkinson's disease led to the first rational therapy of any degenerative neurological disease. The introduction of L-dopa, the precursor of dopamine, to replace the missing neurotransmitter was a therapeutic triumph which has benefitted millions of parkinsonian patients. Dopa treatment, however, is not a panacea and with progression of the underlying degenerative process its efficacy diminished and untoward effects on motor control limit its usefulness. It has become clear that therapies aimed at prevention of the degenerative process or restoration of the damaged neurons may offer substantial additional advantages. One approach has been the use of implants of dopamine-producing tissue into the brains of patients with Parkinson's disease. It is the purpose of this brief review to provide a perspective on the current status of this rapidly advancing frontier.

Rodent models of dopaminergic neuronal deficits

With the availability of sensitive and specific methods for histochemical localization of biogenic amines and immunohistochemical methods for visualizing tyrosine hydroxylase it became possible not only to trace specific neuronal pathways, but to examine tissues which had been implanted into brain. Over a generation ago Anden et al. (1966a) described the large uncrossed nigrostriatal dopaminergic pathway. This was followed by the demonstration that unilateral lesions of this pathway of the caudate-putamen resulted in abnormal motor behavior, particularly after administration of drugs which had effects on dopaminergic mechanisms (Anden et al., 1966b). When haloperidol, reserpine, or chlorpromazine were given to interfere with dopaminergic function, the rats rotated towards the unoperated side whereas after administration of DOPA, they rotated towards the operated side. These effects were attributed to an imbalance of the dopaminergic and cholinergic systems. When a more specific dopaminergic lesion became possible with the application of 6-hydroxydopamine to induce central nervous system neuronal deficits (Ungerstedt, 1968), the importance of involvement of this catecholamine became more evident. Ungerstedt (1971) showed that unilateral chemical lesioning with 6-hydroxydopamine of the substantia nigra resulted in depletion of dopamine from the ipsilateral caudate attended by the development of pharmacological responses consistent with supersensitive dopamine receptors on that side. After administration of L-dopa or the dopamine agonist, apomorphine, the asymetric dopaminergic stimulatory effects of these drugs induced rotational motor activity in a direction contralateral to the lesion. These results suggested that replacement of dopaminergic innervation might reverse the effects of nigrostriatal destruction. Thus, about 15 years ago a satisfactory rat model of dopamine deficiency in the caudate-putamen became available for testing efficacy of neuro-transmitter replacement.

Although the survival of neural implants has been demonstrated, newly developed biochemical and histochemical methods were now applied to examine the ability of dopamine producing cells to survive, grow, produce dopamine, and be functionally important after transplantation into the region of the caudate nucleus. Adrenal medullary tissue and fetal ventral mesencephalic tissue containing dopaminergic neurons had been shown to survive transplantation and to produce catecholamines. It now remained to apply these techniques to 6-hydroxydopamine lesioned rats. This was done by Perlow et al. (1979), Freed et al. (1980) and by Björklund et al. (1980a, 1980b). Although technical details differed, fetal mesencephalic tissue – small pieces or dissociated cells – were implanted in the caudate nucleus or in the lateral ventricle on the surface of the caudate of rats which and been unilaterally administered 6-hydroxydopamine into the substantia nigra. Unoperated rats were demonstrated to rotate in response to dopamine agonist (such as apomorphine) administration. The grafts were shown to survive, mature into dopamine-containing neurons which grew axons at least some of which entered the medial portion denervated caudate nucleus. The grafted tissue remained healthy without the evidence of aging evident in the surrounding tissue and maintained higher concentrations of dopamine than adult substantia nigra. The functional efficacy of the graft was demonstrated by a diminution in the intensity of apomor-phine-induced rotation, suggesting that there was a reduction in dopamine receptor supersen-sitivity. This was presumed to result from exposure to dopamine derived from the grafted dopaminergic tissue.

Adult chromaffin tissue from the adrenal medulla grafted to caudates of 6-hydroxydopamine lesioned rats also was reported to be effective in reducing apomorphine-induced rotation (Freed, 1983).

Implants in patients with Parkinson's disease

These observations in rats led to the first clinical trials employing autologous adrenal medullary brain grafts to treat severe cases of Parkinson's disease (Backlund et al., 1985). Tissue obtained from the patient's own adrenal medula was stereotactically transplanted into the caudate nucleus. There was only transient improvement in parkinsonian symptoms and no direct evidence from measurements of metabolites in cerebrospinal fluid that dopamine was being produced. The feasibiltiy and safety of the procedure, however, was established. In contrast to these initial discouraging results. Madrazo et al. (1987a) reported beneficial effects from direct open microsurgical autografts of adrenal medulla into caudate nucleus of two relatively young patients with Parkinson's disease resistant to medical treatment. Although the implant was on one side only, improvement in motor symptoms was bilateral. This report stimulated a host of similar attempts at treating parkinsonian patients with adrenal autografts. Although some limited improvements have been noted in some patients, there has not been the enthusiastic response that would have been expected if the subsequent attempts in other centers were as successful as the first two cases described by Madrazo et al. (1987a). Additional successes however, were subsequently reported by Madrazo et al. (1987b).

The feasibility of using human fetal tissue for implants was demonstrated by Strömberg et al. (1987). They found that grafting of human fetal substantia nigral tissue (obtained from therapeutic termination of pregnancy in the first trimester) into rats with 6-hydroxydopamine lesions of the substantia nigra reduced apomorphine-induced turning by up to 70-80% after 3-5 months. The animals were treated with an immunosuppressant and the grafts were found to have matured and formed tryosine hydroxylase-containing nerve fibers which penetrated into the host rat striatum. Because of limited therapeutic success of adrenal transplants in humans as well as morbidity and mortality attending the operative procedure required to obtain donor tissue for the adrenal autografts, particularly in elderly patients, Madrazo et al. (1988) attempted to use brain mesencephalic and adrenal medullary tissue from a spontaneously aborted fetus to treat two parkinsonian patients. The tissue was placed in a cavity made in the right caudate nucleus, but in contact with the cerebrospinal fluid. Both patients are maintained on cyclosporine and predisone and have been reported to show objective improvement in their parkinsonian symptoms; again bilateral improvments were noted.

The preliminary results in patients raise several important questions. Why is there bilateral improvement when the implants are unilateral? Are therapeutic benefits related solely to dopamine formation by the grafted tissue? Could the host's dopaminergic neurons play a role or is the graft alone responsible? The answers to these questions are being sought in primate models of Parkinson's disease.

Primate models of Parkinson's disease

As indicated above, clinical trials of autologous adrenal medullary grafts, and most recently fetal mesencephalic and adrenal medullary tissue, appear to have resulted in at least temporary improvement of parkinsonian symptoms in some patients and varying degrees of more lasting efficacy. The procedures followed have paralleled those used in 6-hydroxydopamine lesioned rats where functional improvement was assessed by diminution in apomorphine-induced rotation. Rodents appear to be suitable for studies of replacement of dopamine deficits and have been valuable in assessing various sources of tissue as well as the time course for the expected functional changes. The disorder produced, however, does not closely resemble Parkinson's disease.

Various attempts to produce by brain lesions in primates a suitable model of Parkinson's disease met with only limited success (see review by Poirier, 1971). Although some parkinsonian features can be produced by various lesions of neuronal pathways or brain nuclei, in no case is the complete parkinsonian syndrome obtained. More recently, however the accidentally discovered neurotoxin, 1-methyl-4-phenyl-1,2,3,6-tetrahydropyridine (MPTP), which causes parkinsonism in humans (Davis 1979; Langston, 1983) was demonstrated to provide relatively specific damage to the dopaminergic neurons in the nigrostriatal pathway of monkeys and produces a clinical syndrome which closely resembles Parkinson's disease (Burns et al., 1983).

It is beyond the scope of this presentation to review the extensive work which has been done to establish the mode of MPTP toxicity and to explain the particular vulnerability of primates substantia nigra neurons to this toxin. A recent volume describes much of this work (Markey, 1986). Briefly, it is known that MPTP is rapidly bioactivated to the quarternary derivative MPP+, which is the active toxin. This conversion is mediated by monoamine oxidase type B (MAO-B) and occurs in many tissues. Since MPP+ does not readily enter the brain, in some species conversion of MPTP to MPP+ in the brain capillary endothelium may protect the brain from the toxic actions of MPTP. In primates including humans, brain capillary endothelium does not contain sufficient amounts of MAO-B to prevent entry of MPTP into the brain. Once past the blood-brain barrier MPTP is converted to MPP+ in cells containing MAO-B (most astrocytes). MPP+ is a substrate for catecholamine transport mechanisms and can be concentrated in dopaminergic neurons. Pharmacological blockade of MAO-B or of dopamine uptake mechanisms can prevent MPTP toxicity, indicating that both mechanisms are important. Once in the dopaminergic neurons, MPP+ is further concentrated in mitochondria where sufficiently high levels are attained to interfere with mitochondrial oxidation (at complex I) and deplete ATP. This is believed to be the means by which cells are destroyed. The particular vulnerability of neuromelanin-containing neurons in primates may be due to the binding of MPP+ to the neuromelanin or the presence of neuromelanin may be indicative of diminished oxidative capacity which is associated with MPP+ vulnerability. MPTP-parkinsonism has been produced in many primate species and several groups of investigators have used such animals as models in which to study the efficacy of neural implants. Others at this meeting have described their experience with alterations in motor performance produced by adrenal medullary or fetal nigral implants in the caudate nucleus of MPTP-parkinsonian monkeys.

MPTP hemiparkinsonian monkeys

Because MPTP is rapidly converted to MPP+ in many tissues and because the charged molecule does not rapidly penetrate the blood-brain barrier, by infusing a solution of MPTP into one carotid artery it has been possible to cause unilaterial destruction of the nigrostriatal pathway in monkeys (Bankiewicz et al., 1986). This model appears to have several advantages over monkeys made parkinsonian by repeated systemic administration of MPTP. The parkinsonian symptoms produced are limited to the contralateral side. Such animals are able to feed themselves and are readily maintained in good health without Dopa treatment. Spontaneous motor activity is attended by turning towards the lesioned side. They respond to administration of apomorphine by rotating away from the lesioned side, as do rodents, providing a means for objective quantification of asymetric dopamine receptor supersensitivity. By this criteria, the lesion produced and the functional abnormalities remain stable for up to two years, in contrast to the varying severity of parkinsonism in monkeys treated systemically with MPTP. Unlike rats, in monkeys volitional movements can be tested by presenting the animal with food. Untreated hemiparkinsonian monkeys invariable use the intact hand to retrieve food or handle desired objects. After treatment with L-dopa, the animals are able to use both sides. By use of ^{18}F-DOPA, the formation and storage of ^{18}F-dopamine may be studied using positron emission tomography (PET) scanning. We have been using hemiparkinsonian monkeys to assess changes in volitional movement, parkinsonian status (rigidity, bradykinesia, tremor), responses to dopamine receptor agonists, and ^{18}F-dopamine formation and storage in intact animals. PET studies can also be used to measure local glucose utilization by measuring uptake and accumulation of ^{18}F-2-deoxyglucose and appropriate [18]F- or ^{11}C-labelled ligands may be used to assess regional dopamine or other neurotransmitter receptor density. These latter methods are currently being planned for use in patients as well as in experimental animals.

Other methods requiring direct examination of the brain after death are available for use in experimental animals. After intravenous administration of 14C-2-deoxyglucose, the rates of glucose uptake in various regions of brain can be measured using autoradiography of brain slices (Sokoloff et al., 1977). Changes in local cerebral glucose utilization (LCGU) of MPTP parkinsonian animals have been mapped and the widespread changes before and during treatment with L-Dopa described (Porrino et al., 1987). In hemiparkinsonian animals untreated with DOPA, increased LCGU in the outer portion of the globus pallidus is the most striking abnormality.

Alterations in receptor density can be assessed in post mortem brain slices by specific binding of radiolabelled ligands. In hemiparkinsonian monkeys there is an obvious increase in binding of ^{3}H spiroperidol binding to D_2-type dopamine receptors in the caudate and lateral putamen of the MPTP treated side (Joyce et al., 1985). Dopamine uptake sites are believed to be located exclusively on prejunctional dopaminergic nerve terminals and can be mapped by ^{3}H-mazindole binding. Using this technique, Joyce et al. (1985) found almost complete loss of binding sites in the caudate and putamen of the MPTP-treated side.

After implants of adrenal medullary or fetal mesencephalic tissue, evidence of functional recovery is sought by repeatedly examining the volitional responses of the animal when food is presented and by recording apomorphine-induced rotation. ^{18}F-dopamine formation in the implant on other areas can also be assessed to determine if there is significant, measurable enhancement of dopamine formation, but if supersensitive receptors are present, the magnitudes of changes may be too small to become apparent by in vivo methods.

Direct examination of the brains of animals which had received adrenal or fetal implants provides the only definitive evidence for determining if the transplanted tissue survives, if there have been ingrowths of fibres from the implant to the host, or if sprouting of the hosts surviving dopaminergic neurons play a significant role in the recovery process. At this time only a limited number of studies in these animals have been completed and only a few results are available to provide direction to the various studies which we have planned.

The results obtained thus far are summarized in Table 1. Control operations without implants or implants of fat, adrenal cortex, etc. result in only transient decreases in apomorphine turning (Bankiewicz et al., 1987). After adrenal medullary tissue is implanted, volitional movements transiently return, but after a month improvements do not appear significantly greater than in the control animals. After fetal mesencephalic tissue implants, however, the improvements in motor function persist as long as these animals have been studied. It seems that diminished apomorphine induced rotation is a more sensitive (but perhaps less relevant) index of dopaminergic reinnervation than return of volitional responses.

The few studies which have been done using ^{18}F-DOPA or LCGU show improvement only after fetal mesencephalic tissue implants. In the brains of animals which have been studied, TH positive cells were found to survive only after fetal mesencephalic implants. In no case, however, have fibres been found to grow from the implant into the host caudate. With either adrenal or control tissue implants, some sprouting of fibres from the hosts surviving dopaminergic neurons was found, but not to the extend found when fetal tissue was used.

These preliminary findings suggest that growth enhancing factors may play an important

Table 1

Evaluation of Tissue Implants in Brains of MPTP Hemiparkinsonian Monkeys

	Adrenal	Control	Fetal DA Cells
Volitional movement of affected side	Transient	No	Yes
Decrease in apomorphine-induced turning	Transient	Transient	Yes
Formation of ^{18}F-DA by implant (PET)	No	—	Yes
Return to normal of LCGU in outer GP	No	No	Yes
Decrease in elevated 3H-spiperone binding	—	—	Yes
Survival of TH-positive cells	No	—	Yes
Ingrowth of fibres from implant to host	No	—	No
Sprouting of host's dopaminergic neurons	+ +	+ +	+ + + +

role in the effects of fetal (or other) implants. They are consistent with the observations of Bohn et al. (1987) showing that adrenal medullary grafts enhance recovery of striatal dopaminergic figures and explain bilateral improvement with unilateral grafts in Parkinson's disease (Madrazo et al., 1987). There is, of course, much work to be done in this exciting field. The possibilities to reverse, by chemical means, the degenerative process causing Parkinson's disease and or produce conditions under which fetal tissue may be made to develop functional fibres which grow into the caudate (as has been found in rodents) are being actively pursued.

References

Andén, N.E., Dahlström, Fuxe, K., Larsson, et al., Ascending monoamine neurons to the telecephalon and diencephalon. Acta Physiol. Scand. 67:313-326, 1966.

Andén, N.E. Dahlström, Fuxe, K., et al., Functional role of the nigrostriatal dopamine neurons. Acta Pharmacol. 24:263-274, 1966.

Backlund, E.O., Granberg, P.O., Hamberger, B., et al., Transplantation of adrenal medullary tissue to striatum in parkinsonism. J. Neurosurg. 62:169-173, 1985.

Bankiewicz, K.S., Oldfield, E.H., Chiueh, C.C., et al., Hemiparkinsonism in monkeys after unilateral internal carotid artery infusion of 1-methyl-4-phenyl-1,2,3,6-tetrahydropyridine (MTP), Life Sci, 39:7-16, 1986.

Bankiewicz, K.S., Plunkett, R.J., Kopin, I.J., et al., Transient and long term functional improvement by adrenal and fetal mesencephalic implants into caudate nuclei of MPTP parkinsonian monkeys, presented at: Schmitt Neurological Sciences Symposium, Rochester, New York, 1987.

Björklund, A., Dunnett, S., Stenevi, et al., Reinnervation of the denervated striatum by substantia nigra transplants: functional consequences as revealed by pharmacological and sensorimotor testing. Brain Res. 199:307-333, 1980a.

Björklund, A., Schmidt, R., and Stenevi, U., Functional reinnervation of the neostriatum in the adult rat by use of intraparenchymal grafting of dissociated cell suspensions from the substantia nigra. Cell Tissue Res. 212:39-45, 1980b.

Bohn, M.C., Cupit, L., Marciano, F. and Gash D.M., Adrenal medulla grafts enhance recovery of striatal dopaminergic fibers. Science 237:913-915, 1987.

Burns, R.S., Chiueh, C.C., Markey, S.P., et al., A primate model for parkinsonism: selective destruction of dopaminergic neurons in the pars compacta of the substantia nigra by N-methyl-4-phenyl-1,2,3,6-tetra-hydropyridine, Proc. Natl. Acad. Sci. 80:4546-4550, 1983.

Davis, G.C., Williams, A.C., Markey, S.P., et al., Chronic Parkinsonism secondary to intravenous injection of meperidine analogues. Psychiatry Res. 1:249-54, 1979.

Fletcher H., McDowell, M.D., and Markham, C.H., (Eds.) Recent Advances in Parkinson's Disease, Contemporary Neurology Series, F.A. Davis, Company, Philadelphia, Pennsylvania, 1971.

Freed, W.J., Perlow, M.J., Karoum, F., et al., Restoration of dopaminergic function by grafting of fetal rat substantia nigra to the caudate nucleus: long-term behavioral, biochemical and histochemical studies. Ann. Neurol. 8:510-519, 1980.

Freed, W.J., Functional brain tissue transplantation: reversal of lesion-induced rotation by intraventricular substantia nigra and adrenal medulla grafts with a note on intro-cranial retinal grafts. Biol. Psychiatry 18: 1205-1267, 1983.

Joyce, J.N., Marshall, J.F., Bankiewicz, et al., Hemiparkinsonism in a monkey after unilateral internal carotid artery infusion of 1-methyl-4-phenyl-1,2,3,6-tetrahydropyridine (MPTP) is associated with regional ipsilateral changes in striatal dopamine D_2 receptor density. Brain Res. 382:360-364, 1985.

Langston, J.W., Ballard, P., Tetrud, JW., et al., Chronic Parkinsonism in humans due to a product of meperidine-analog synthesis. Science 219: 979-980, 1983.

Madrazo, I., Drucker-Colin, R., Diaz, V., et al., Open microsurgical autograft of adrenal medulla to the right caudate nucleus in two patients with intractable Parkinson's disease. N. Engl. J. Med. 316:831-834, 1987a.

Madrazo, I., Drucker-Colin, R., Leon V., et al., Adrenal medullar transplanted to caudate nucleus of treatment of Parkinson's disease: report of 10 cases, Surg. Forum, 38:510-511, 1987b.

Madrazo, I., Leon, V., Torres, C., et al., Transplantation of fetal substantia nigra and adrenal medulla to the caudate nucleus in two patients with Parkinson's disease, N. Eng. J. Med., 318:51, 1988.

Markey, S.P., Castignoli, N., Jr., Trevor, A.J., Kopin, I.J., (Eds.) MPTP: A Neurotoxin Producing a Parkinsonian Syndrome, Academic Press, Orlando, Florida, 1986.

Perlow, M.J., Freed, W.J., Hoffer, B.J. et al., Brain grafts reduce motor abnormalities produced by destruction of nigrostriatal dopamine systems. Science 204:643-647, 1979.

Poirier, L.J., The development of animal models for studies in Parkinson's disease, in Recent Advances in Parkinson's Disease, McDowell, F.H., and Markham, D.H. (Eds.), F.A. Davis Company, Philadelphia, Pennsylvania, 1971, pp 83-117.

Porrino, L.J., Burns, R.S., Crane, et al., Local cerebral metabolic effects of L-dopa therapy in 1-methyl-4-phenyl-1,2,3,6-tetrahydropyridine-induced parkinsonism in monkeys. Proc. Natl. Acad. Sci. 84:595-5999, 1987.

Sokoloff, L., Reivich, M., Kennedy, C. et al., The [^{14}C] deoxyglucose method for the measurement of local cerebral glucose utilization: theory, procedure, and normal values in the conscious and anesthetized albino rat. J. Neurochem, 28:897-916, 1977.

Stromberg, I., Bygdeman, M., Goldstein, M., et al., Human fetal substantia nigra grafted to the dopamine-denervated striatum of immunosuppressed rats: evidence for functional reinnervation, Neurosc. Lett. 71:271-276, 1986.

Ungerstedt, U., 6-Hydroxydopamine induced degeneration of central monoamine neurons. Eur. J. Pharmacol. 5:107-110, 1968.

Ungerstedt, U. Postsynaptic supersensitivity after 6-hydroxydopamine induced degeneration of the nigro-striatal dopamine system. Acta Physiol. Scand. 367:69-93, 1971.

27

REVERSAL OF EXPERIMENTAL PARKINSONISM
IN AFRICAN GREEN MONKEYS
FOLLOWING FETAL DOPAMINE NEURON TRANSPLANTATION

T.J. Collier, D.E. Redmond Jr.*, R.H. Roth**,
J.D. Elsworth** and J.R. Sladek Jr.

Department of Neurobiology and Anatomy, University of Rochester
School of Medicine, Rochester, NY 14642; and Departments of
Psychiatry* and Pharmacology**, Yale University School of Medicine,
New Haven, CT 06510

INTRODUCTION

The motor abnormalities and progressive debilitation associated with Parkinson's disease has been linked to depletion of the neurotransmitter dopamine in the nigrostriatal system of the brain (12). Supplementation of brain dopamine by administration of the precursor substance L-dopa significantly improves the motor symptoms of the disease, and has been the preferred treatment for the past 20 years (16). However, chronic L-dopa administration often yields negative side-effects (3), and the benefit of the drug diminishes as the disease progresses (7,24).

Research on improved therapies has taken two main directions. The first is represented by the ongoing clinical tests of the efficacy of pharmacological treatment with the antioxidants deprenyl-and-tocopherol; substances that may retard progression of the syndrome in patients identified early in the disease process (20). The second, involves replacement of striatal dopamine through intracerebral grafting of a new tissue source of the neurotransmitter: fetal dopamine neurons (2,4,18,21), cells of the adrenal medulla (9,17), and potentially, other paraneural tissues (10) and cultured cell lines (11,13). This approach may be of greatest benefit to patients late in the disease process that have too few of their own nigral neurons remaining to benefit from L-dopa therapy, and consequently may require replacement of cells to provide a new source of dopamine. These severely parkinsonian patients would be less likely to benefit from antioxidative therapy. Indeed, dopamine replacement via autografts of adrenal cells has been performed on an increasing number of Parkinson's patients to date, with highly variable results (1,14,15). The increasing skepticism surrounding the general applicability of adrenal cell grafts as supplementation therapy has encouraged continued study of grafted neuronal tissue in non-human primates.

We have been studying the efficacy of grafted fetal dopamine neurons as an experimental supplementation therapy in the non-human primate model of parkinsonism produced by administration of the neurotoxin N-methyl-4-phenyl-1,2,3,6-tetrahydropyridine (MPTP). Treatment of primates with MPTP produces a pattern of morphological damage (6, also, see Langston et al., this volume) and behavioral symptoms strikingly similar to Parkinson's disease in humans. Utilizing this animal model, we (19), and others (2), have reported that implantation of dopamine-rich neural tissue into the striatum of symptomatic monkeys ameliorates the deficits in motor function and is related to the survival and growth of grafted neurons. In the first year of our studies we established that fetal primate dopamine neurons could be successfully transplanted into MPTP-treated adult monkey hosts, and that after a 70 day survival interval these grafted neurons could be identified in the host brain (19,21). Grafted cells continued to express an enzyme marker for dopamine synthesis (tyrosine hydroxylase), and the mild to moderate motor deficits exhibited by these monkeys recovered in proportion to the number of grafted neurons identified immunohistochemically (22,23). Importantly, control animals receiving neural grafts that did not contain dopamine, or animals receiving dopamine-containing grafts into an incorrect brain target, did not exhibit behavioral recovery. While these initial results were exciting, many questions remain.

In the second year of our studies we began to address the following issues: 1) Would grafts of fetal dopamine neurons survive and function for extended periods of time in the host brain? 2) Would extended survival times yield signs of continued integration between the graft and the host brain? 3) Is behavioral recovery following fetal tissue implantation related to replacement of dopamine or can it be accounted for by supplementation of some other fetal tissue factor? 4) Would fetal neuronal grafts ameliorate the behavioral symptoms of severely debilitated parkinsonian monkeys? In this report we will highlight the evidence provided by a pair of severely parkinsonian monkeys, one receiving multiple implants of fetal midbrain tissue rich in developing dopamine neurons and one receiving control implants of non-dopaminergic fetal cerebellum, with particular reference to the issues outlined above. This pair of cases has been the most extensively studied to date, and provide the clearest examples of the efficacy of neural grafts. However, they represent our working hypotheses based on an ongoing study of over 30 transplanted monkeys that are in various stages of data collection and analysis.

METHODS

Monkeys receiving neural grafts were young adult male African green monkeys (Cercopithecus aethiops sabaeus) from the St. Kitts colony, treated with MPTP (0.4mg/kg, intramuscular injection, five times over 5 days). In the pair of animals decribed here, S089 and S241, MPTP treatment yielded a progressive parkinsonian syndrome characterized by head and limb tremors, bradykinesia and hypokinesia. At the time of transplant surgery, these animals had declined to a state of near immobility.

Grafting of fetal neural tissues was accomplished utilizing a protocol described previously (21). Monkey S098 was implanted with four 1 cubic millimeter blocks of fetal substantia nigra tissue (donor: 11.0cm crown-rump length), rich in developing dopamine neurons, into the striatum of each hemisphere. Monkey S241 received intrastriatal implantation of a comparable volume of fetal cerebellar tissue, a brain region that does not contain dopamine neurons. After transplantation these animals were behaviorally monitored for 5 (S241) and 7.5 (S089) months.

At the conclusion of the study, brain tissue was prepared for combined biochemical determination of levels of dopamine and its metabolite homovanillic acid in tissue punches proximal and distal to neural grafts, and immunocytochemical analysis of the dopamine marker tyrosine hydroxylase to identify grafted neurons and evaluate their growth and integration with the host brain.

RESULTS

Transplantation of fetal dopaminergic substantia nigra neurons bilaterally into the striatum of monkey S089 resulted in a gradual, progressive improvement of motor deficits over a period of 4 months following surgery. At the conclusion of the experiment, 7.5 months after grafting of nigral tissue, monkey S089 remained fully functional in its spontaneous behavior. Monkey S241, equally debilitated and immobile at the time of surgery, showed no behavioral improvement following implantation of cerebellar tissue that did not contain dopamine neurons. Five months after transplant surgery this animal exhibited no change in its debilitated condition and the experiment was terminated.

An analysis of the brains of these animals yielded several interesting findings. Tyrosine hydroxylase (TH) immunocytochemistry in conjunction with nissl staining confirmed that these severely parkinsonian monkeys indeed did have the expected profound loss of substantia nigra neurons and the accompanying marked depletion of TH in the striatum. Thus, there was no anatomical evidence that the behavioral recovery in monkey S089 could be attributed to minor damage and subsequent recovery in the host nigrostriatal system.

TH immunocytochemistry also revealed that the behavorial results observed could be related to the presence or absence of grafted dopamine neurons in the striatum of the host animals. Monkey S089 had well-developed clusters of TH-positive neurons, typical in morphology of nigral dopamine cells, in the striatum of both hemispheres (Figure 1). These grafted neurons sent their processes into the surrounding denervated striatum, suggesting a physical basis for integration with the host brain (Figure 1). Monkey S241 exhibited no TH-positive

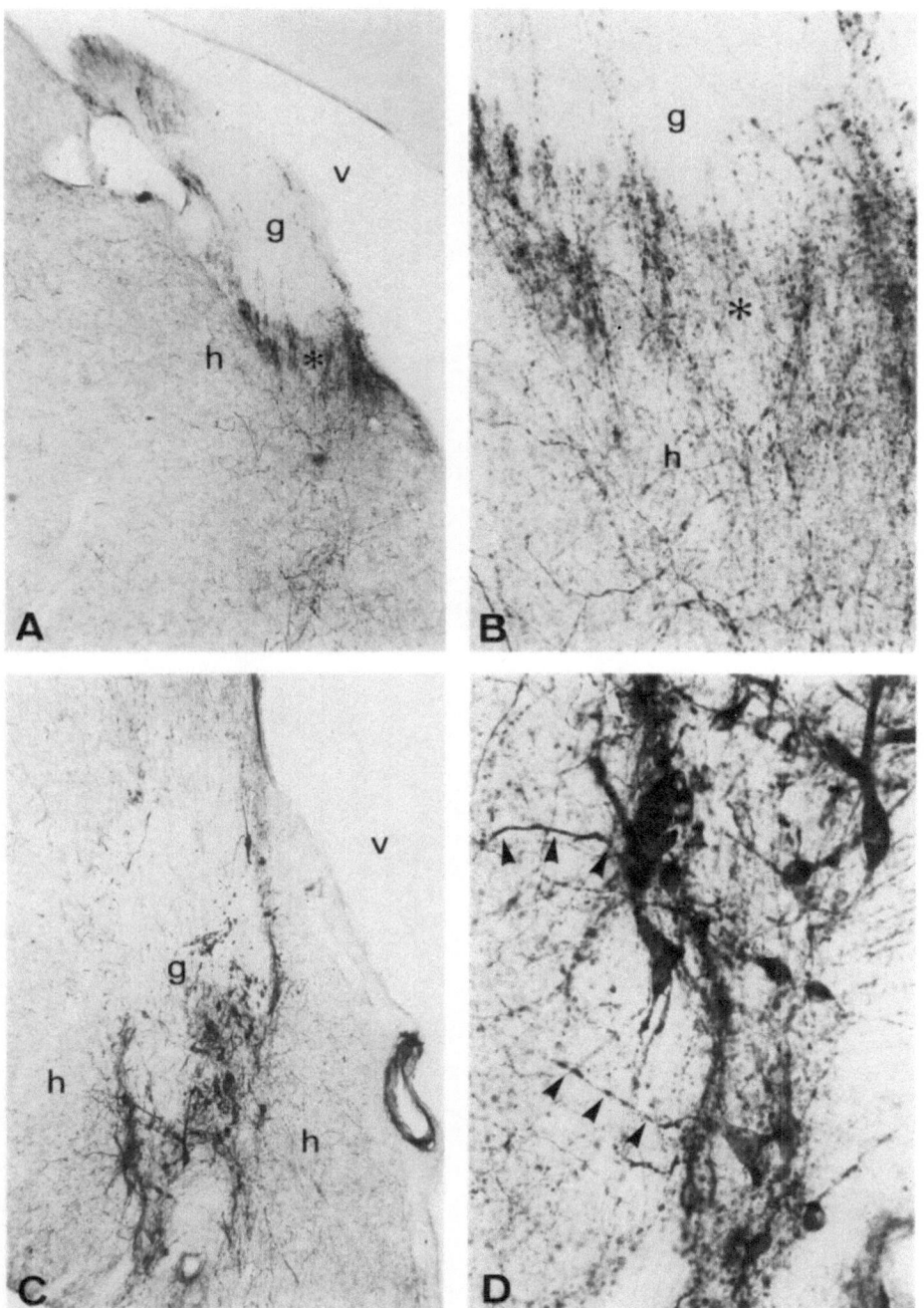

Figure 1. Grafted fetal primate dopamine neurons. Panel A, C and E are low magnification (50X) views of rostral, middle and caudal levels through a single nigral graft in monkey S089. Panel G is a similar view of another implant in this animal. Panels B,D,F and H show high magnification (150X) details of these nigral neuron grants. Tissue has been processed for visualization of tyrosine hydroxylase (TH), a marker enzyme of dopamine neurons. A). Rostrally, this fetal graft (g) appears as a peninsula of tissue extending into the lateral ventricle (v). In all four low magnification views the implanted tissue appears paler than the surrounding host striatum (h). A dense sweep of TH-positive fibers (*) at the interface of the graft and the host is seen to advantage in panel B. C.). At its middle level, this graft (g) is seen as a wedge of tissue deep in the medial part of the host

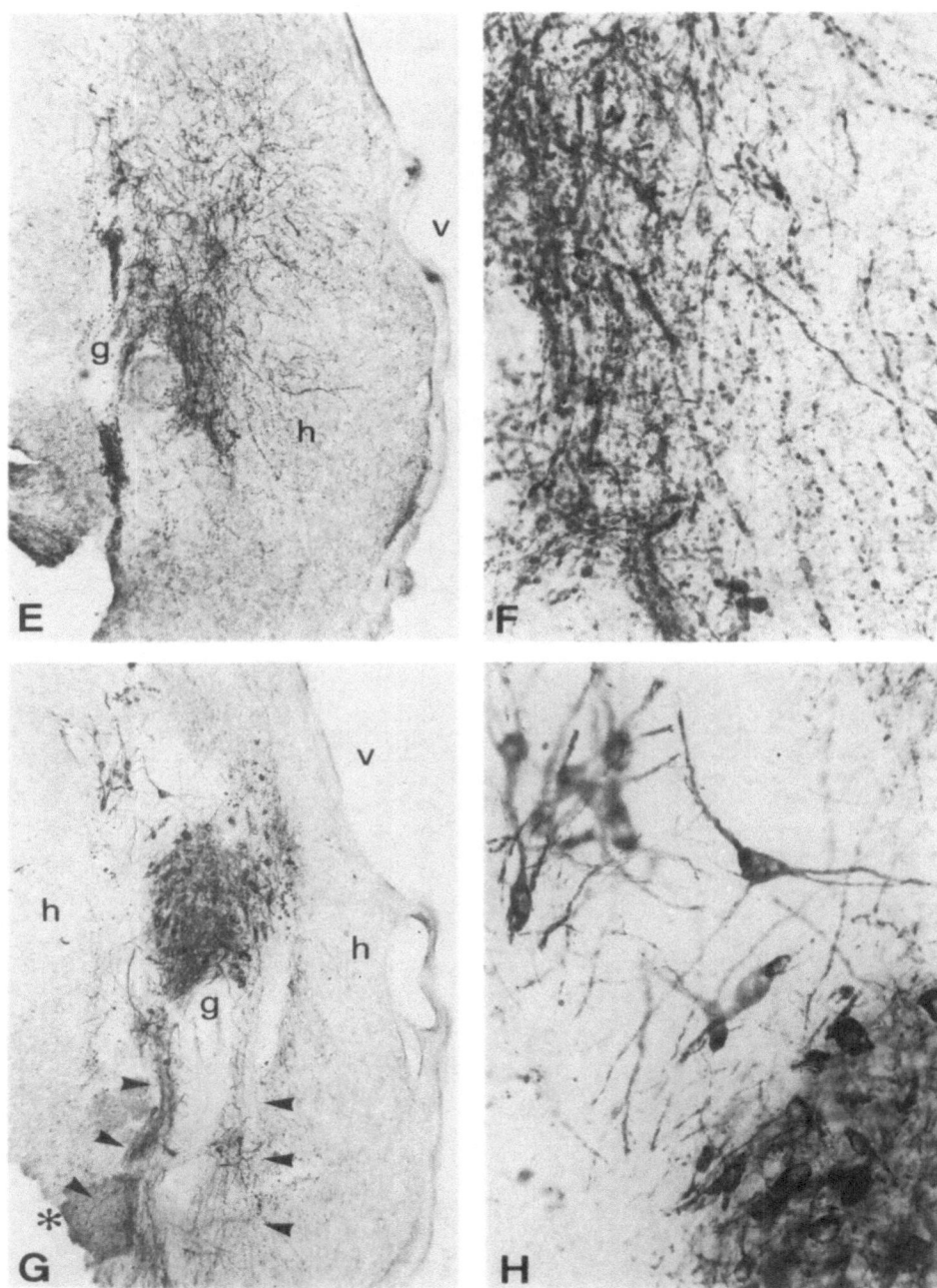

striatum (h). Clusters of TH-positive neurons line the edges of this implant, sending clear fiber projections into the surrounding host brain (panel D, arrows). E). At its caudal border, the implant is seen as a sliver of tissue no wider than the cannula track (g), but exhibits a dense spray of TH-positive fibers innervating the host stritum (h). The beaded appearance of these fibers is typical of dopamine neurons (panel F). G). A second cluster of grafted neurons in monkey S089 is shown here. Similar to the other graft shown, an aggregate of transplanted TH-positive neurons (g) gives rise to a bundle of fibers (arrows) entering the host striatum (h) ventrally. The asterisk marks the edge of a hole where tissue was punched for biochemical measurement of dopamine levels, as discussed in the text. Grafted neurons are seen to advantage in panel H.

neurons in its intrastriatal grafts, consistent with transplantation of non-dopaminergic cerebellar tissue in this control animal. While the cerebellar tissue yielded clearly identifiable transplants in this monkey, there was no indication of enhanced TH staining in the striatum (Figure 2). Thus, the presence of grafted fetal substantia nigra neurons in monkey S089 is consistent with supplementation of dopamine activity in this animal yielding the marked behavioral improvement detected. The absence of TH-positive cells in the cerebellar grafts of monkey S241 is consistent with no effect on the dopamine system of this animal and the observed continued behavioral debilitation.

Biochemical assessment of dopamine (DA) and homovanillic acid (HVA) levels obtained from tissue punches suggested that grafted fetal nigral neurons in monkey S089 did exert a modest, local, normalizing influence on the host striatum. Our evidence (8) indicates that MPTP treatment reduces DA levels, in the part of the striatum studied here, from 150-165ng/mg protein to 1-3ng/mg protein, and elevates the HVA/DA ratio from approximately 1.0 to 8.0-12.0. Monkey S089, in tissue punched from the striatum adjacent to the fetal nigral neuron grafts, had DA levels of 8.2ng/mg protein rostrally and 7.5ng/mg protein caudally, and an HVA/DA ratio of 5.1 rostrally and 3.6 caudally: measures comparable to those of asymptomatic MPTP-treated monkeys and in the direction of normalization. Measures of DA and HVA levels were not improved in striatal tissue punched at a distance from the nigral grafts.

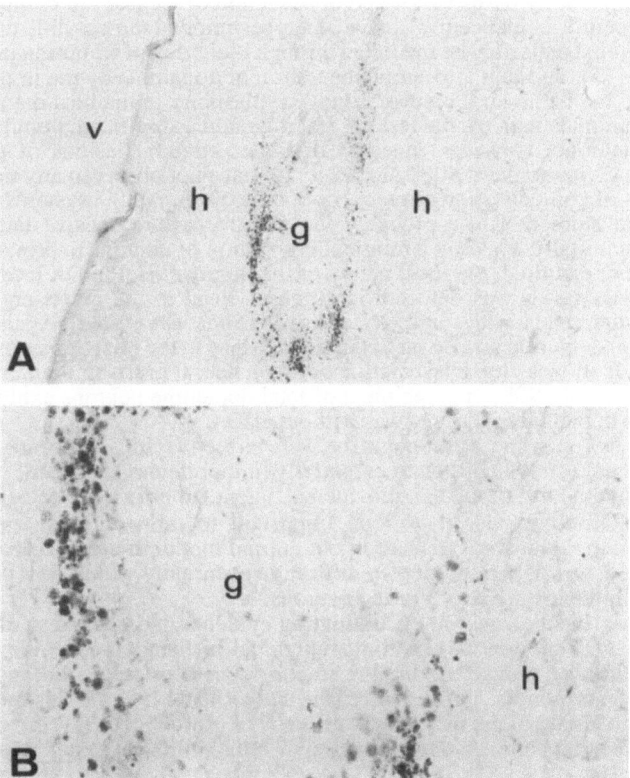

Figure 2. Grafted fetal cerebellar tissue. A). Low magnification (40X) view of a control cerebellar tissue implant in monkey S241. This wedge of implanted tissue (g) is negative for tyrosine hydroxylase staining, consistent with the absence of dopamine neurons in this donor brain region. As seen to advantage in panel B (160X), cerebellar grafts are associated with no enhancement of TH staining over the background levels of the dopamine-depleted host striatum (h). This implant is partially outlined with cells that may be reacting to tissue damage associated with the implantation procedure. v = lateral ventricle.

DISCUSSION

The case studies of monkeys S089 and S241 support the view that behaviorally-significant dopamine replacement, over a several month period, can be provided by instrastriatal implantation of fetal dopamine neurons. Implantation of fetal brain tissue that did not contain dopamine neurons was not sufficient to provide a therapeutic effect over the five month time interval studied.

This ongoing study of long-term survival and efficacy of fetal primate neural grafts provides some initial answers to the questions posed at the outset of this report. First, we now have evidence that grafted fetal dopamine neurons survive, grow, continue to express the marker enzyme TH, supplement dopamine in the host striatum, and ameliorate parkinsonian symptoms in MPTP-treated monkeys for up to 7.5 months following transplantation. Ongoing studies are examining an even more extended timecourse, but the current data argues that neural grafts provide an enduring influence on the host brain.

Second, study of extended survival intervals has provided a clearer indication of integration between the grafted neural tissue and the host brain. Our morphological evidence demonstrates large numbers of TH-positive fibers coursing from clusters of grafted neurons into the surrounding denervated host striatum. Biochemical assays indicate local changes in the direction of normalization of dopamine levels and the homovanillic acid/dopamine ratio in areas of the host striatum associated with neural grafts. Both types of evidence suggest a structural and functional interaction between grafted neurons and the host brain.

Third, study of severely parkinsonian monkeys, like S089 and S241, begin to support the view that behavioral recovery is specifically related to dopamine replacement rather than supplementation of some other tissue factor and/or induction of recovery or compensation in host brain systems. Recently, evidence has appeared suggesting that the possible therapeutic influence of grafted adrenal cells may be mediated through induction of sprouting in remaining host dopamine neurons (5), and that any supplementation of dopamine by the implant may not be causally related to the behavioral recovery. Indeed, plasticity in the host dopamine system is likely to occur in animals with partial lesions that maintain a significant population of undamaged dopamine neurons. However, in cases of large, bilateral lesions of the nigrostriatal dopamine system, as in monkeys S089 and S241, we have not observed any morphological or biochemical signs of compensation or recovery in the host dopamine systems over the 5.0-7.5 month time interval studied. This is probably due, in part, to the extensive damage done to the dopamine systems initially, yielding a minimal population of neurons to provide any compensatory response. For example, the local nature of the improvement in DA levels - no improvement in DA or HVA levels was detected in striatum distal to the grafts-argues for a graft-mediated effect rather than a widespread recovery of the host DA system. Against the backdrop of severe dopamine depletion and no appreciable recovery in the host dopamine system, monkeys S089 and S241 show a clear dissociation between neural grafts of dopamine neurons and control brain tissue: only S089, the recipient of fetal dopamine neurons exhibited behavioral recovery. The non-dopaminergic tissue was without effect.

Finally, fetal dopamine neuron grafts were effective in ameliorating parkinsonian symptoms even in a severely debilitated, essentially immobile monkey. While this encourages the hope that grafted neural tissue can provide sufficient support to be of benefit to severely dopamine depleted individuals, it will be important to more closely simulate the long timecourse of human Parkinson's disease in our animal model in order to determine whether long-term changes in the dopamine-depleted striatum eventually yield a host environment that is less responsive to the influence of grafted neurons.

In conclusion, we believe our accumulating evidence indicates that transplanted fetal dopamine neurons can provide an enduring functional influence on the dopamine-depleted striatum of MPTP-treated monkeys, yielding amelioration of impaired motor performance for at least a period of 7.5 months. Furthermore, the study of severely parkinsonian animals provides the clearest example of the therapeutic efficacy of grafted neurons, minimizing the contribution of any potential compensation provided by remaining host systems.

ACKNOWLEDGMENTS

The authors thank Jeffrey P. Blount, Daniel R. Feikin, Joseph P. Fenerty, Paul N. Foster, Valerae Lewis, Lawrence M. Salzer, Lori Kaplowitz, David Solomon, Bettina Steffen and Ashley Mears for care of the treated monkeys, for behavioral observations and for assistance in neural grafting. Skilled technical assistance was provided by Brian Daley, Barbara Blanchard and Judith Van Lare and the staff of the St. Kitts Biomedical Research Foundation. Supported by USPHS Grants NS24032, NS15816, MH14092, AG00847, MH25642, MH14276 and core support from the St. Kitts Biomedical Research Foundation, the Axion Research Foundation, and the Pew Charitable Trust. TJC is an Alzheimer's Disease and Related Disorders Association Faculty Scholar.

REFERENCES

1. Backlund, E.-O., Grandberg, P.-O., Hamberger, B., Knutson, E., Martensson, A., Sedvall, G., Seiger, A., and Olson, L. (1985) Transplantation of adrenal medullary tissue to striatum in parkinsonism: First clinical trials. J. Neurosurg., 62:169-173.

2. Bakay, R.A.E., Barrow, D.L., Fiandaca, M.S., Iuvone, P.M., Schiff, A., and Collins, D.C. (1987) Biochemical and behavioral correction of MPTP parkinson-like syndrome by fetal cell transplantation. Annals N.Y. Acad. Sci., 495:623-640.

3. Bergmann, K.J., Mendoza, M.R., and Yahr, M.D.(1986) Parkinson's disease and long-term levodopa therapy. In: Advances in Neurolgy Volume 45, M.D. Yahr and K.J. Bergmann (Eds.). Raven Press, New York, pp. 463-467.

4. Bjorklund, A., and Stenevi, U. (1979) Reconstruction of the nigrostriatal dopamine pathway by intracerebral nigral transplant. Brain Res., 177:555-560.

5. Bohn, M.C., Cupit, L., Marciano, F., and Gash, D.M. (1987) Adrenal medulla grafts enhance recovery of striatal dopaminergic fibers. Science, 237:913-916.

6. Burns, R.S., Chiueh, C.C., Markey, S.P., Ebert, M.H., Jacobowitz, D.M., and Kopin, I.J. (1983) A primate model of parkinsonism: Selective destruction of dopaminergic neurons in the pars compacta of the substantia nigra by N-methyl-4-phenyl-1,2,3,6-tetrahydropyridine. PNAS, 80:4546-4550.

7. Cedarbaum, J.M., and McDowell, F.H. (1986) Sixteen year follow-up of 100 patients begun on levodopa in 1968: Emerging problems. In: Advances in Neurology Volume 45, M.D. Yahr and K.J. Bergmann (Eds). Raven Press, New York, pp. 469-472.

8. Elsworth, J.D., Deutch, A.Y., Redmond, D.E. Jr., Sladek, J.R. Jr., and Roth, R.H. (1987) Effects of 1-methyl-4-phenyl-1,2,3,6-tetrahydropyridine (MPTP) on catecholamines and metabolites in primate brain and CSF. Brain Res., 415:293-299.

9. Freed, W.J., Morihisa, J.M., Spoor, E., Hoffer, B.J., Olson, L., Seiger, A., and Wyatt, R.J. (1981) Transplanted adrenal chromaffin cells in rat brain reduce lesion-induced rotational behavior. Nature, 292:351-351.

10. Gash, D.M., Collier, T.J., and Sladek, J.R. Jr. (1985) Neural transplantation: A review of recent developments and potential applications to the aged brain. Neurobiol. Aging, 6:131-150.

11. Gash, D.M., Notter, M.F.D., Okawara, S.H., Kraus, A.L., and Joynt, R.J. (1986). Amitotic neuroblastoma cells used for neural implants in monkeys. Science, 233:1420-1422.

12. Hornykiewicz, O., and Kish, S.J. (1986) Biochemical pathophysiology of Parkinson's disease In: Advances in Neurology Volume 45, M.D. Yahr and K.J. Bergmann (Eds.). Raven Press, New York, pp.19-34.

13. Kordower, J.H., Notter, M.F.D., Yeh, H.H., and Gash, D.M. (1987) An in vivo and in vitro assessment of differentiated neuroblastoma cells as a source of donor tissue for transplantation. Annals N.Y. Acad.Sci., 495:606-622.

14. Lindvall, O., Backlund, E.O., Farde L., Sedvall, G., Freedman, R., Hoffer, B., Nobin, A., Seiger, A., and Olson, L. (1987) Transplantation in parkinson's disease: Two cases of adrenal medullary grafts to the putamen. Ann. Neurol., 22:457-468.

15. Madreazo, I., Drucker-Colin, R., Diaz, V., Martinez-Mata, J., Torres, C., and Becerril, J.J. (1987) Open microsurgical autograft of adrenal medulla to the right caudate nucleus in two patients with intractable Parkinson's disease. N. Engl. J. Med., 316:831-834.

16. McDowell, F.H., Lee, J.E., Swift, T., Ogsbury, J.S., and Kessler, J. (1970) Treatment of Parkinson's syndrome with L-dihydroxyphenylalaine (levodopa). Ann. Int. Med., 72:29-35.

17. Morihisa, J.M., Nakamura, R.K., Freed, W.J., Mishkin, M., and Wyatt, R.J. (1987) Transplantation techniques and the survival of adrenal medulla autografts in the primate brain. Annals N.Y. Acad. Sci., 495:599-605.

18. Perlow, M.J., Freed, W.J., Hoffer, B.J., Seiger, A., Olson, L., and Wyatt, R.J. (1979) Brain grafts reduce motor abnormalities produced by destruction of the nigrostriatal dopamine system. Science, 204:643-647.

19. Redmond, D.E., Jr., Sladek, J.R. Jr., Roth, R.H., Collier, T.J., Elsworth, J.D, Deutch A.Y., and Haber, S. (1986) Fetal neuronal grafts in monkeys given methylphenyltetrahydropyridine. Lancet, I #8490:1125-1127.

20. Shoulson, I. (1988) Experimental therapeutics directed at the pathogenesis of Parkinson's disease. In: Handbook of Experimental Pharmacology: Drugs for the Treatment of Parkinson's Disease, D.B. Calne (Ed.). Springer-Verlag, New York, in press.

21. Sladek, J.R. Jr., Collier, T.J., Haber, S.N., Roth, R.H., and Redmond, D.E. Jr. (1986) Survival and growth of fetal catecholamine neurons transplanted into primate brain. Brain Res. Bull., 17:809-818.

22. Sladek, J.R., Jr., Redmond, D.E. Jr., Collier, T.J., Haber, S.N., Elsworth, JD., Deutch, A.Y., and Roth, R.H. (1987) Transplantation of fetal dopamine neurons in primate brain reverses MPTP induced parkinsonism. In: Progress in Brain Research Volume 71, F.J. Seil, E. Herbert and B.M. Carlson (Eds.) Elsevier, New York, pp. 309-323.

23. Sladek, J.R. Jr., Collier, T.J., Haber, S.N., Deutch, A.Y., Elsworth, J.D., Roth, R.H., and Redmond, D.E. Jr. (1987) Reversal of parkinsonism by fetal nerve cell transplants in primate brain. Annals N.Y. Acad. Sci., 495:641-657.

24. Yahr, M.D. (1984) Limitations of long-term use of anti-parkinsonian drugs. Can. J. Neurol. Sci., 11:191-194.

28

TRANSPLANTATION OF HUMAN DOPAMINERGIC NEURONS IN PARKINSONISM: EXPERIMENTAL REALITY AND FUTURE CLINICAL FEASIBILITY

Ake Seiger, Lars Olson[1], Ingrid Stromberg[1], Marc Bygdeman[2], Menek Goldstein[3], Barry Hoffer[4]

Department of Neurological Surgery, University of Miami, School of Medicine, Miami, Florida, USA, [1]Department of Histology and Neurobiology, Karolinska Institute, Stockholm, Sweden, [2]Department of Obstetrics and Gynecology, Karolinska Institute, Stockholm, Sweden, [3]Department of Psychiatry, New York University Medical Center, New York, New York, USA, [4]Department of Pharmacology, University of Colorado Medical Center, Denver, Colorado, USA.

The extensive degeneration of mesostriatal dopamine (DA) neurons in patients with Parkinson's disease lead to severe motor deficits with tremor, rigidity and hypokinesia. Levodopa treatment counteracts many of these symptoms for a few years, but patients invariably go into an end stage of severe "on-off" oscillations irrespective of further modulations of the medication (Marsden 1980). The need for a new and more effective long-term treatment of Parkinson's disease is therefore urgent. After unilateral electrolytic lesions of the nigrostriatal pathway that cause degeneration of striatal DA terminals, rats become asymmetric and display a rotational behavior when given drugs that interfere with DA neurotransmission (Anden et al 1966). Animal models of Parkinson's disease based on these observations have been available for almost 20 years. The first one was the 6-hydroxy-dopamine (6-OH-DA) lesion of the nigrostriatal pathway in rats (Ungerstedt 1971). By specific unilateral lesions of the DA system, rats have been produced whose asymmetric motor behavior can be quantified by a rotometer (Ungerstedt and Arbuthnott 1970). It was later shown that the transplantation of fetal syngeneic DA neuroblasts from substantia nigra to animals with such lesions resulted in a partial structural and functional restoration of the DA system in striatum (Perlow et al., 1979, Bjorklund and Stenevi 1979). Later modification of the procedure has improved reinnervation of striatum from the grafted DA neuroblasts leading to extensive functional restitution of the motor performance of the lesioned animals, (reviews see: Olson et al., 1985, Brundin et al., 1987).

The second animal model was developed following reports of a "Parkinson-like" syndrome in drug addicts traced to the drug "impurity" l-methyl-4-phenyl-1,2,3,6-tetrahydropyridine (MPTP), (Langston et al 1983). This chemical produced a syndrome in these young individuals indistinguishable from Parkinson's disease. MPTP was later shown to have similar effects in monkeys (Burns et al,. 1984) and in certain mouse strains (Hallman et al., 1984, Heikkila et al., 1984). Especially the MPTP lesioned monkeys have been used for transplantation of homotypic fetal DA neuroblasts from substantia nigra into the DA denervated basal ganglia (Backay et al., 1985, Redmond et al., 1986).

Another paradigm to counteract experimental parkinsonism in rodents and monkeys is transplantation of chromaffin cells to striatum. The first report on chromaffin cell transplantation described morphological transformation in intraocular grafts of adrenal medullary cells (Olson 1970). It was later shown that sequential double intraocular grafts of adrenal medulla and CNS regions such as cerebral cortex resulted in "adrenergic" innervation of the CNS graft from the transformed chromaffin cells (Olson et al., 1980). Freed and coworkers (1981) showed that chromaffin cells transplanted into the DA denervated striatum could counteract partially the motor asymmetry elicited by apomorphine. The first patient with Parkinson's disease was transplanted with chromaffin autografts in the spring of 1982 in Stockholm (Backlund et al.,

1985). It was subsequently shown both in the eye chamber model (Stromberg et al., 1985a) and in intrastriatal grafts (Stromberg et al., 1985b) that the addition of nerve growth factor (NGF) markedly enhanced the in vivo morphological transformation of chromaffin cells to form catecholamine-containing nerve fibers. In the case of NGF added to intraparenchymal chromaffin cells, also the functional restoration was enhanced by chronic infusion of NGF (Stromberg et al., 1985b).

From 1986 to date, more than two hundred Parkinsonian patients have been transplanted with chromaffin autografts. Most patients have been operated on in Mexico (Madrazo et al., 1987), China (Shou-shu Jiao et al., 1987), USA (Allen et al., 1987) and Sweden (Backlund et al., 1985; Lindvall et al., 1987). Taken together, highly variable results have been obtained with the most positive reports coming from Mexico and China. It has not been possible so far to tie long-term neurological improvements after the transplantation to definite chromaffin cell survival in the basal ganglia of the patients. Judged on the animal experiments, it is possible that chromaffin tissue autografting to Parkinsonian patients with chronic infusion of NGF locally could improve the survival and transformation of the chromaffin cells and thereby, possibly the neurological status of the patient (cf. Stromberg et al., 1985b).

In a direct comparison between auto-transplantation of chromaffin cells and homo-transplantation of fetal nigral cells it becomes clear that the long-term structural and functional restoration of the DA fiber system is more extensive with the nigral cells. Although the chromaffin cell transplantation to Parkinsonian patients might eventually become beneficial, it would, in the light of this comparison between catecholamine cell sources, be an interesting and possibly very important improvement to use fetal brain cells for transplantation purposes in parkinsonian patients. An absolute requirement for such a future development would be to show convincingly in animal experiments that human fetal DA neuroblasts obtained from elective abortions could serve the purpose to replace structurally and functionally the degenerated striatal DA fiber system. Some of the obvious additional problems that emerge from such xenograft experiments, are the ethical considerations of access to the fetal material; the immunological problems likely to require immunosuppression; and practical problems such as sterility, yield, transfer time and others.

It has been shown recently that human fetal nigral DA neuroblasts from first trimester elective abortions can be obtained and transplanted into DA denervated striatum of recipient rats (Stromberg et al., 1986, Brundin et al., 1986, 1987, Seiger et al., 1987). Dopaminergic neuroblasts in small fragments of brain stem survive in cavities (Stromberg et al., 1986), as well as intraparenchymally injected in a cell suspension (Brundin et al., 1986). In both of these studies, the morphological differentiation of the human DA neuroblasts was strikingly slower than for the corresponding neuroblasts from a rat embryo. Morphological evaluation after 1-2 months showed very immature, relatively small neuroblasts with large nuclei and a thin rim of cytoplasm. The cells showed few signs of forming processes and only short scattered fibers left the transplant area to enter the DA denervated target striatum of the host animal (Stromberg et al., 1986, Brundin et al., 1986). At this time there was no functional restitution of the nigro-striatal DA system as tested with rotational behavior during apomorphine (Stromberg et al., 1986) or amphetamine (Brundin et al., 1986) treatment. 4-6 months after transplantation morphological evaluation revealed substantial structural differentiation of the grafted human DA neuroblasts. The cell bodies were larger, with prominent dendrites and visible dendritic spines and smooth, thin neurites extending far into the host rats' striatal neuropil. The distribution of terminal branches in the target region was considerably more even than what is seen around grafted rat DA neuroblasts in a similar situation (Stromberg et al., 1986, Brundin et al., 1986, cf. Bjorklund et al., 1983, Stromberg et al., 1985c). A significant functional restoration of the DA fiber system was seen at these longer post grafting stages. Both apomorphine- (Stromberg et al., 1986) and amphetamine-(Brundin et al., 1986)-induced rotational behavior was reduced after 4-6 months. There was a good correlation between density of DA fiber reinnervation and degree of functional restoration (Stromberg et al., 1986). When donor material was used from second trimester elective abortuses (up to 19 weeks) insignificant DA cell survival was seen and with no functional recovery (Brundin et al., 1986). Thus, it is clear that human fetal DA neuroblasts obtained from elective first trimester abortions can survive for at least six months in immunocompromized rodent hosts with experimental Parkinsonism, and that these human neurons can functionally counteract some of the neurological deficits encountered by the lesion of the nigrostriatal DA system. These results certainly increases the hope for a future similar use of such DA neuroblasts for transplantation therapy of Parkinsonian patients. A number of additional considerations such as ethical, immunological, and practical problems have to be taken into account, however, before such a procedure could become routine treatment in patients.

The procedure that has been developed in the only two research centers to date, using human fetal cells for xenografts on a larger scale (Stockholm and Lund), includes the following considerations: the aborted fetus can only be used after a written informed consent from the abortion-seeking woman. Only elective first trimester abortions have been utilized, and there has been no significant change of the abortion procedure imposed on the patient. The use of the material has been entirely passive with regard to the justification of the abortion, and entirely after the fact with regard to getting access to the cells from the dead aborted fragmented fetus. It has been made impossible to track a particular set of cells to any given donor, thereby securing the anonymity of the woman. No pre-, per- or post-abortion treatment or procedure has been altered besides the informed consent process. The recipient animals have all been immunocompromised, either by immunosuppression using Cyclosporin A in combination with prophylactic antibiotics or using immunodeficient athymic (nude) rats incapable of rejecting xenografts.

The yield of DA neuroblasts from substantia nigra of the aborted fetus has improved gradually and in 30-60% of the abortuses these cells can now be found. The location of the nigral cell complex in the human embryo has been known for many years (Olson et al., 1973, Nobin and Bjorklund 1973). Although the routine abortion technique entails suction to empty the uterine cavity, thereby partly destroying the integrity of the embryo proper, our previous neuroanatomical knowledge has made the location of the DA neuroblasts in the aborted fetal CNS possible in a large number of cases. The time from abortion to transplantation has been kept to a minimum, usually less than three hours. No obvious differences in DA cell survival upon grafting within that time has been noted. Additional screening for human DA cell yield, survival and capacity to grow has been carried out in intra-ocular transplantation experiments to similar hosts (Olson et al., 1987, Seiger et al., 1988). It seems as if the optimal donor stage for DA neuroblasts is up to 11 weeks of gestation, where the tissue yield is better from 10-11 week pregnancies and the taking seems to be better from 8-9 week pregnancies. Judging from these xenograft procedures developed in Sweden, it is quite possible that human fetal cells could be used on a larger scale for clinical trials with transplantation into the basal ganglia of Parkinsonian patients within the next few years.

Supported by the Swedish Medical Research Council, The Miami Project Foundation, National Institute of Health, The "Expressen" Prenatal Research Foundation, Soderberg's Research Foundations, and "Magnus Bergvalls Stiftelse."

References

Anden N-E, Dahlstrom A., Fuxe K., Larsson K.: Functional role of the nigro-striatal dopamine neurons. Acta Pharmacol. Toxicol. 24: 263-274, 1966.

Allen, G.S., Burns, R.S. and Tulipan, N.B.: Human adrenal autografts as a potential therapy for Parkinson's disease. Abstract. Transplantation into the mammalian CNS. Rochester, June 1987.

Backay, R.A.E., Fiandaca, M.S., Barrow, D.L. et al: Preliminary report on the use of fetal tissue transplantation to correct MPTP-induced parkinson-like syndrome in primates. Appl. Neurophysiol. 48: 358-361, 1985.

Backlund, E.O., Granberg, P.O., Hamberger, et al: Transplantation of adrenal medullary tissue to striatum in parkinsonism. First clinical trials. J. Neurosurg. 62: 169-173, 1985.

Bjorklund, A. and Stenevi, Reconstruction of the nigrostriatal dopamine pathway by intracerebral nigral transplants. Brain Res. 177: 555-560, 1979.

Bjorklund, A., Stenevi, U., Schmidt, R.H., Dunnett, S.B., and Gage, F.H.: Intracerebral grafting of neuronal cell suspensions. II. Survival and growth of nigral cell suspensions implanted in different brain sites. Acta Physiol. Scand. [Suppl] 522: 9-18, 1983.

Brundin, P., Nilsson, O.G., Strecker, R.E., Lindvall, O., Astedt, B., Bjorklund, A.: Behavioral effects of human fetal dopamine neurons grafted in a rat model of Parkinson's disease. Exp. Brain Res. 221: 235-240, 1986.

Brundin, P., Strecker, R.E., Lindvall, O., et al: Intracerebral grafting of dopamine neurons: experimental basis for clinical trials in patients with Parkinson's disease. In: Azmitia E, Bjorklund A (eds): Cell and tissue transplantation into the adult brain, New York, New York Academy of Sciences, pp. 473-495, 1987.

Burns, R.A., Markey, S.P., Phillips, J.M. and Chieuh, C.C.: The neurotoxicity of l-methyl-4-phenyl-1,2,3,6-tetrahydropyridine in the monkey and man. Can. J. Neurol. Sci. 11: 166-168, 1984.

Freed, W., Morihisa, J., Spoor, E., Hoffer, B., Olson, L., Seiger, A., and Wyatt, R.: Transplanted adrenal chromaffine cells in rat brain reduce lesion-induced rotational behavior. Nature 292: 351-352, 1981.

Hallman, H., Olson, L., and Jonsson, G.: Neurotoxicity of the meperidine analogue N-methyl-4-phenyl-1,2,3,6-tetrahydropyridine on brain catecholamine neurons in the mouse. Eur. J. Pharmacol. 97: 133-137, 1984.

Heikkila, R.E., Hess, A., and Duvoisin, R.C.: Dopaminergic neurotoxicity of l-methyl-4-phenyl-1,2,3,6-tetrahydropyridine in mice. Science, 224: 1451-1453, 1984.

Jiao, Shou-shu, Zhang, W.C., Ding, M.C. and Sun, J.B.: The clinical study of adrenal medullary tissue transplantation to striatum in Parkinsonism. Abstract. Transplantation into the mammalian CNS, Rochester, June 1987.

Langston, J.W., Ballard, P., Tetrud, J.W. and Irwin, I.: Chronic parkinsonism in humans due to a product of meperidine-analog synthesis. Science 219: 979-980, 1983.

Lindvall, 0., Backlund, E.-O., Farde, L., Sedvall, G., Freedman, R., Hoffer, B., Nobin, A., Seiger, A and Olson, L., Transplantation in Parkinson's disease: two cases of adrenal grafts to the putamen. Ann. Neurol. 22. 457-468, 1987.

Madrazo, I., Drucker-Colin, R., Diaz, V., et al: Open microsurgical autograft of adrenal medulla to the right caudate nucleus in two patients with intractable Parkinson's disease. N. Eng. J. Med. 316: 831-834, 1987.

Marsden, C.D.: "On-off" phenomena in Parkinson's disease. In: Rinne UK, Klingler M,, Stamm, G. (eds.): Parkinson's Disease-Current Progress, Problems and Management. Amsterdam, Elsevier/NorthHolland, pp. 241-254, 1980.

Nobin, A. and Bjorklund, A.: Topography of the monoamine neuron systems in the human brain as revealed in fetuses. Acta Physiol. Scand. [Suppl.] 388: 1-40, 1973.

Olson, L.: Fluorescence histochemical evidence for axonal growth and secretion from transplanted adrenal medullary tissue. Histochemie 22: 1-7, 1970.

Olson, L., Backlund, E.-O, Freed, W., Herrera-Marschitz, M., Hoffer, B., Seiger, A., Stromberg, I.: Transplantation of monoamine-producing cell systems in oculo and intracranially: Experiments in search of a treatment for Parkinson's disease. In: Hope for a New Neurology (F. Nottebohm, ed.) Ann. N.Y. Acad. Sci. 1985, pp. 105-126.

Olson, L., Backlund, E.O., Gerhardt, G., Hoffer, B., Lindvall, O., Rose, G., Seiger, A. and Stromberg I.: Nigral and adrenal grafts in parkinsonism: recent basic and clinical studies. In: Yahr M.D., Bergmann K.J. (eds.) Advances in neurology, Vol 45, Raven Press, New York, pp. 85-94, 1986.

Olson, L., Boreus, L.O. and Seiger, A., Histochemical demonstration and mapping of 5-Hydroxytryptamine- and catecholamine-containing neuron systems in the fetal brain. Z. Anat. Entwickl. Gesch. 139: 259-282, 1973.

Olson, L., Seiger, A., Freedman, R., and Hoffer, B.: Chromaffine cells can innervate brain tissue: Evidence from intraocular double grafts. Exp. Neurol., 70: 414-426, 1980.

Olson, L., Stromberg, I., Bygdeman, M., Granholm, A.-Ch., Hoffer, B., Freedman, R. and Seiger, A.: Human fetal tissues grafted to rodent hosts: structural and functional observations of brain, adrenal and heart tissues in oculo. Exp. Brain Res. 67: 163-178, 1987.

Perlow, M., Freed, W., Hoffer, B., Seiger, A., Olson, L. and Wyatt, R.: Brain grafts reduce motor abnormalities produced by destruction of nigrostriatal dopamine systems. Science 204: 643-647, 1979.

Redmond, D.E., Sladek, J.R. Jr., Roth, R.H., et al: Fetal neuronal grafts in monkeys given methylphenyltetrahydropyridine. Lancet 1: 1125-1127, 1986.

Seiger, A., Bygdeman, M., Goldstein, M., Almqvist, P., Hoffer, Stromberg, I. and Olson, L.: Human fetal catecholamine-containing tissues grafted intraocularly and intracranially to immunocompromized rodent hosts. In: Transplantation into the mammalian CNS. Progr. Brain Res. (Sladek, JR and Gash DM, eds.), 1988, in press.

Seiger, A., Stromberg, I. and Olson, L: Aspects of embryonic CNS transplantation. Int. Pediatr. 2: 79-82, 1987.

Stromberg, I., Bygdeman, M., Goldstein, M., Seiger, A. and Olson, Human fetal substantia nigra grafted to the dopaminedenervated striatum of immunosuppressed rats: evidence for functional reinnervation. Neurosci. Lett. 71: 271-276, 1986.

Stromberg, I., Ebendal, T., Seiger, A., and Olson, L.,: Nerve fiber production by intraocular adrenal medullary grafts: stimulation by nerve growth factor or sympathetic denervation of the host iris. Cell Tissue Res. 241: 241-249, 1985a.

Stromberg, I., Herrera-Marschitz, M., Ungerstedt, U., Ebendal, T. and Olson, L.: Chronic implants of chromaffin tissue into the dopamine-denervated striatum. Effects of NGF on survival, fiber growth and rotational behavior. Exp. Brain Res. 60: 335-349, 1985b.

Stromberg, Johnson, Hoffer, B.J. and Olson, Reinnervation of the dopamine-denervated striatum by substantia nigra transplants: Immunohistochemical and electrophysiological correlates. Neuroscience 14: 981-990, 1985c.

Ungerstedt, Histochemical studies on the effects of intracerebral and intraventricular injections of 6-hydroxydopamine on monoamine neurons in the rat brain. In: 6-Hydroxydopamine and catecholamine neurons, eds. T. Malmfors and H. Thoenen. North Holland Publishing Company, 1971.

Ungerstedt, U. and Arbuthnott, G.W.: Quantitative recording of rotational behavior in rats after 6-hydroxydopamine lesions of the nigrostriatal dopamine system. Brain Res. 24: 485-493, 1970.

INDEX